Apollo
The Lost and Forgotten Missions

Springer

London
Berlin
Heidelberg
New York
Barcelona
Hong Kong
Milan
Paris
Santa Clara
Singapore
Tokyo

David J. Shayler

Apollo

The Lost and Forgotten Missions

Springer

Published in association with

Praxis Publishing
Chichester, UK

David J. Shayler
Astronautical Historian
Astro Info Service
Halesowen
West Midlands
UK

SPRINGER–PRAXIS BOOKS IN ASTRONOMY AND SPACE SCIENCES
SUBJECT *ADVISORY EDITOR*: John Mason B.Sc., Ph.D.

ISBN 1-85233-575-0 Springer-Verlag Berlin Heidelberg New York

British Library Cataloguing-in-Publication Data
Shayler, David, 1955–
   Apollo: the lost and forgotten missions. – (Springer–Praxis
   books in astronomy and space sciences)
   1. Manned space flight   2. Space flight to the moon
   I. Title
   629.4'54

   ISBN 1-85233-575-0

A catalogue record for this book is available from the Library of Congress

Printed by MPG Books Ltd, Bodmin, Cornwall, UK

Copy editing and graphics processing: R.A. Marriott
Cover design: Jim Wilkie
Typesetting: BookEns Ltd, Royston, Herts., UK

Printed on acid-free paper supplied by Precision Publishing Papers Ltd, UK

Dedicated to the inspiration of Dr Wernher von Braun
and the commitment of President John F. Kennedy

To the efforts of the NASA astronauts who participated in the
pioneering American manned spaceflight programme of 1961–1975

To the memory of the Apollo 1 crew and the courage of the Apollo 13 astronauts

To the thousands of Americans for whom Apollo was to forever
become part of their heritage, career, youth, and family life

To the small group of talented men who could have been remembered
as Apollo crew-members but who lost the opportunity

To those who still ask the question: *When will we go back?*

On 19 November 1969, Apollo 12 astronaut Al Bean became the fourth man to walk on
the surface of the Moon. At that time at least another sixteen Apollo astronauts were
expecting to leave their boot-prints in the lunar soil. Eight of them achieved that goal,
but the other eight would not do so. Of the twelve men who have lived and worked on a
world other than our home planet, three of them are no longer with us. In 2002 – the
thirtieth anniversary of the departure from the Moon of the last astronauts of Apollo –
the world still awaits the thirteenth human to make another small step to resume the
adventure.

Table of contents

Foreword

Of all the great adventures that mankind has pursued during the ages, the exploration of space must surely rank among the top few on anyone's list. And for sheer excitement and thrill, the Apollo programme – when man first walked on the Moon – is, so far, the high point.

As an astronaut I was deeply involved in that Apollo programme as we prepared for and finally landed on our nearest celestial neighbour. I helped design and test all of the equipment that was to be used on the early lunar landings. I helped write and evaluate all of the procedures that would be carried out between touch-down and lift-off. For the three-and-a-half years before we landed and while we were exploring the Moon, the Apollo programme was the centre of my professional life. Because of this preparation, I had every reason to expect that I, too, would walk on the Moon. The way the flight crews were being formed, I expected to be the sixteenth or possibly the eighteenth man to leave my footprints on another astronomical body. However, as history records, only twelve Apollo astronauts ever walked on the Moon. Was I disappointed? Absolutely. Was I discouraged? Yes, I was. But even with that personal frustration I am still convinced that I had the exceptional privilege of participating in one of the greatest human adventures.

Those of us who were, and still are, involved in the exploration of space wish that more lunar flights had been made; that the programme had not been terminated so abruptly; and in general that this marvellous adventure was currently being pursued more enthusiastically. I am anxious to see my fellow men walking on Mars. Why cannot we go faster? We certainly know what we want to do.

It is probably useful to try to put this space effort in historical perspective. For comparison, a good example is the exploration and travel by Western countries into the Pacific basin. From 1519 to 1522 Magellan sailed through the Pacific Ocean; but it was a generation and more later, in 1580, when Sir Francis Drake next arrived; and Captain Cook did not arrive on the scene until 1774. For more than 250 years we made only intermittent visits to what for Westerners was a new world. I surely hope that it will not be another 250 years before we return to the Moon! However, I am afraid that some of us as space enthusiasts will have to develop a little patience and persistence.

(*Left*) Onboard a KC-135 aircraft, Don Lind participates in simulated lunar gravity deployment of the ALSEP. Lind had been selected as an astronaut by NASA in April 1966, and was working on LM and ALSEP development issues. (*Right*) Lind might have walked on the Moon as the second scientist-astronaut in the early 1970s, but the Apollo programme was cut back, and he was transferred to Skylab (AAP) as a back-up crew-member for the second and third missions. He finally flew into space in 1985, nineteen years after selection as MS-1 onboard STS-51B (*Challenger*) carrying Spacelab 3. He left NASA in 1986.

In the exploration of the Pacific we hear only about the voyages that were actually launched. The history books that I read do not even provide a hint of any worthwhile voyages that were seriously discussed but were not approved by someone in authority, or of some sailings which had to turn back before they accomplished their mission, or that funds simply could not be raised to support some brilliant ideas. For explorations of the past, the plans that were never quite realised were not recorded. But that is not the case for space exploration in our day. The space programme – particularly the attempt to land on the Moon – was so well reported that even programmes that were never realised were open to public view. Some of these programmes did not materialise because technical problems remained to be solved. Some of the funds went to other more demanding or promising approaches. Development was so fast that the goals of some of these programmes could be met more efficiently by other strategies. There was certainly no shortage of ideas.

David Shayler has done us a service that was not done in past ages. In this book he performs the very useful and interesting task of preserving the history of the wonderful ideas of space exploration that never were covered by the evening news because they never happened. They were great concepts that died at birth. Many of these programme goals have since been realised, but many have not. I personally, of course, wish that there had been many more flights to the Moon. But all of these plans – the missions that did not fly as well as those that did – are part of the fabric of man's first faltering steps into the Universe. Even the aborted programmes and Apollo 13 that did not land on the Moon are a part of one of mankind's greatest adventures. Our journey into the Solar System will continue just as surely as our previous attempts at exploration were finally carried to their completion. The difference this time is that we cannot at the moment imagine what could properly be called the 'completion' of space exploration. We will surely see a lunar base, and men walking on Mars – but what will our children's children live to see? It is exciting just to try to imagine what it will be like.

Don L. Lind   November 2001
NASA astronaut, 1966–1986

Author's preface

I will always remember the early hours of 21 July 1969, when I was sitting on the floor in my grandparents' home at Paignton, South Devon, in south-west England, so enthralled with the activities portrayed on the small black-and-white television screen in front of me. As a 14-year-old, the thrill of staying up all night was itself a personal 'small step' to adulthood, but with my grandfather I was a witness to a 'giant leap' in history, watching the ghostly images of astronauts Armstrong and Aldrin working on the surface of the Moon. As a young boy I had always been interested in rockets and space adventures, but it was Apollo 8's mission to the Moon during Christmas 1968 that was to spark the passion for the real space programme that has remained with me for more than three decades. In the run-up to Apollo 11, I had read articles and heard reports of what the mission was to attempt and what lay beyond, and I was hooked.

Mixed in with the accurate accounts of Apollo hardware and mission profiles were accounts of the science fiction stories of lunar exploration from the 1950s and what would follow the first landing. There would be at least ten initial landings across the Moon, with astronauts riding on Moon-buggies or using a personal flying unit to explore the mountains, valleys and craters. On later missions, astronauts could spend up to two weeks on the surface by using a Mobile Laboratory which would also carry their living quarters, to conduct extended expeditions in more remote regions. Later Apollo missions would include three or four Lunar Modules landing at the same site, with teams of astronauts constructing the first lunar base. At the same time, other astronauts would use Apollo hardware to construct the first space stations in Earth orbit, and from there, build the spacecraft to take the first men to Mars – and all in the next ten years. Incredible! It was also a time when the activities of the Soviet Union were unpredictable, and this uncertainty and hope that the Russians would also soon send cosmonauts to the Moon fuelled the imagination even further. Unfolding events in space during 1969 provided a story to follow, and I eagerly soaked it all up, tolerated by my parents who expected me to eventually grow out of such a strange hobby – but I never did.

However, it was not long before I became aware that if you really want to follow the space programme then the first thing you have to learn is not the technology or terminology, but patience and disappointment. I was frustrated that the television pictures from Apollo 12 were lost, and I had to rely on listening to the astronauts work instead of watching them. By the time of the next flight I had my cuttings file and books ready to follow the mission to the highlands of the Moon. That mission was Apollo 13, and the disappointment of the aborted landing is one I will always recall, but one that was soon replaced with concern over bringing the astronauts safely home. The coverage of Apollo 13 after the accident was comprehensive, and I hoped that it would continue when Apollo 14 went back to the Moon. I was wrong. By late 1970, three missions had been scrapped and the others stretched out, and talk of losing a second Skylab space station made for gloomy reading.

It was difficult to understand how, after so much effort in reaching the Moon, NASA seemed to be giving up lunar exploration at a time when it should be expanded, as the stories had indicated a little over a year before. With an element of sadness I followed Apollo 17 to the Moon, was frustrated at the lack of coverage of the Skylab mission, and was not looking forward to a lull after Apollo–Soyuz. The Space Shuttle was an exciting, challenging and important programme – but it was no Apollo. A generation later I continue to maintain a keen interest in developments of humans in space with the Soviet/Russian space station era, the International Space Station, and the developing Chinese manned space programme. But I still recall those exciting, pioneering days of the late 1960s, and wonder just what could have been.

This book is the result of three decades of collecting lost dreams and old articles. It began by trying to ascertain what the Apollo 18, 19 and 20 missions would have done had they flown, and who the crews would have been – which of the astronauts had lost the Moon. Memories of the school service after the Apollo 1 pad fire and the dramatic incidents of Apollo 13 were recalled while reading numerous accounts of what happened in both accidents. But what those accounts hardly ever mention is what Grissom and his crew would have accomplished had they flown their mission, or what Lovell and Haise would have done at Fra Mauro.

From my interest in Skylab, I learned how the programme evolved from the idea of flying telescopes on the Service Module and then converting the Lunar Module to carry the instruments. Over the years, further research unearthed more details of what was involved in the Apollo Applications Program, both in Earth orbit and on or around the Moon. More amazing were the details of plans to use Apollo/Saturn hardware to send astronauts on a flight around Mars, 'Apollo 8-style', leading to the first landings on the Red Planet.

This book, therefore, is a recollection of the Apollo missions that never happened. Although discussion is presented on how they evolved (and how they were eventually cancelled or were unable to complete the mission assigned to them), this is not intended as a definitive history of unrealised Apollo missions. Indeed, it is doubtful whether such a work could be written. What is intended here is a supplementary account of the story of Apollo, and of missions that are no longer lost or forgotten, providing an insight into what the Apollo program could have accomplished in addition to the achievements for which it will always be remembered.

David J. Shayler   December 2001
West Midlands, England
www.astroinfoservice.co.uk

Acknowledgements

This book is the result of more than thirty years' interest in studying what the Apollo programme and hardware offered but never delivered. With such a long-term project, a great number of key people offered their assistance, cooperation, comments, suggestions, criticisms and guidance.

The origins of this book began in a series of articles written in the late 1970s for the British Interplanetary Society's magazine *Spaceflight* – articles which examined astronaut assignments in the early American manned space programmes. In particular, what could have been the outcome of later missions had the Apollo 1 crew flown their mission, if Apollo 13 had landed, and if the missions of Apollo 18–20 had been retained on the launch manifest? My thanks therefore go to the BIS – notably Ken Gatland and Len Carter – who encouraged such investigative articles during the 1970s, and to a trio of other 'space sleuths' who were the co-authors in this research: Curtis Peebles in the USA, and Phil Snowden and Andy Wilson in the UK. The staff of the BIS – notably Shirley Jones, Suszann Parry and Ben Jones – have also generously supported my research for this book by allowing me access to the Society's extensive library and photographic archive. This archive remains a valuable resource for UK-based space research.

For the past twenty years, Mike Cassutt in the US, Rex Hall in England, and Bert Vis in Holland have continued to share unique and important information on aspects of human spaceflight history. These three constitute the best 'space sleuthing' prime crew in the business, and I am honoured to be included in their circle of friends.

A book of this kind would not be possible without the generous help and assistance of the NASA History Department in Washington and at JSC, Houston. Since this volume mostly concerns missions and plans that never reached the launch pad, let alone be launched, access to the NASA files to uncover this intriguing story would not have been possible without the cooperation of Roger Launius and Lee Saegesser in Washington, Janet Kovacevich, Joey (Pellerin) Kuhlman at JSC, and Joan Ferry at Rice University. Information on KSC facilities was obtained from its History Office and Library staff: notably Ken Nail, Elaine Liston and Kay Grinter. Illustrative material has been obtained with the help of Mike Gentry and staff (notably Lisa Vasquez, Debbie Dodds, Jody Russell and Mary Wilkinson) at JSC, and Mary Persinger at KSC. Throughout this book, illustrations are from the Astro Info Service collection and, unless otherwise stated, are reproduced courtesy of NASA.

Two individuals who have conducted extensive personal research into aspects of this story of unflown Apollo missions are John Charles of NASA JSC, who provided invaluable information on the plans for Apollo 1 and Apollo 2, and writer David Portree, who has extensively researched the often frustrated efforts of fifty years of lunar and martian manned exploration, as detailed on his web site *Romance to Reality* and in his publication *Humans to Mars* (see Bibliography). Also, Lois Lovisolo supplied valuable information from the Grumman archives on designs for adapting the original LM Apollo lunar landing configuration for other objectives.

Several other key figures played prominent roles in supporting my research: notably former NASA astronauts Jerry Carr, Fred Haise and Bill Pogue (who would have been the prime crew of Apollo 19), Jack Lousma and Paul Weitz (often tipped as LMP and CMP respectively for Apollo 20). And, of course, my sincere thanks are due to Don Lind – who was expecting to be the second scientist on the Moon – for his help, and for writing the Foreword for this book.

Sincere thanks are again extended to my brother Mike, who continues to provide excellent initial reading and editing skills – and yet still talks to me after each exhausting project; to my project editor Bob Marriott who, besides editing the text, once again devoted many hours of his 'quality ale time' to scanning, processing and cleaning very old images and presentation illustrations to the standard reproduced here; to Arthur and Tina Foulser and staff at BookEns, for their typesetting skills; to Jim Wilkie, for undertaking the challenge of interpreting photographs to design a cover depicting real images of Apollo activities for events that never occurred; to Clive Horwood, Chairman of Praxis who, along with his staff, consistently provide enthusiasm and encouragement for each of my book projects; and to David Anderson, Springer's Northern Europe Sales Director, and all the members of the Springer promotion, sales and marketing teams for producing such fine end results and for unfailing support after publication.

List of illustrations and tables

Block I missions

Small steps

Lunar logistics

Exploration planning

Changing the mission

Beyond the Moon

Tables

Front cover
The main image is taken from the 1971 Apollo 15 mission to the Hadley Apennine landing site, near the original primary landing site intended for the cancelled Apollo 19 mission. In the foreground photograph (taken during training for Apollo 13), 'Apollo 19 Commander' Fred Haise carries an ALSEP experiment package in a scene that could have taken place during the Apollo 19 mission. Had this been the cancelled mission, then the astronaut saluting the flag would have been Apollo 19 LMP Jerry Carr, and not Apollo 15 LMP Jim Irwin.

Back cover
A frame from a movie film of the first walk on the Moon in 1969, taken through the LM ascent stage window, and showing Armstrong and Aldrin deploying the first flag on the lunar surface. This was planned as the first of at least ten such events during the Apollo programme, but there were only six deployments. The effect of this photograph is of a distant yet familiar memory of the triumph of Apollo. In 2002, thirty years after the end of the Apollo programme, we are no closer to planning the next landing or to identifying from which country or organisation will originate the thirteenth human to walk on the Moon.

Acronyms, abbreviations and notes

AP   Apollo Applications Program
AES   Apollo Extension System
ALS   Apollo (lunar) Landing Site
ALLS   Advanced Lunar Logistics System
ALSEP   Apollo Lunar Surface Experiment Package
ATM   Apollo Telescope Mount
BP   Boilerplate
CapCom   Capsule Communicator
CB   Astronaut Office Directorate Code
CDR   Commander
CM   Command Module
CMP   Command Module Pilot
CSM   Command and Service Module
EASEP   Early Apollo Scientific Experiment Package
EOR   Earth Orbital Rendezvous mode
EVA   Extravehicular Activity (space walk)
FD   Flight Director *or* Flight Day
GET   Ground Elapsed Time
IVA   Intravehicular Activity
KSC   Kennedy Space Center, Florida
LBNP   Lower Body Negative Pressure (device)
LC   Launch Complex
LEC   Lunar Equipment Conveyor
LES   Launch Escape System
LFV   Lunar Flying Vehicle
LH   Liquid Hydrogen
LM   Lunar Module
LMP   Lunar Module Pilot
LOX   Liquid Oxygen
LOR   Lunar Orbital Rendezvous
LRV   Lunar Roving Vehicle

LSS	Life Support System
LSSM	Local Scientific Survey Module
MMM	Manned Mars Mission
MCC	Mission Control Center (Houston)
MESA	Modular Equipment Stowage Assembly
MET	Modular Equipment Transporter
MEV	Mars Excursion Vehicle
MOLAB	Mobile Laboratory
MOR	Mars Orbital Rendezvous
MSC	Manned Spacecraft Center, Houston, Texas
MSFC	Marshall Spaceflight Center, Huntsville, Alabama
NAA	North American Aviation
NERVA	Nuclear Engine for Rocket Vehicle Application
OMSF	Office of Manned Spaceflight
OWS	Orbital Workshop
PLSS	Portable life Support System
RCS	Reaction Control System
SIM	Scientific Instrument Module
SLA	Spacecraft LM Adapter
SM	Service Module
SPS	Service Propulsion System
VAB	Vehicle Assembly Building

Saturn stages

Saturn	1st stage	Engine	2nd stage	Engine	3rd stage	Engine
1	S-I	H-1 (8)	S-IV	Rl10 (6)	–	–
1B	S-1B	H-1 (8)	S-IVB	J-2 (1)	–	–
V	I-C	F-1 (5)	S-II	J-2 (5)	S-IVB	J-1 (1)

SA	Saturn–Apollo development flights
AS	Apollo–Saturn operational (?) flights

Prologue

In December 1960, an *ad hoc* scientific committee reported to President Eisenhower on the goals, missions and costs of sustaining a man-in-space programme. The report reviewed the status of the Project Mercury series (in preparation for the first manned sub-orbital flights), the capabilities of the Saturn booster and the proposed Apollo spacecraft, and possible future directions. The Report stated that the main purpose of Apollo was to gain experience in manned spaceflight and, by utilising the larger Saturn rockets, an 'orbiting laboratory' could provide valuable experience for future steps. The report suggested that although 'the Apollo program in itself does not reach what might be considered to be the next major goal in manned spaceflight, *i.e.* manned landing on the Moon... it does, however, appear to represent a logical approach to that goal.' Such an approach might have been the successful completion of more elliptical orbits carrying it further into space, which could have culminated in a manned circumlunar flight around 1970.

During the first months of 1961 a new President, John F. Kennedy, was in the Oval Office reviewing this and other reports on his national space policy and programmes. Then, following the Soviet achievement of placing the first man in space, on 12 April 1961, week-long discussions were held in the White House, whereupon the President issued a memo, dated 20 April, to Vice President Lyndon B. Johnson (who was also the Chairman of the Space Council) to ask 'other responsible officials' if America had a chance of beating the Soviets in space by orbiting a space laboratory, by sending astronauts to the Moon, or with any other suitable programme. The subsequent recommendations strongly indicated that the only clear option that met the criteria set out by the President for dramatic results, and which the US could win, was an all-out effort to put Americans on the Moon.

During space review meetings early in May 1961, there appeared to be tension between Johnson, who wanted to push that strong recommendation to Kennedy, and the NASA Administrator Jim Webb, who was not convinced that such a major commitment was technically feasible or politically sound. But the resulting document from Webb to the Vice President on 8 May put forward recommendations to change the national space effort to include the objective of 'landing a man on the Moon and returning him to Earth in the latter part of the current decade.' This document also stated that the initial flights of the 'Apollo multi-manned orbiting laboratory' would qualify both the spacecraft and manned flights around the Moon, prior to any actual attempt to land on the lunar surface.

On 25 May 1961, Kennedy delivered his historic 'Urgent National Needs' speech to the Joint Session of Congress, in which he committed America to placing a man on the Moon, and returning him safely to Earth, by 31 December 1969. Kennedy stated that no single space project in this period would be more exciting or impressive, or 'more important to the long range exploration of space,' and that none would be as difficult or expensive to accomplish. The speech also mentioned additional funds to develop a nuclear rocket, which could one day 'provide the means for even more exciting and ambitious explorations of space, perhaps beyond the Moon... to the very end of the Solar System itself.' Kennedy emphasised that money alone would not achieve the goal, and that the whole nation would need to answer the commitment. And if the effort wavered, then it would be better not to begin the journey at all. It was a journey that Kennedy would not see fulfilled, as he was assassinated on 22 November 1963, the mantle of leadership falling to Vice President Lyndon B. Johnson.

Five years later, and halfway through the decade, Jim Webb was becoming increasingly irritated by the White House (Johnson) and the Democrat-led Congress over the lack of financial support committed to the programme each year. Despite the fact that Kennedy's deadline was only little more than three years away, the funds to achieve that goal were far from certain. Indeed, for NASA, whose budget peaked in 1966, the future of the agency once Apollo had achieved the lunar goal was still very unclear. The 1967 budget request had already received a $712 million reduction, and there was the threat of a further $1 billion cut.

In attempting to establish the security of Apollo, the future of NASA and the American space programme, a frustrated Webb wrote to Johnson on 26 August 1966: 'Almost six years ago, when you urged me to accept the responsibilities which devolve upon the Administrator of NASA, I asked if my task would be to carry out a preconceived program or to figure out what needed to be done and do it. You said, 'the latter', and, of course, this was on the basis that President Kennedy would have to approve whatever you and I worked out. You will remember that in the sessions you had in 1961 with your advisors and Congressional Leaders, I was quite reluctant to undertake the responsibility of building a transportation system to the Moon, and that you almost drive me to make the recommendation which you sent on to President Kennedy.'[1]

Barely five months after Webb's letter was written, the Science Advisory Committee issued a report to the President entitled *The Space Programme in the Post Apollo Period*, published in January 1967. In stark contrast to Webb's difficulties in securing funding for the mainline Apollo programme, the Committee was tasked with identifying potential programmes (and budgets) for the 1970s.[2] In its conclusions and recommendations the report stated that 'with the termination of the basic Apollo project now only three years away, it has become urgent to begin planning for the post-Apollo space program', and asked the question: 'Where do we go in space from here?' One option, according to the Panel, was the selection of a single dominating project for NASA, such as a manned Mars fly-by in 1975 or a manned orbiter or lander on Mars in the late 1970s.

This new proposal for a national space commitment came at the time of the fatal Apollo 1 flash fire – well before any Apollo hardware had flown with a crew on board, let alone take anyone to the Moon. In the late 1960s, NASA was already struggling to find enough funded hardware to support the Apollo programme; and perhaps in a desire to ensure its future beyond 1970 it publicly supported the suggestions of the Advisory Committee to promote expanded Apollo operations in Earth orbit (Apollo Applications) and on the Moon, and the genesis of manned planetary missions a decade after Apollo had achieved its lunar goal.

By 1969, in the wake of Apollo 8's historic flight around the Moon during the previous Christmas, and in the build-up to sending the first crew to attempt the landing on the Moon that summer, NASA appeared to have a strong future. Its space infrastructure included committed manned lunar exploration, the creation of orbital space stations, regular access to space by a new vehicle called a space shuttle, a lunar base, and plans to reach Mars by 1980. All of this appeared very impressive – and expensive – to the public and members of Congress. Despite in-house discussions to identify the flight schedule for the remaining Apollo hardware, to define the configuration of the Saturn Orbital Workshop, and to present serious post-Apollo plans to Congress that would maintain the lead and momentum that Apollo had created, the reality was that the golden era of NASA was coming to a close.

This, then, is the setting for the lost and forgotten missions of Apollo: from the dreams and plans of space scientists and engineers to push the boundary of human and space exploration, to the limits of the budgetary reigns of Congress and the White House, while trying to convince the public of the benefits and value of continuing the programme. It is a story of what could have been, of changing directions and unclear goals after achieving the first landing on the Moon, and of the dedication and frustration of a talented group of individuals who, despite all adversities, still believed in NASA and the commitment to the manned exploration of space.

1   *Exploring the Unknown*. Volume 1: Organising for Exploration, Chapter 3, The Evolution of US Space Policy and Plans, by John M. Logsdon, 408–494; NASA SP-4407, 1995.
2   *The Space Program in the Post Apollo Period*, A Report of the President's Science Advisory Committee, prepared by the Joint Space Panels, The White House, February 1967. From the archives of the British Interplanetary Society, London.

Extending the capabilities

The huge cost of any programme of space exploration ensures that the designers, engineers, scientists, managers and those who provide the funds are always looking to achieve maximum returns for the investment.

On 29 July 1960, the Deputy Administrator of NASA, Hugh L. Dryden, announced plans of the follow-on programme to the one-man Mercury series, under the name of Apollo. He stated that the new series would include not just manned missions in Earth orbit, but also circumlunar missions before 1970 and the creation of a temporary orbital scientific research platform (known more commonly as a space station). In the decade of the 1970s there would be a programme of manned lunar landings, a permanent space station, and advanced hardware, based on the Apollo spacecraft design and the Saturn and Nova family of launch vehicles. It was also expected that this technology would lead to manned planetary expeditions sending the first Americans to Mars. Dryden's announcement appeared a full year before President Kennedy publicly committed America to a manned lunar landing by the end of 1969, and before any human, let alone any American, had journeyed into space.

DEFINING THE HARDWARE

In 1960 the design of the 'Apollo' was still under discussion, and no prime contractor had been selected. As the programme developed through 1961, the configuration of the spacecraft evolved into a three-man conical command spacecraft mounted on top of an unmanned cylindrical Equipment-Propulsion Module, which attached to the top of the upper stage of the Saturn booster. This, of course, soon became known as the Apollo Command and Service Module (CSM), the parent spacecraft of the Apollo system. The contract for construction of the CSM was awarded to the Space and Information Systems Division of North American Aviation Inc on 28 November 1961.

By then, President Kennedy had delivered his historic speech to Congress, and Apollo became the programme to ensure that the lunar challenge was met. In order

An artist's impression depicting the main components that were to support actual and planned Apollo operations in Earth orbit and to the Moon, as well as providing the genesis for studies towards initial manned spaceflight to Mars. At left is the Saturn V S-IVB third stage that would form the basis of the Skylab orbital workshop. At right the Apollo CSM retrieves the LM for a mission to the Moon. The LM design was to feature in extended lunar and Earth orbital exploration plans, and the CSM was planned as the primary space transportation system and manned logistics spacecraft for US astronauts from the 1960s to the 1980s. (Courtesy NASA via the British Interplanetary Society.)

to achieve its focused role of landing Americans on the Moon and returning them safely to Earth by 1970, the Apollo 'programme' first needed to accrue experience in extended spaceflight, the techniques of orbital rendezvous, precision launching, and working outside the spacecraft in pressure suits. To supply this much needed experience, while still developing Apollo hardware, NASA and McDonnell Douglas, primary contractor of the Mercury spacecraft, had designed a two-man Mercury Mark II design. On 7 December 1961 this became the official second American manned spaceflight programme, to be completed after Mercury and prior to any Apollo manned missions. On 3 January 1962, Mercury Mark II officially became known as Gemini.[1]

On 11 July 1962, NASA decided that the most suitable method of sending Apollo to the Moon and achieving the manned landing by 1970 was Lunar Orbital Rendezvous (LOR). This method was chosen over that of assembling the lunar spacecraft in Earth orbit (Earth Orbital Rendezvous), and relying on multiple launch

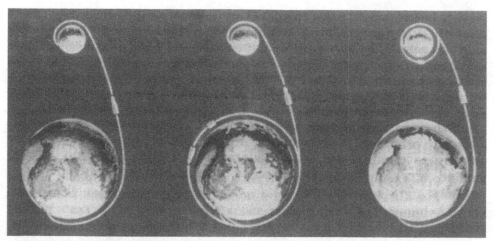

The three methods short-listed to send astronauts to the Moon by the end of the 1960s. (*Left*) Direct Ascent, (*centre*) Earth Orbital Rendezvous (EOR), (*right*) Lunar Orbital Rendezvous (LOR) – the option finally chosen for Apollo.

techniques or the more direct approach (Direct Ascent) with rockets so large that it would probably take another decade to develop them. Under the LOR profile, all elements of the Apollo spacecraft would be launched into Earth orbit by a single three-stage Saturn, and after an orbit or two of systems checks would be sent on its way towards the Moon by re-starting the third-stage engine. The trip to and from the Moon would be completed in the Apollo CSM, but the 'excursion' from lunar orbit to the surface and back would be achieved by two of the crew in a separate landing module. The Lunar Excursion Module (LEM) became the second major spacecraft in the Apollo programme, and on 7 November 1962 Grumman Aircraft Engineering Corporation was selected to build the lunar landing vehicle, though contracts still had to be negotiated and were not signed until March 1963. Its name became shortened to Lunar Module (LM), but throughout the programme it was still pronounced 'LEM'.

Building blocks
Due to the obvious complexity of the lunar programme it was early on decided to follow a multi-phase development approach, with each phase serving as a qualification before progressing to the next step. Following unmanned launch vehicle and boilerplate (mock-up) spacecraft flight tests, the three main manned phases would begin initially with Earth orbital flights using the smaller Saturn launchers, to be followed by a series of high-apogee, circumlunar and lunar-orbital flight tests using the larger Saturn V to further develop the spacecraft and operational techniques. The programme would conclude with the manned lunar landing mission using the Saturn V.

The selection of LOR in 1962 altered the approach to preparing for the lunar mission that now included a second spacecraft, the LM, and deletion of the Nova vehicle. In addition a change from land to water recovery resulted in a further complexity to the design of the spacecraft. In 1964 a programme definition study was initiated in order to identify the basic requirements for the CSM to fly a lunar mission which would result in a version of the parent spacecraft that could be tested and used in Earth orbit without an LM, and thus speed up qualification of basic CSM systems and procedures prior to introducing the more advanced lunar mission CSM and LM combination. Therefore a multi-phased programme of spacecraft development evolved for Apollo.

Mock-ups   These were usually some of the first 'spacecraft' constructed by NASA and its contractors to establish the design, dimensions and general layout of what was proposed to evolve into the final version. The mock-ups included full and scale models constructed of wood or metal, and included a variety of test rigs, structural test articles, training devices, system evaluation devices, and fit and function units.

Boilerplates (BP)   These were vehicles used in a variety of test and qualification programmes that were pre-production spacecraft, similar to the full production versions in the size, shape and centre of gravity, but lacking many of the flight systems and incapable of supporting a human crew in space. These boilerplates were used for a wide variety of research and development test programmes as launch vehicle payloads, spacecraft recovery systems, vibration and stability tests, and system verification. Several boilerplates were employed in a series of unmanned Pad Abort tests. These simulated emergency ejections of the spacecraft by means of the launch escape tower were launched by the Little Joe II (340,000-lb thrust) rocket. This was powered by four *Recruit* and two *Algol* rocket motors, and could boost the capsules several thousand feet into the air from the White Sands Missile Range (WSMR) in New Mexico. These tests duplicated conditions of a CM leaving a rogue Saturn at speed and altitude in tests of the abort and parachute recovery systems. The Pad Abort tests were as follows:

1963 November 7   PA-1   WSMR   LES   BP-6
Pad Abort test of the launch escape system was completed successfully. The flight lasted 165.1 sec, and resulted in a successful demonstration of ballistic recovery of a boilerplate CM. The only small problem was that exhaust from the escape system coated the spacecraft.

1964 May 13   A-001   WSMR   Little Joe II   BP-12
The first flight of an Apollo spacecraft on a Little Joe II vehicle. The test was successful except for one parachute failure, though the two remaining parachutes allowed spacecraft recovery within human tolerances at 26 fps instead of the planned 24 fps. Flight time was 5 min 50 sec, with the spacecraft landing 22,400 feet downrange.

1964 December 8   A-002   WSMR   Little Joe II   BP-23
This test carried canards on the front of the LES to help orientate the blunt end of

Boilerplate CM No.22 atop a Little Joe II launch vehicle for an escape system test at White Sands Missile Range, New Mexico, during May 1965.

the CM forward after engine burn-out, and a boost protective cover on the CM to present deposits of soot on the spacecraft, especially the crew observation windows. It also tested the abort capability under maximum dynamic pressure (Max Q) regions to simulated limits of the emergency detection sub-system.

1965 May 19   A-003   WSMR   Little Joe II   BP-22
A planned high-altitude Abort Test. Approximately 25 sec after launch the LJ-II completed an unprogrammed and violent roll, resulting in the structural break-up of the launcher. However, the emergency detection system performed perfectly, and initiated the LES to pull the spacecraft clear of the disintegrating Little Joe II. Although the prime objective of a high-altitude Abort Test was not met, the LES demonstrated its effectiveness in an emergency situation.

1965 June 29   PA-2   WSMR   Little Joe II   BP-23A
A successful test of the LES carrying the CM to 5,000 ft above the pad. Similar to the PA-1 profile with a successful operation of the Earth Landing System (ELS – the parachutes).

1966 January 20   A-004   WSMR   Little Joe II   CM-002 (Block I)
The sixth and final test of the CM abort development programme, and the first to use a Block I production spacecraft. The spacecraft was boosted to a 15-mile altitude in a demonstration of a launch abort using a power-on spacecraft configuration. The launch vehicle began to tumble at altitude, and the escape system, sensing this motion, activated the LES to pull the CM clear and initiated the ELS sequence.

Block I A series of limited-production light-weight spacecraft designed to qualify systems and procedures initially in the unmanned role but later with a flight crew. These spacecraft were limited to Earth orbital activity without the LM, and were launched by the smaller members of the Saturn rocket family.

Block II The lunar mission class CSM manned spacecraft, which during the course of the Apollo programme underwent a number of modifications (discussed later) but which could be launched by the larger Saturns on a variety of missions.

Block III This configuration was intended for operations in what became the Apollo Applications Program (see p. 47) but which was not fully developed as a separate series.

Powering Apollo
To launch the Apollo hardware into Earth orbit and out to the Moon, the NASA Marshall Spaceflight Center in Huntsville, Alabama, under the direction of the former German V-2 engineer Dr Wernher von Braun, had, since the late 1950s, evolved a family of major launch vehicles under the US Army's Advanced Research Projects Agency (ARPA). Originally called Juno, these launch vehicles would become the workhorse of the Apollo programme for the ensuing fifteen years.[2]

On 3 February 1959 ARPA renamed the project Saturn, and this was followed by NASA assuming the technical direction of the programme on 18 November that year. Formal approval of the programme was received on 18 January 1960, with the Saturn being given the highest national priority for development. The official transfer of the Saturn programme from ARPA to NASA occurred on 16 March 1960.[3]

The Saturn launch vehicle relied on the technology of cryogenics – the use of liquefied (below approximately –240° F – also termed 'super-cold') gases in various propellant combinations. This required less volume to store the propellants than it did in the conventional gaseous state, which would have increased the size and weight of launch vehicle to beyond the capacity of the engines then available or under development to lift the rocket off the ground. All the first stage engines of the Saturn family used a non-cryogenic fuel derived from kerosene and designated RP-1, with the upper stages using liquid hydrogen (LH). Liquid oxygen (LOX) was used in all engines as the oxidisers.

Five types of main engine were used on the different configurations of Saturn rockets proposed, and the grouping of them together (clustering) offered an increase in the combined thrust for each stage. The H-1 engine, developed from Jupiter and

Redstone missile technology, used LOX and RP-1, and was capable of delivering approximately 200,000 lbs of thrust at sea level, while the RL-10 engine used LOX and LH and delivered about 15,000 lbs at altitude. The LR-119, derived from the Centaur upper stage engine, also used LH and LOX to produce about 17,500 lbs of thrust at altitude, as did the J-2 engine, which could provide a thrust (at altitude) of approximately 200,000 lbs. Finally, the F-1 (LOX and RP-1) had a sea-level thrust of approximately 1.5 million lbs.

During the early development of the series there were several configurations proposed, and this led to some confusion in the designations assigned to each type of design. Essentially, the Saturn 'family' included the following variants:

Saturn C-1 The first stage (designated S-1) consisted of eight H-1 engines clustered to provide 1.5 million pounds of lift-off thrust; on the S-IV second stage were four LR-119 engines, offering about 80,000 lbs of combined thrust. As a result of development problems with the LR-119 engine configuration, in March 1961 NASA authorised a change in the configuration on the S-IV second stage to a six-engine

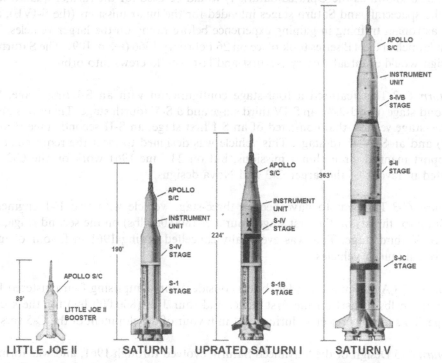

LITTLE JOE II   SATURN I   UPRATED SATURN I   SATURN V

Comparative sizes of the four launch vehicles of the US Apollo programme. (*Left–right*) Little Joe II, used for ballistic tests of the emergency escape tower; the Saturn 1, used for the first sub-orbital and orbital flights of Apollo hardware; the Uprated Saturn 1 (later termed the Saturn 1B), used for the first manned Apollo Earth orbital flights and later for Skylab and the ASTP; and the Saturn V (or Saturn 5), designed for the manned lunar missions, and used to launch the unmanned Skylab workshop.

design using a less powerful unit (the RL-10), which delivered 90,000 lbs thrust capability. There was originally a third stage planned for the C-1, designated the S-V (two LR-119s) and totalling 40,000 lbs thrust, but in June 1961 the need for a three-stage C-1 was dropped, as this requirement would be more efficiently met on later vehicles. The first C-1 launch, using a dummy second stage, occurred on 27 October 1961, and a series of ten launches was conducted for qualification tests of the Apollo CSM (see p. 104). In February 1963 the C-1 was renamed Saturn 1. The programme originally included several manned launches, but in October of that year the manned mission requirement was dropped from the programme (see p. 102), resulting in the deletion of six Saturn 1 vehicles.

Saturn C-1B The S-IB had eight clustered and improved (uprated) H-1 engines (1.6 million lbs thrust) and an S-IVB second stage with one J-2 engine (200,000 lbs thrust). On 11 July 1962 NASA announced a requirement for a two-stage launch vehicle to conduct Earth orbital missions of both unmanned and manned Apollo spacecraft. In February 1963 the C-1B was renamed Saturn 1B. This vehicle (which during 1966–67 was also known as the Uprated Saturn 1) would be used for the further qualification of the spacecraft and Saturn stages intended for the lunar missions (the S-IVB), and for astronaut training in gaining experience before riding on the larger vehicles. The first launch in the 1B series took place on 26 February 1966 (see p. 109). The Saturn 1B design would eventually carry the first and last Apollo crews into orbit

Saturn C-2 This featured a four-stage configuration with an S-I first stage, S-II second stage (two J-2s?), an S-IV third stage and a S-V fourth stage. There was also a three-stage version that consisted of an S-I first stage, an S-II second stage (four J-2s?) and an S-IV third stage. This vehicle was designed to meet the requirement to support manned circumlunar missions, but on 23 June 1961 work on the C-2 was halted in favour of the larger C-3 and Nova designs.

Saturn C-3 This was to have been a three-stage vehicle with two F-1 engines (3 million lbs thrust) on the first stage, four J-2s (800,000 lbs) on the second stage, and an S-IV third stage. This was eventually cancelled during 1961 in favour of more powerful launch vehicles.

Saturn C-4 A design also under (brief) consideration comprising four clustered F-1s (6 million lbs thrust) on the first stage, and four J-2s (800,000 lbs) for the second stage. This too was rejected during 1961 in favour of development of the C-5 version.

Saturn C-5 Design of the 'advanced Saturn' evolved through 1961, and was formally approved on 25 January 1962 as a three-stage launch vehicle for the Apollo lunar programme. The first S-IC stage featured five F-1s (7.5 million pounds), an S-II second stage with five J-2s (1 million pounds), and an S-IVB third stage. Estimates forecast that dual Saturn V launches could meet the Earth Orbital Rendezvous mode, if selected, but with the decision to adopt the LOR mode a single Saturn V launch would adequately meet that goal. In February 1963 the C-5 was renamed the Saturn

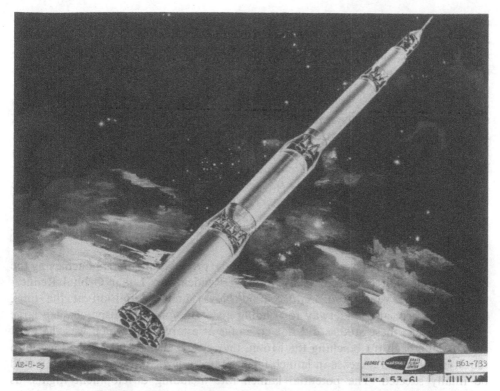

A four-stage design for the Nova/Apollo concept proposed for Direct Ascent flights to the Moon. In this 1961 concept there were to be eight engines in the first stage, eight engines in the second stage, four engines in the third stage, and six engines in the fourth stage, with the Apollo CM and Earth return lunAr landing stage and escape tower completing the huge vehicle. (Courtesy NASA via the British Interplanetary Society.)

V (also called Saturn 5). A test vehicle (Saturn Apollo-500 series SA-500-F) was used for launch pad qualification in 1966, leading to the first unmanned launch on 9 November 1967 (see p. 168). The final Saturn V was launched (unmanned and with two stages supporting the Skylab space station) on 14 May 1973.

Saturn C-8 This was the largest Saturn design proposed, and was in the Nova class of a 'super-booster'. The C-8 configuration featured eight F-1s (12 million lbs) on the first stage, a second stage with five J-2s (1.6 million lbs), and a third stage with a single J-2 (200,000 lbs). On paper this vehicle had the capacity to perform the Direct Ascent approach to the Moon, but during 1962 it was rejected in favour of the C-5.

Nova This was as an early proposal, frequently changed for a Direct Ascent launch vehicle using Saturn-based technology with up to ten or twelve F-1 engines clustered in its first stage. Some of the Nova designs also featured nuclear-powered engines for the upper stage. It was developed at the same time as the C-5, but NASA recognised that a conflict of interest could develop with the decision to adapt the Saturn or Nova for the Apollo programme. Study contracts were issued in July 1962, but the selection of the Saturn C-5 for LOR the same month eliminated the immediate requirement for any further development of the Nova. Some conceptual studies continued for the next two years under the NASA Future Projects Office, with the idea of a post-Saturn V, to be phased in during the early 1970s and to be used to support manned planetary missions. However, these designs were never finalised, and no hardware was ever developed. From the various configurations proposed, two options suggested were a first-stage cluster of up to ten F-1s delivering 15 million lbs, a second stage with four M-1 engines (LOX/LH) and a third stage with a single J-2, or by clustering twelve F-1s (a staggering 18 million pounds – more than twice that of the Saturn V) in the first stage, ten J-2s (2 million lbs thrust) in the second stage, and a nuclear engine in the third stage.

Handling the hardware

In 1961, as the studies continued into each of the proposed methods of achieving the lunar landing (Direct Ascent, Earth Orbital Rendezvous, Lunar Orbital Rendezvous) with the versions of Saturn (and Nova) under definition at the time, discussions were also underway for finding a suitable place to dispatch such large vehicles. While these studies did not go as far as establishing which was the most suitable approach for reaching the Moon, they clearly demonstrated the long-term requirement for an increase in launch facilities to support whichever method would be finally selected. The potential launch rate including development flights, as indicated in 1961, estimated that as many as 28 C-1 and 38 C-4 flights would be required to support a Lunar Orbital Rendezvous (LOR) approach. For Earth Orbital Rendezvous (EOR) it was estimated that 32 C-1 and 53 C-4 launch would be needed, and a programme of 22 C-1 and 38 Nova launches would be required if the Direct Ascent approach were to be adopted.

Additional studies and discussions for the lunar launch complex focused on safety issues such as the minimal ground safety zones for clearing personnel in the event of pad explosions, protection from blast shock waves or falling debris including spent rocket stages, launch abort flight trajectories, and crew recovery and the boundary where it was safe to witness a launch without any damage to hearing.

The location chosen to construct the facilities to support the Apollo launch vehicle programme centred on an area to the north of Cape Canaveral, home of the Air Force Missile Test Center on the Atlantic coastline of Florida. In 1958 the DOD Missile Firing Laboratory (from 1960, NASA's Launch Operations Directorate) following several weeks of surveying, discussion with the Air Force and evaluation of other sites, selected an area half a mile north of the existing Titan I Launch Complex (LC 20) site for the first Saturn launch complex. Designated LC 34, this was to become the launch site for the first unmanned Saturn 1 and Saturn 1B missions, and the site from where the first astronauts to fly Apollo were planned to begin their mission.

The second Apollo launch pad constructed was LC 37, located about three-quarters of a mile north of LC 34. The decision to construct a second launch site

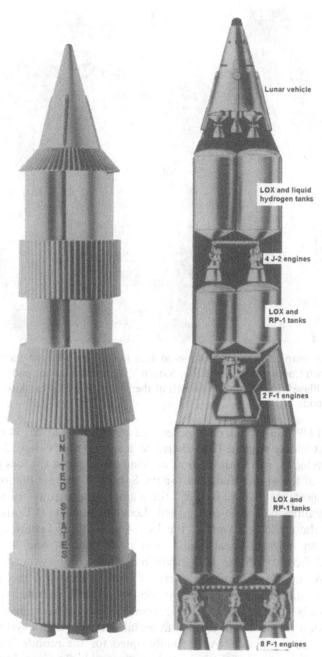

NASA pre-1961 artwork, revealing an early design of the Nova launch vehicle towering over 350 feet. The accompanying data stated a capacity of placing 300,000 lbs in low Earth orbit or 100,000 lbs on a trajectory to the Moon, with a 50,000-lb capability for an Earth return lunar capability. There were also plans for four or five stages that were dependent on the mission to be performed.

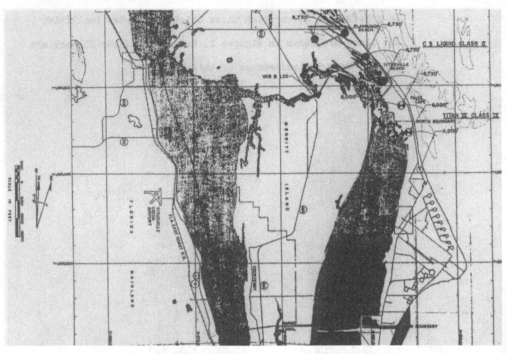

An early 1960s map showing the proposed location of the Vertical (later Vehicle) Building, Launch Control Center and four Saturn V/Nova class launch pads around the Titusville and Playalinda beach areas north of the 'Cape' launch area. Also visible are two Titan III pads.

resulted from a 1959 study which determined that an explosion at LC 34 would render it unusable for at least a year. In 1960, therefore, a second pad, designed for Saturn 1 and Saturn 1B vehicles, was authorised as an insurance against the loss of LC 34.

The creation of the launch facilities for the Saturn V became one of the largest construction projects in history. By 1962 the major considerations of where to site the Saturn V facilities had been established. Located on Merritt Island in an area immediately to the north and west of the LC 34 and LC 37 facilities, NASA had negotiated 130 sq miles of scrub land, most of it marshy, and 90 sq miles of submerged land from the State of Florida on which to construct the Saturn V facilities. NASA had considered a variety of launch sites and methods for the preparation and launching of the Moon rocket which had included off-shore and remote-area launch facilities. In addition, several methods of horizontal and vertical vehicle preparation featured rail, barge and mobile transportation systems from the preparation area to the pads. NASA finally opted for the mobile launch facility, resulting in the construction of the huge Vehicle Assembly Building, Mobile Launch Platforms, Crawler-way, Crawler Transporter system, and two launch pads (Pad A and Pad B). This huge area was designated Launch Complex 39, and was located on what was initially designated the Merritt Island Launch Area (MILA). From November 1963, in honour of the assassinated President the area was renamed the

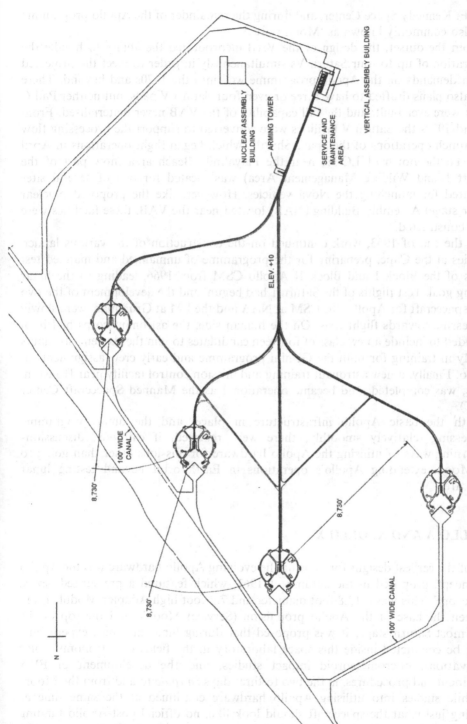

An early crawler access system from the VAB to the planned four pads of LC 39. Note the single-file six-bay VAB and the nearby Nuclear Assembly Building. A canal system also serviced each pad.

John F. Kennedy Space Center, and during the remainder of the Apollo programme was also commonly known as 'Moon port'.[4]

From the outset, the design of the VAB incorporated the ability to handle the preparation of up to four Saturn Vs simultaneously in order to meet the projected launch demands on the Apollo programme well into the 1970s and beyond. There were also plans drafted to have three or even four Saturn V pads, but neither Pad C nor D were ever built, and the full capability of the VAB never materialised. From the mid-1970s the Saturn V facilities were converted to support the processing flow and launch operations of the Space Shuttle, which began flight operations in April 1981. To the north of LC-39, near the Playalinda Beach area (now part of the Merritt Island Wildlife Management Area) was located for one of several sites evaluated for launching the Nova vehicles. However, like the proposed Nuclear (upper stage) Assembly Building (NAB) located near the VAB, these facilities were never constructed.[5]

By the end of 1963, work continued on the construction of the various launch facilities at the Cape, preparing for the programme of unmanned and manned test flights of the Block I and Block II Apollo CSM from 1966, leading to the lunar landing goal. Test flights of the Saturn 1 had begun, and the development of the two main spacecraft for Apollo, the CSM at NAA and the LM at Grumman, were slowly progressing towards flight tests. On the human side, the astronaut corps had been expanded to include a new class of fourteen candidates to join the sixteen astronauts already in training for both the Gemini programme and early crew assignments in Apollo. Finally, a new astronaut training and mission control facility near Houston, Texas, was completed, and became operational as the Manned Spacecraft Center (MSC).

With the basic Apollo infrastructure in place, and the lunar programme progressing relatively smoothly, there were renewed, if cautious, discussions concerning ways of utilising the Apollo hardware for missions other than going to the Moon, extending Apollo's operations in Earth orbit beyond testing lunar hardware.

APOLLO A AND APOLLO X

One of the earliest designs for utilising the evolving Apollo hardware was the Apollo A concept, proposed in the autumn of 1961, which featured a pressurised 'space laboratory'. This was a 12.8-foot diameter and 7.9-foot high Adapter Module fitted between the base of the Apollo propulsion (Service) Module and the top of the uppermost Saturn stage. It was proposed that during lunar missions, experiments could be conducted inside this space laboratory in the fields of astronomy, solar observations, micrometeoroid impact studies, and the development of EVA equipment and procedures, in the two to three days *en route* to and from the Moon.

While studies into utilising Apollo hardware continued at the same time as defining just what the spacecraft should look like, no official post-Apollo funding had actually been authorised. Mindful of depleting Apollo funds, NASA was careful not to detract from the national goal of landing a man on the Moon before 1970, but was also keen to develop additional programmes evolved from the hardware designed to achieve the lunar mission.

Throughout the 1960s the space agency carefully balanced the need to secure funding for Apollo to reach the Moon along with encouragement to develop other 'extended' objectives under the generic name of Apollo, so as not to create a brand new programme that might result in restrictions on the main lunar goal. The further complications of an escalating war in south-east Asia, domestic welfare issues, and a growing awareness of global environmental issues, all meant that promoting any new

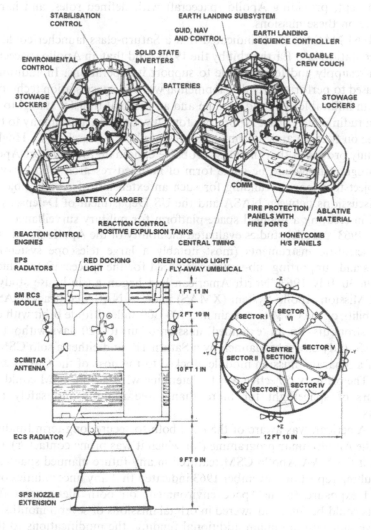

Cutaways of the Apollo Block II CM and SM. Note the vacant bay in Sector I of the Service Module used to locate scientific instruments.

programme would pose serious threats to funding for the main Apollo programme. So although non-lunar missions began to be studied in-house at NASA and with aerospace contractors, most would feature Apollo-derived hardware and even identification under the Apollo programme name. A space station plan could be developed, but it would have to be an application for Apollo hardware already developed and provided for in the budget.

Initially, these Apollo design studies were termed Apollo X (experimental), but this also became the term for the series of Apollo missions designed to 'extend' the scope of the lunar programme. Thus Apollo X also stood for Apollo 'Extended' missions. Many of these paper studies involved the early evolution of the space station concept, providing Apollo spacecraft with defined roles, and hardware to incorporate on these missions.

Early 1963 NASA reports indicated that a Saturn-class launcher could deliver a large space station into Earth orbit by the 1970s, and that an Apollo spacecraft could serve as a resupply and ferry vehicle to support its operations. In addition, Apollo could be used to perform early developmental work in space station techniques, such as fitting an inflatable spheroid module and an internal transfer tunnel to the CSM to create a rudimentary space laboratory for a few days. In a testimony to the House Committee on Science and Astronautics presented on 4 March 1963, Hugh Dryden had carefully proposed that the most probable candidate for any post-Apollo lunar landing programme would be some form of manned research laboratory in Earth orbit, subject to adequate funding for such an extension of Apollo objectives and further discussions between NASA and the US Department of Defense, which was interested in creating a manned space platform for military surveillance.

During 1963, parallel studies evaluated the ability of the Apollo parent spacecraft to carry scientific instruments (most notably a large telescope system) and the techniques and supporting sub-systems required for the spacecraft to achieve useful results. On 30 July 1963, North American had begun an in-house study, entitled Extended Mission Apollo System (XMAS) under a NASA contract (NAS 9-1963), into the ability of the CSM to remain in a 100–300-mile altitude orbit with either two or three astronauts, on an 'extended' mission of up to 120 days without resupply. The XMAS study envisaged launch by a Saturn 1B, and either a solo CSM flight or the use of a separate 'mission module' docked to the nose of the CM as extra living quarters. The primary objective was to determine whether the CSM could withstand the rigours of spaceflight for more than three months and safely return the astronauts.

North American was aware of the need both to secure long-term funding for the CSM in the Apollo lunar programme (for which it was prime contactor) and also to ensure that the NAA Apollo CSM featured in any future manned spacecraft plans. Their resulting report in November 1963 indicated that any uncertainties over such a prolonged exposure to the space environment on both the spacecraft and the astronauts could be finally answered in orbital missions of several months' duration. To achieve this with minimum additional funding, the modifications to the Apollo CSM, in their opinion, would be most the suitable, as opposed to the difficult, lengthy and costly development of totally new hardware. The report proposed a range of Apollo system modifications from the basic CSM to separate laboratory and research modules with self-contained life support and sub-systems. Based on the 'favourable' results from the 100-day study, it was suggested that each of the proposed concepts were both technically sound and fully achievable with minimal costs and within attainable time frames.

By the autumn of 1963, NASA was progressing on the development of the hardware for the lunar mission. The same technology was also being applied to a range of science-based objectives aimed at the development of an Apollo-based space station, to support and complement that of the USAF-developed Manned Orbiting Laboratory (which had been announced on 10 December 1963 upon the cancellation of the X-20 space-plane). Two weeks later, on 26 December, the Director of the Marshall Space Flight Center, (MSFC), Wernher von Braun, in a letter to Apollo Spacecraft Manager Joseph F. Shea, had suggested a series of Apollo missions that extended the basic Apollo lunar system, as well as expand lunar exploration to include a programme of short-term lunar shelters under an Integrated Lunar Exploration System. Although not initially receiving great support, this proposal was the basis for several detailed studies for using Apollo hardware both for an expanded lunar programme after the initial landings and for the creation of an Apollo-based space station.[6]

Development of Apollo X

To explore the possibility of using Apollo technology beyond the initial lunar effort, in April 1964 NASA created the Apollo Logistics Support System Office. Their studies included the use of both the CSM and LM for other objectives in Earth orbit, deep space and on the Moon. Three areas emerged that considered the expansion of the lunar surface and lunar orbital scope of Apollo beyond that of the initial goal of landing men on the surface and ensuring a safe return. First, the use of the spacecraft for a series of independent scientific research flights; second, the evolution of Apollo and Saturn hardware to create a preliminary manned space station; and third, the creation of a lunar logistics system to support extended lunar exploration operations after the initial landings had been accomplished. At no point would Apollo's primary lunar landing target be compromised, and so all other programmes would be secondary in long-term planning. They were also, of course, not deemed 'new-start' programmes, but further applications of the Apollo–Saturn system, and would be subject to funding (or as it would transpire, lack of funding) from the Apollo budget.

The lunar and space station applications of basic Apollo hardware are discussed in other sections in this book. Here we concentrate on the series of independent scientific missions proposed for solo-CSM-type missions in the mid-1960s. However, following on from their 1963 studies NAA had begun a second series of NASA-contracted studies (NAS 9-3140) that were based on an Apollo CSM with minimum modifications for missions up to 45 days in duration. Parallel to those studies, NAA conducted additional studies into evaluating what modifications, additions and redesigns would be required to adapt the Apollo vehicle to become 'essentially useful for manned Mars or Venus flight with mission durations on the order of 600 days' (see p. 317).

A report issued on 14 August 1964 by the Spacecraft Integration Branch at the NASA Manned Spacecraft Center (which from 17 February 1973 became the Lyndon B. Johnson Spacecraft Center in honour of the recently deceased former President) proposed the development of a series of Apollo X missions which also featured a separate laboratory module launched by Saturn 1B, on missions of 14–45 days in Earth orbit at 230 miles. A programme of biomedical and scientific experiments and research would be linked to studies of extended-duration spaceflight. The series would build upon the experiences and achievements of each previous flight, and would eventually lead to a 120-day mission using an independent laboratory module with self-contained systems and life support independent of the Apollo spacecraft – a space station.

Four configurations were suggested. The first featured an initial mission (Configuration A) on a flight of 14–45 days, with a two-man crew but with no laboratory module. Configuration B would see three astronauts flying a 45-day mission, with a single laboratory module to support the extended flight. The module would be carried aloft in the Spacecraft Lunar Adapter, and the crew would perform the transpiration and docking manoeuvre to retrieve it in Earth orbit – similar to extracting the LM from the top of the third stage on the lunar missions. Use of a single or dual launch Saturn 1B would depend upon the launch weight of the module (although the diagrams indicated that the Laboratory Module would be launched on a manned Saturn 1B).

Configuration C was a duplicate three-man 45-day mission, but with a double Laboratory Module with systems dependent on the Apollo. Finally, Configuration D envisaged a three-man mission of 120 days, using an independent Laboratory Module, presumably launched separately from the Apollo CSM. In releasing this information, NASA was quick to point out that this synopsis was in no way competing with the developing USAF MOL programme (which had clear military objectives), and focused on technical and scientific objectives more suited to the civilian agency's long-term goals. By the end of 1964, these studies suggested that such missions, though still undefined, could begin flying by 1969, but only after the initial landings had been achieved.

Some on Capitol Hill in Washington voiced concern that MOL and Apollo X were trying to create two national space station programmes at the taxpayer's expense, when only one was really needed. However, with clearly defined military and civilian objectives, both NASA and the USAF argued the need for both programmes, and for some years both survived. But it would not be too long before both began to suffer funding difficulties in light of other, more Earthly concerns.

Apollo Extension System (AES)

By 1965, Apollo X was being termed the Apollo Extension System (AES), and on 18 February, NASA Associate Administrator for Manned Spaceflight, George Mueller – already a keen supporter of expanding the potential of Apollo resources into other areas – outlined NASA's post-Apollo objectives in a testimony to the house Committee on Science and Astronautics. During his speech, Mueller advocated the advanced capabilities of Apollo, and 'sold' the idea that the development of

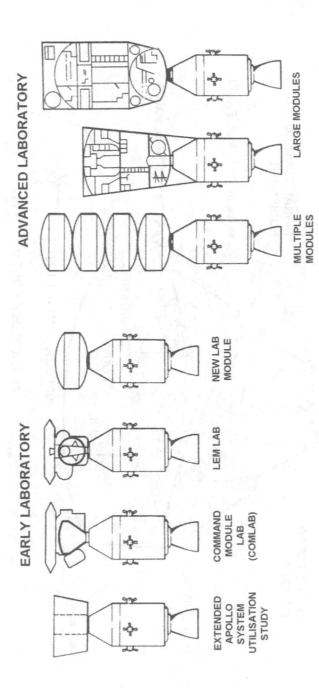

The various configurations of the Comlab (Command Module/Laboratory) system of scientific laboratories flown in conjunction with the Apollo CSM in Earth orbit under the Apollo Extension System.

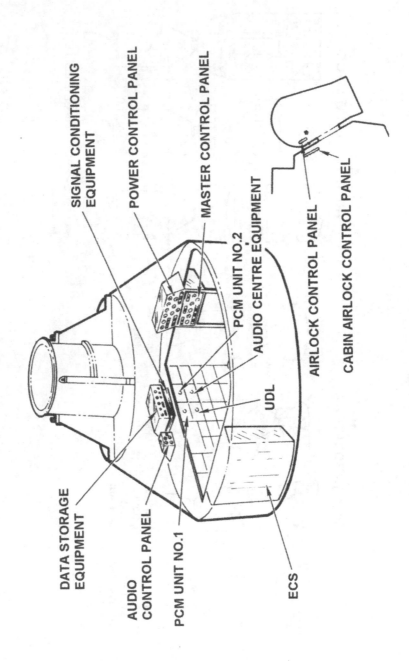

SIGNAL CONDITIONING EQUIPMENT

POWER CONTROL PANEL

MASTER CONTROL PANEL

DATA STORAGE EQUIPMENT

AUDIO CONTROL PANEL

PCM UNIT NO.1

PCM UNIT NO.2

AUDIO CENTRE EQUIPMENT

UDL

ECS

AIRLOCK CONTROL PANEL

CABIN AIRLOCK CONTROL PANEL

The internal arrangement of the CM flown in conjunction with the Comlab Laboratory Module.

hardware to support the initial lunar programme would enable the agency to continue to use hardware on future missions at a small fraction of the development cost. 'This is the basic concept of the Apollo Extension System now under consideration,' he stated. Cleverly, Mueller had promoted the AES as a 'new' development (but not a programme) within the existing Apollo project that could incorporate a whole series of additional flights for minimal additional cost. The additional series would feature both lunar and Earth orbital objectives, including missions lasting between four and six weeks in orbit around the Moon and up to two weeks, on its surface (lunar daytime), and Earth orbital missions lasting up to three months, or even towards the 120–180-day (six-month) goals.

In order to support such an expanded programme in co-operation with each of the manned spacecraft field centres (MSC, MSFC and KSC), the prime contractors would be asked to provided study contracts to identify potential missions and detailed objectives. On 12 May 1965, a formal presentation of the early Apollo Extension System Experiment Program, presented by the Advanced Spacecraft Technology Division (Space Station Study Office) at MSC, revealed the scope of the proposed programme.[7]

Spacecraft configuration

Early in the definition studies only the Apollo CSM was considered as a primary AES spacecraft, but with various configurations of attached Laboratory Modules

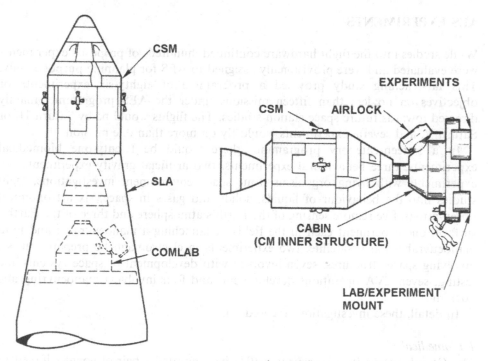

The launch and flight configurations of the Apollo CSM/Comlab design.

(as shown in the diagrams on p. 19. Later, the use of the Lunar Module ascent stage also featured in designs of some AES configurations (see p. 19). Once again, as prime contractor for the Apollo CSM it was NAA that received the contract (NA59-5017) from NASA at the beginning of Fiscal 1966, to study the AES spacecraft concept.

The studies reviewed a number of options, but soon determined that Block II configurations of the CSM could support an extended-duration orbital mission with an improved life support system and additional supplies of propellants. NASA conducted studies on the LSS of Block II spacecraft, and determined that the most efficient method to extend the duration of the existing system was to replace the lithium hydroxide/carbon dioxide ($LiOH/CO_2$) removal system, which was only capable of supporting three men for up to fourteen days. This original Apollo sub-system required the storage of a number of replacement and spare LiOH canisters and disposal of used units in the CM that increased as the mission progressed. The installation of a 'lighter' regenerative molecular sieve system would maintain a relatively constant 'weight' as the mission progressed, although this saving would be counteracted to some degree by an increase in complexity and power requirements. It would allow missions in excess of thirty days to be planned, and using a more advantageous dual gas atmosphere (see p. 82) over the mostly pure oxygen environment on the Apollo lunar missions.

AES EXPERIMENTS

While studies into the flight hardware continued, hundreds of proposed experiments were evaluated and were provisionally assigned to AES for planning purposes only. This far-reaching study provided a programme of eighty-six experiments or objectives on no less than fifteen missions under the AES programme, mainly directed towards future space station studies. The flights would be by Saturn 1B or Saturn V, and several experiments would fly on more than one mission.

In a sixteen-category programme, there would be twenty-one biomedical experiments, three behavioural experiments, two artificial gravity experiments, six experiments with living organisms, four space environment investigations, eight studies into the behaviour of liquids, solids and gases in space, six astronomical observations, five remote sensing of the Earth's atmosphere and three of the Earth's surface, one experiment each in the fields of launching a manoeuvrable and non-manoeuvrable Earth satellite, two experiments in electromagnetic preparation, six involving space structures, seven involved with development of space systems and testing, seven EVA operations development, and four involving manoeuvring and docking.

In detail, these investigations focused on:

1 Biomedical
The *Otolith Sensitivity Experiment (0101)* incorporated a pair of goggles linked by fibre optics to a television camera to record the increase of otolith sensitivity in zero

g and to determine any operational hazards. A similar pair of goggles and hardware was used to record the *Head Rotation Experiment (0102)* data, evaluating the thresholds and maximum tolerances during the rotation of the subject's head. Lower Body Negative Pressure (LBNP) measurements were taken for the *Circulatory Response Experiment (0103)*, using an early model of a device which has since flown on Skylab and more recently on the Shuttle. The Russians have also used it on the Salyut and Mir space stations. This experiment would also record cardiac output, take body mass measurements, and record data from standard bioinstrumentation and clinical observations. The *Work Capacity and Circulatory Response Experiment (0104)* had the primary objective of determining the effect of prolonged zero g on the human body by measuring cardiac output, the use of a bicycle and hand/arm ergometer, the automatic recording of blood pressures, and the use of the standard bioinstrumentation and clinical observation methods.

A 472-lb, 9-cubic foot device was used to measure the changes in blood and other body fluids as part of the *Body Fluid Volume Experiment (0105)*. Key techniques included isotope injection, blood sampling, and analysis on orbit. The device featured a micro-centrifuge, micrometer, spectrocolorimeter and a microscope, and had a supply of syringes, slides, test tubes and reagents. The *Cardiovascular Reflexes Experiment (0106)* was to determine the effects of prolonged spaceflight by carotid baroreceptor forcing, featuring the use of the LBNP device. The experiment would be used to determine adequate countermeasures in addition to gathering basic scientific knowledge of this phenomenon. The LBNP device was also used in the *Venous Compliance Experiment (0107)* to determine alterations to the veins, while evaluation of countermeasures for *Circulatory Reflex Changes (0108)* used intermittent venous cuffs, salt- and water-evading techniques, the use of salt-retaining steroids, and a programme of isometric or isotonic exercises.

By taking lung compartment volume measurements, gas flow rate measurements and the recording of the rate of clearance of a dilutant gas, the evaluation of changes and detection of trends formed the basis of the *Pulmonary Function Experiment (0109)*, with the subject wearing an oxygen mask connected to a mass flow transducer and gas analyser and also wearing an ear oximeter. Periodic atmospheric sampling for O_2 and CO_2 concentrations was the technique employed by the *Ventilatory Gas Exchange Experiment (0110)* to measure the changes and identify any trends. For the *Muscle Mass/Strength Experiment (0111)*, a tape measure, electromyography, dynameter, urinary and creatine measurements were used to determine and qualify any deleterious effects. The experiment would again be used to develop countermeasures. The *Mineral Metabolism Experiment (112)* utilised the same hardware as Experiment 0105 and was used to assess the electrolyte balance, bone density, urinalysis and levels of serum alkaline phosphates in the body, to determine and quantify any adverse effects and to help develop corrective measures.

The astronauts would measure the intake of O_2 and outpour of CO_2, conduct blood lactic acid tests, a nitrogen balance test, body mass and volume measurements (by using a mass measurement device capable of recording the mass of a human body in zero g), and a serum protein test as part of the *Nutritional Status Experiment (0113)*. This was in order to determine and quantify the nutritional value of operational rations supplied to the crews for the duration of the mission, and would help to develop corrective dietary measures. For the *Gastrointestinal Motion and pH Experiment (0114)*, minute transmitters were to be retained in the astronauts' gastric antral area and sensors attached to the body. Skin and core temperature measurements would be taken for the *Thermal Regulation Experiment (0115)*, and a correlation was to be made with the endocrine and cardiovascular experiments – again to determine and quantify any adverse effects and develop corrective measures. In this experiment, the crew would make use of the LBNP device, a telethermometer, an ear canal thermistor, and a *Water Cooled Suit* which was itself an AES experiment *(0116)*.

The objective of the Water Cooled Suit on AES was to evaluate the device for standard and emergency use and to support the cardiovascular experiment programme. Cabin temperatures would be varied while the astronaut was wearing the suit, allowing him to raise or lower the suit's own level to maintain a comfortable temperature irrespective of the cabin temperature. This experiment would weigh 60 lbs and require 5.5 kW per hour, and would be evaluated over 5.5 man-hours on a mission. This type of suit would evolve into the Liquid Cooled Undergarment used during EVAs on Apollo and Shuttle missions, and of the type employed by Soviet cosmonauts during Salyut and Mir EVAs, becoming a standard element of EVA operations.

The *Neuro-Endocrine Function Experiment (0117)* was to determine and quantify changes in samples of endocrine and neuro-endrocrine functions by collecting urine and blood samples in flight, as well as in pre- and post-flight conditions. Three blood study experiments included the *Haemic Cell Study (0118)* to detect haematological cellular changes and to determine the cause and effect. The crew-member would use the preferred technique of in-flight analysis to gather a reticule count, hematocrit measurements, red blood cell count and mass determinations, the survival of red blood cells, and studies of cytogenic effects. In addition, pre- and post-flight measurements would be taken, to provide comparable samples. In the *Haematological Defences Experiment (0119)*, changes in the human immunological system and the leukocyte functions would be assessed. In-flight observation of leukocyte mobility and determination of phagocyte activity would be supplemented by additional post-flight analysis. The *Haemostasis Experiment (0120)* would determine blood coagulation and haemostasis parameters, and study the fundamental process of clotting. To achieve this, techniques used would include coagulation time, bleeding time, platelet count, clot retraction, and platelet adhesion, as well as additional post-flight analysis.

The final biomedical investigation, *Microbiological Evaluation (0121)*, would identify indigenous microganisms and determine dominance and pathogency. Key techniques to be used included standard microbiological techniques, by using glassware, a microscope, incubator and refrigerator, as well as chemicals and containers.

2 Behavioural

The *Sensory and Perceptual Processes Experiment (0201)* was to determine the relationship between physiological condition and performance, as well as detecting changes in performance capability. To achieve this, an 81-lb Acceleration Detection Unit, with a volume of 3.7 cubic feet and power requirement of 410 W, would have been used. Associated equipment included a camera, a tape recorder audiometer, an orthorator, an acceleration chair, a mass determination device, and a performance test console (PTC). Techniques to be used included visual and audio measurement, vestibule, kinaesthetic and orientation measurements.

The *Psychomotor Functioning Experiment (0202)* also made use of the PTC with similar objectives to 0201, but using the key techniques of gross and fine motor performance testing and continuous and discrete processes. The *Higher Mental Process Experiment (0203)*, while having the same objectives as 0201 and 0202, relied on vigilance and attention testing, memory and problem solving tests, as well as a speech, reading, perception and comprehension tests.

3 Artificial gravity

There were two investigations under this category in the AES programme. The *Rotating Station Experiment (0301)* was to determine the radius and rotational speed for a simulation of artificial gravity of a spacecraft and counterbalance. Using the CSM, it was suggested that the counterbalance could have been an S-IVB stage, LM, or experiment rack, using telescopic or cable connections. Variations in the rotational radius and rate of spin would have been attempted. Tethered experiments were performed on Gemini 11 and 12 in 1966, and met with varying success. Thirty

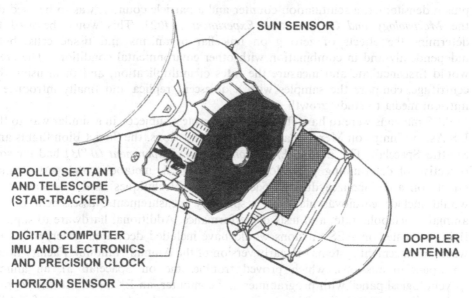

SUN SENSOR

APOLLO SEXTANT
AND TELESCOPE
(STAR-TRACKER)

DIGITAL COMPUTER
IMU AND ELECTRONICS
AND PRECISION CLOCK

HORIZON SENSOR

DOPPLER
ANTENNA

Weight (lbs) 400 / Volume (cu ft) 10 / Crew time (m/h) 20

An artist's impression of the use of dual Comlab modules for an extended mission in Earth orbit. Detailed are space navigation experiments.

years later, the difficulties of using tethers in space were demonstrated by the Tethered Satellite System experiments on STS-46 and STS-75. What was proposed for the AES experiment was far more complex than anything that has since been attempted.

The second investigation was the *On-board Centrifuge (0302)*, which was to support a human test subject for periodic g conditioning using varied levels of time, position and the levels of g experienced. The subject's tolerance would be correlated with 'other factors.' The stated objective was to reduce or prevent debilitation and to determine g loading, or the exercise that would be required to maintain physical fitness in flight. The volume needed to support this experiment was stated as 390 cubic feet, and would probably have taken up most of the diameter of a mission module. (When this proposal is reviewed, visions of the Skylab astronauts running around the storage lockers of the Upper Workshop area performing gymnastics come to mind, as this proposal would have provided a similar level of g forces.)

4 Living organisms

The objectives of the *Genetic Effects in Microorganisms Experiment (0401)* were to determine the effects of zero g on replication and recombination of DNA, and to test synergistic effects on mutation rates and phage production. Using the refrigerator, incubator, lyophiser, a microscope, chemostat and a medical centrifuge, the crew would maintain viable cultural transfers, and by using the microorganisms to cultures technique and then infecting the cultures with phage strains, enable them to halt infection at timed intervals and then observe the results. The same hardware, plus a densitomer, a scintillation counter and a particle counter, was to be used on the *Morphology and Growth Rate Experiment (0402)*. This would be used to determine the effects of zero g on unicellar organisms and tissue cells, both independently and in combination with other environmental conditions. The crew would first incubate and measure the rates of multiplication, and then, using the centrifuge, compare the samples with a quiescent replica and finally introduce a nutrient media to study growth rate.

AES missions were to have flown primates as test subjects, in a similar way to the US Astro Chimps on Mercury and Biosatellite missions, the Soviet Bion flights and Shuttle–Spacelab. The *Zero G Effects on Primates Experiment (0403)* had the sole objective of determining physiological, neurological, metabolic and performance effects on a chimpanzee during spaceflight. Key techniques used to obtain data would include cardiovascular and respiratory measurements, assessment of the animal's metabolic rate, and tests of performance. Additional hardware to support flying primates on AES missions would have included dedicated water, food and waste management systems (an early version of the Animal Handling Facility flown on Spacelab missions, which proved troublesome on Spacelab 3), an animal 'psychological panel' with programmer and computer, an EKG, ballistocardiogram and cardiac output device. This equipment was quoted as having a mass of 394 lbs and 40 cubic feet of volume, with a power requirement of 35 W and minimal crew time (0.5 man-hours). (The thirty minutes of crew time planned for AES was later proven to be inadequate on Spacelab 3 in 1985, when additional and unplanned crew time was crucial to support the primate's comfort and survival in order to supply usable experiment results.)

The *Tissue Regeneration Experiment (0404)* was to investigate the changes in capacity and rate in newts and planara by means of pre-flight surgical preparation, and in-flight growth and morphology observations and measurements, as well as in-flight periodic sample freezing. Once again, dedicated life support systems would have been required to supply food and deal with waste, as well as aeration equipment to deal with aquatic test subjects. The *Drug Effects Experiment (0405)* was to determine the effects of the drugs on the behaviour of rats in a zero g environment. The rodents were to be placed in flight cages supplied with feeders, a water supply, and observation cameras. The associated equipment to support this investigation was quoted at a mass of about 668 lbs, with a volume of 24 cubic feet, a power requirement of 56 W, and crew time of 5 man-hours. (Twenty years later, experiences on Spacelab 3 once again illustrated the 'difficulties' of flying animal cages and their supporting systems. In this case, they never worked as designed, and contaminated the main crew atmosphere with particles of food and droppings!) On the AES experiment in 1965, it was proposed that pre-flight activities would include training isolated groups of rats, injecting drugs into chosen subjects, observing their behaviour and then dissecting and analysing the animals to arrive at control results. In flight, a repeat of the pre-flight procedures would be attempted, including, apparently, surgical dissection in orbit.

The final topic in this field was proposed as the *Space Borne Microorganism Experiment (0406)* using Mylax-covered screens, silicon grease-covered impaction surfaces, and electrostatic precipitation and field capture devices to perfect techniques for capturing such microorganisms – if, of course, they existed in the first place.

5 Space environment
Apollo lunar missions would have to traverse the Van Allen radiation belts and go further into the magnetic fields that surround the Earth than any other manned spacecraft had thus far penetrated. Unmanned probes and satellites had begun to investigate the new space environment over the previous few years, but AES was to take these investigations a stage further with studies from manned spacecraft in high Earth orbits and, in time, lunar distance flights.

The *Radiation Environment Monitoring Experiment (501)* was a pioneering investigation that has continued to the current Shuttle and space station missions, gathering a record of radiation received by the crew inside the spacecraft and recording levels of the external radiation environment being traversed. For AES, this data was to be gathered by a proton-alpha spectrometer, an electron spectrometer, ionisation chambers, and film packs.

The *Study of Magnetic Field Line Experiment (0502)* was designed to study the geometry of lines of force of the terrestrial magnetic field. To do this, a 'remote manoeuvring unit' would be released from an Apollo, which would measure these paths followed using electrons injected into the field from a linear accelerator. This would create artificial aurorae that could have been observed and tracked either on the ground or by aircraft.

A *Comet-Like Particulate Cloud Experiment (0503)* was to include photographic, photometric, and crew visual observations of an artificially generated comet cloud (created by the release of chemicals from an experiment module) in order to determine the composition and physical characteristics of comets.

The AES series of experiments produced a wide-ranging programme of investigations and the configurations to achieve them. One of the more interesting spacecraft configurations was for the *Micrometeoroid Collection Experiment (0504)*, designed to determine the physical and chemical properties of the dust clouds that surround the Earth. In addition to measuring particle velocity and flux, particles were to have been trapped in collector panels deployed from the LM and SM, to be returned to Earth (presumably via EVA retrieval) for post-flight analysis.

6 Liquid/gas/solids behaviour
In order to aid understanding of the interactions of matter at simultaneous solid/ liquid and vapour interfaces under conditions of zero g, the *Capillary Studies Experiment (0601)* was to take static measurements of wetting angles and dynamic measurements of capillary flow rates and accelerations, using two separate sets of equipment, for either static or dynamic tests.

Experiment 0602 consisted of two separate investigations and chambers. *Dynamics of Vapour/Gas Bubbles (0602A)* and *Liquid Drop Dynamics (0602B)*. The first investigation was to be performed in a controlled chamber with transparent walls, and the results recorded on film. It was a study of free boiling, bubble growth and collapse, and the dynamics of vapour/liquid interfaces in zero g. The second investigation would employ stereoscopic cameras to observe the effects of induced perturbations in various types and size of drops, to determine the fluid dynamic behaviour of freely suspended liquid drops.

Experiment 0603 also consisted of two investigations: *Pool Boiling in Zero g (0603A)* and *Nucleate Condensation (0603B)*. The objective of the first experiment was to gain knowledge of bubble growth rates, the interaction of bubbles, hydrodynamic stability of bubble columns and nucleation criteria by obtaining direct images and shadowgraphs of both bubbles and thermal layers from a heated surface. The other part of this experiment was to study the initiation of condensation of a pure substance in zero g. To do this, a pure fluid would have been cooled in a sealed vessel, and the condensation observed photographically by means of scattered light.

In the *Fluid at a Near Critical State Experiment (0604)*, the objective was to study the Photo-Voltaic-Thermal (PVT) behaviour of a near-critical fluid under zero g. To do this, a test cell would have been loaded with a predetermined amount of liquid and vapour, and then heated. Resulting density profiles would then have been measured by diffraction of light. The *Crystallization Studies Experiment (0605)* was designed to determine the relationship between the characteristics of crystalline materials and their modes of formation. The techniques designed to achieve this would have included the heating of various samples to a molten state in a test-chamber using a 3-foot-diameter solar concentrator. After cooling, the samples would be removed from the quartz test chamber for return to Earth. The *Cosmic Ray*

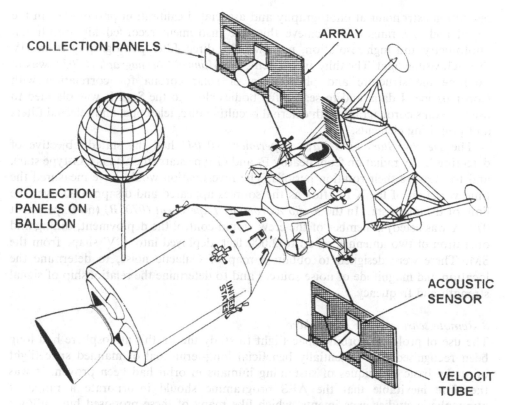

COLLECTION PANELS

ARRAY

COLLECTION
PANELS ON
BALLOON

ACOUSTIC
SENSOR

VELOCIT
TUBE

UNITED
STATES

Weight (lbs) 135 / Volume (cu ft) 22 / Power (kWh) / Crew time (m/h) 1 per week

This design for an experiment (0504) to collect micrometeoroids during a trip towards the Moon was one of many proposed for early Apollo Extension System missions.

Emulsion Experiment (0606) was designed to obtain a charge spectrum of the primary cosmic radiation and a measure of intensity and angular distribution. A three-dimensional photographic emulsion would have been exposed to radiation. The crew were to terminate the experiment by partially developing the emulsion in-flight, with post-flight development and examination of the results.

7 Astronomical observations

Three investigations under this category would have been incorporated into one set of instruments, and were once part of the early development of the Apollo Telescope Mount that was launched on Skylab in 1973 (see p. 52).

A Michelson interferometer operating in the infrared region was to be used to determine the distribution of patterns and densities of interstellar gases and to record the output on magnetic tape under the *Emission Line Radiometry Experiment (0701)*. The *Intermediate Size Telescope (0702)* was designed to perform high-resolution astronomical photography and associated calibration photometry in the visual and UV ranges. To achieve this, the instrument recorded high-resolution photometry and high-resolution, narrow-field, direct-filter photography from UV through visible light. The third instrument, the *Manned Coronagraph (0703)*, was to examine the structure and phenomena of solar corona for correlation with Coronascope II data, and to search for bodies close to the Sun. It was planned to use an Evans coronagraph with external occulting disc, telescope, and selected filters and polarising elements.

The *Nearby Stars in X-Rays Experiment (0704)* had the stated objective of detecting X-ray radiation from nearby F- and G-type stars (the Sun is a G-type star), and to measure their rotation periods. The investigation would have measured the time variation of the X-ray flux as the sources appeared and disappeared over the limb of the target star. In the *Radio Astronomy Experiment (0705B)* (no Experiment 0705A was listed) a member of the crew would control the deployment, tuning and operation of two antennae of up to 2,000 feet, deployed into a 'V' shape from the SM. These were designed to obtain a map of Galactic noise, to determine the location and magnitude of noise sources and to determine the relationship of signal strength and frequency.

8 Remote sensing of the atmosphere

The use of prolonged orbital spaceflight to study the Earth's atmosphere had long been recognised as a potentially beneficial long-term goal of manned spaceflight once the basic techniques of sustaining humans in orbit had been proven. It was therefore inevitable that the AES programme should incorporate a range of atmospheric studies experiments, which like many of these proposed but unflown investigations, were the genesis of later flight-proven investigations. (The remote sensing atmospheric experiments evolved through many guises to studies from Skylab in 1973 and 1974 and from several Shuttle missions – most notably the ATLAS series of missions in the 1990s. The Soviets also conducted similar studies from Salyut and Mir, and it is a research field that will continue aboard the ISS.)

The *Conjugate Aurora and Airglow Experiment (0801)* was to investigate the detailed nature of the conjugate auroral forms and geometries in both the Arctic and Antarctic regions at the same time. To do this, the astronauts would have conducted a programme of visual, optical and photometric measurements, in addition to data gathering by particle detectors and a magnetometer, in conjunction with coordinated ground-based data collection, if possible. Simultaneous twin-pole data collection would have necessitated Apollo flying in a synchronous orbit – probably over continental USA.

Experiment 0802 was a four-part investigation. In the *High-Resolution IR Radiometer Experiment (0802A)*, a detector would be manually pointed through mirrors at areas of meteorological interest, such as cloud cover phenomena, with an auxiliary camera taking documentary photographs. The second element was an *IR Scanning Spectrometer (0802B)* designed to evaluate the test spectrometer band and to obtain measurements of the spectral radiances of Earth and its atmosphere in the 0.2–40 µm band. It was expected that detailed examination of the chosen bandwidth would determine which portion produced the highest-accuracy measurement. The *Microwave Spectrometer (0802C)* would also test the spectrum, in the range of 0.5–3 cm. To accomplish this, an astronaut would have aimed the spectrometer at clouds and studied their emissions, making adjustments to compensate for instrument degradation. The fourth instrument in this set was to be the *Prototype Star-Tracker (0802D)*. This experiment was to demonstrate the ability of a prototype star-tracker to lock on and track stars of the fourth magnitude when viewed near the horizon, beneath the airglow layer. The astronaut would have searched for, acquired and tracked designated stars under those conditions.

9 Remote sensing of Earth

As with the previous field of research, remote sensing of Earth from space was a rapidly developing science and a stated benefit for manned platforms to supplement observations from unmanned spacecraft and satellites. Preliminary photographic work had been assigned to the Gemini programme, and with AES, the Earth resources survey and inventory included studies of agriculture and forestry, geology, oceanography, hydrology, topography and biogeography. This was part of the genesis of the Earth Resources programme that continued into Skylab, the Shuttle and Soviet space stations.

In the *Multi-Spectral Target Characteristics Experiment (0901)*, the objective was to establish how characteristic spectral signatures gathered from space could be used to identify and measure Earth and atmospheric properties. To do this, a wide variety of targets would be used, under different conditions, to gather multispectral imagery and spectral response from between near-UV and radio wavelengths. By obtaining photographs of land areas of the Earth, basic imagery could be used to construct topographical maps to provide up-to-date details of the various uses of the land and the network of transportation systems. Scientists could then determine if the current use was efficient, and construct future plans to ensure that the most economical application of land was achieved. The investigation on AES was termed *Synoptic Earth Mapping (0902)*, in which the astronauts were to obtain black-and-white stereoscopic images with a resolution of 20 feet at the red end of the spectrum, along with photographs in the blue, yellow and near-infrared areas of the spectrum, to a resolution of 200 feet. In addition, broadband VHF radar-gathered data as a radiometer provided emission characteristics data of surface moisture conditions. Finally in this category, the *Multi-Frequency Radar Imagery Experiment (0903)* was to provide continued radar observation of the Earth on a seasonal synoptic basis, using a variety of scientific disciplines to collect complementary data from the same target areas over the progress of a year or more. In addition to a radar altimeter, the array of instruments planned to support this investigation included intermediate-resolution side-looking ground radar, a high-resolution coherent side-looking radar, three high-resolution systems, and panoramic cameras. (Developments of this type of survey continued after the AES programme, and would evolve into what became known as the Skylab Earth Resources Experiment Package, the Apollo lunar Scientific Instrument Module package, the Landsat series of Earth Resources Technology Satellites, and as a variety of experiment packages flown on the Space Shuttle in the 1980s and 1990s.)

10 Manoeuvring satellite
The *Small Manoeuvrable Satellite (SMS) Experiment (1001)* was to be launched from the Apollo 'mother ship' (presumably the Service Module), and would remain under remote control (it is not clear whether this would be from the Apollo or from ground control) for the duration of the experiment. This was planned for 40 man-hours during the mission. The mass was quoted at around 371 lbs, with dimensions of 30 × 30 × 12 inches. It would include scientific instruments to study the wake of the Apollo as it passed through its orbit, as well as hydromagnetic waves, antenna patterns, cosmic radio sources at low frequencies, the extent of X-ray sources, the structure of the solar corona, tug-deployed antenna and hazardous radiation belt areas. This was quite a comprehensive programme for such a small package. A sub-satellite was subsequently deployed from Apollo 15 and Apollo 16 in lunar orbit, and developments of free-flying pallet type structures – SPAS or Spartan – have been frequently flown on Shuttle missions. The AES structure would probably have been spring-ejected and abandoned after the end of the investigation period.

11 Launch of unmanned satellite
One of the more optimistic experiments planned for inclusion in an AES mission was the proposed launch an unmanned Orbiting Geophysical Observatory (OGO) satellite with a mass of 1,300 lbs and volume of 20 cubic feet. Crew time estimated to complete the *Launch of an Unmanned Satellite (OGO) Experiment (1101)* was quoted as 15 man-hours, and in light of actual activities of this type, they would probably have needed every minute – if not more. The presentation suggested that the unmanned satellite would be 'unpacked, checked out, deployed, adjusted and launched from the manned vehicle in orbit' (Apollo SM). The objective was to 'develop procedures and operations for checkout, activation and launch of OGO from a manned vehicle'. It was further stated that it might also have been recovered, re-serviced and re-launched.

Experiences on EVA during the Gemini missions in 1966 demonstrated how difficult even the most straightforward EVA task can become if the astronauts are not securely tethered and properly cooled. It was only after underwater training and the installation of adequate foot- and hand-holds that the skill required for EVA became apparent during the final Gemini flight. Skylab provided more EVA experience in Earth orbit, but it was not until the advent of the Shuttle and the satellite repair and retrieval missions (such as Solar Max and the Hubble Space Telescope) that a full demonstration of what was proposed in 1965 was achieved, with spectacular effect.

12 Electromagnetic propagation
The *Radiation Frequency Radiation Experiment (1201)* had the objective of determining the radio frequency interference level and background noise environ-

ment that could be experienced in communication with Earth. With the Apollo missions designed to journey to the Moon, reliable and clear communication between the spacecraft and mission control was essential for a safe flight. This experiment was to investigate this field by using wide and narrow bandwidth antennae on the Experiment Module attached to the Apollo. They would have been directed towards Earth to record radiation in the 100 mc and 100 kmc region. At this time Apollo was to be used not just for a few landings on the Moon, but also for prolonged expeditions of up to two weeks on the surface and a month in lunar orbit. Plans being formulated for using Apollo-derived hardware to send astronauts on flights to Mars made this experiment even more important.

In connection with the previous investigation, the *Wide Bandwidth Transmission Experiment (1202)* was to develop a high-power spacecraft transmitting system capable of sending high-rate data back to Earth and to other spacecraft. To achieve this, an Experiment Module would have been provided, with a common power supply and three transmitters in the high-ferquency, UHF and millimetre bands. Omnidirectional antennae would also have been provided on the configuration.

13 Space structures

In order to determine criteria for the development of advanced expandable, erectable and/or inflatable space structures, the *Environment Effects of Structures Experiment (1301)* was designed to evaluate the effects of vibration, damping, pressure, temperature and leakage. The experiment structure would have had a mass of about 480 lbs and a volume of 60 cubic feet, which the crew would access through an internal transfer tunnel. The effects were to be monitored by telescope, camera and other remote instrumentation, and presumably by the astronauts themselves. (Over the past four decades, studies of this type of space structure have continued with potential applications in space, on the Moon, and at Mars. The most recent configuration was the TransHab structure planned for the ISS in the late 1990s.)

Some of the apparently straightforward experiments assigned to the AES series have had related applications in later programmes, with mixed results. The *Deploy RF Reflective Structure (1302)* would have seen the crew observe and describe the deployment of a large RF reflective structure (similar to the early Echo passive communication satellite balloons) in the environment of space. The crew was to release a tethered structure from the spacecraft and initiate its deployment. They would then observe and photograph its deployment and be ready to assist (presumably by EVA) in the event of a malfunction. (This has since been a standard contingency activity for each Shuttle flight. Here, a team of EVA astronauts are trained to perform unplanned or contingency EVAs, such as manually closing the payload bay doors, or jettisoning the robot arm or other spacecraft payload appendages such as solar arrays. The two Shuttle flights that deployed the Tethered Satellite System in 1992 (STS-46) and 1996 (STS-75) demonstrated the difficulty of using tethers in space, and the crew had trained for the possibility of separating from a tangled mass of tethers – hopefully not from around the orbiter.)

In a second experiment of this type, the crew was to have observed and described

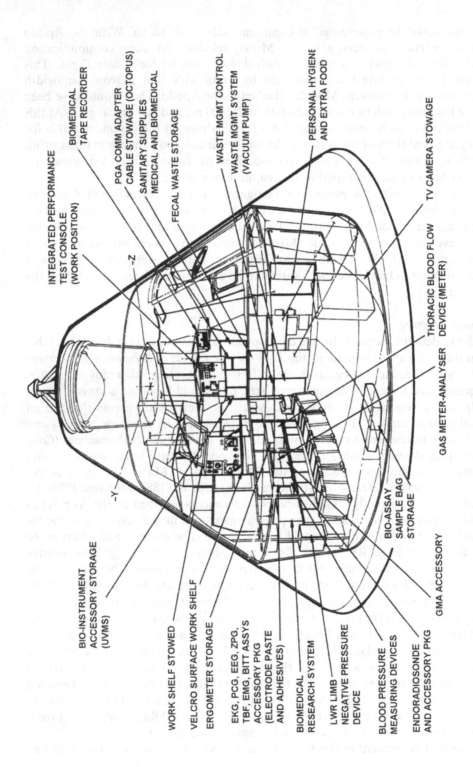

A 'typical' internal configuration for a Block II CM on an AES mission.

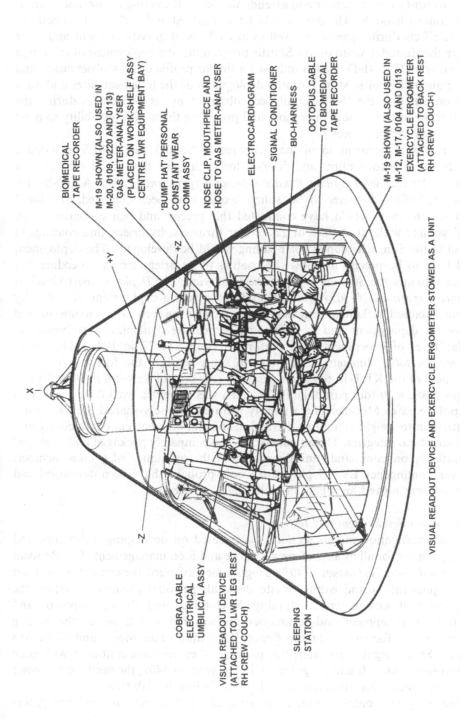

BIOMEDICAL
TAPE RECORDER

M-19 SHOWN (ALSO USED IN
M-20, 0109, 0220 AND 0113)
GAS METER-ANALYSER
(PLACED ON WORK-SHELF ASSY
CENTRE LWR EQUIPMENT BAY)

'BUMP HAT' PERSONAL
CONSTANT WEAR
COMM ASSY

NOSE CLIP, MOUTHPIECE AND
HOSE TO GAS METER-ANALYSER

ELECTROCARDIOGRAM

SIGNAL CONDITIONER

BIO-HARNESS

OCTOPUS CABLE
TO BIOMEDICAL
TAPE RECORDER

M-19 SHOWN (ALSO USED IN
M-12, M-17, 0104 AND 0113)
EXERCYCLE ERGOMETER
(ATTACHED TO BACK REST
RH CREW COUCH)

+Z

+Y

−X

−Z

VISUAL READOUT DEVICE AND EXERCYCLE ERGOMETER STOWED AS A UNIT

COBRA CABLE
ELECTRICAL
UMBILICAL ASSY

VISUAL READOUT DEVICE
(ATTACHED TO LWR LEG REST
RH CREW COUCH)

SLEEPING
STATION

Experiment locations on an AES Block II CM.

the deployment of long, slender and extendable rods in the zero-g environment, from an Experiment Module. The crew would have recorded and verbally described the dynamic effects during payout, as well as any effects of gravity gradient and solar pressure they found. (Again on the Shuttle programme, the deployment of the Large Solar Array on STS 41-D in 1984 followed a similar profile of crew observation and interpretation of results. More recently, deployment of the tether satellites and huge arrays on Mir and the ISS have all been observed by crew-members during the procedure, to ensure smooth operation, and providing them with the ability to react to unforeseen circumstances.)

In an early demonstration of the technique of 'solar sailing,' AES included a research objective to confirm and develop techniques for the orbital control of a satellite under the *Solar Sailing Passive Comsat Experiment (1304)*. A 100-lb, 10-cubic-foot, 50-foot diameter Mylar sphere would have been deployed, and for 3 man-hours the crew would have conducted the 'precise and gentle control of an orbital satellite without the use of gas jets or thrusters, by orientating coatings in relation to the Sun, whereupon solar sailing would be developed'. The deployment would have to be precise to allow the coating to be orientated (a procedure not explained) to face the Sun and to catch the solar wind. The *Deploy Gravity Gradient Structure Experiment (1305)* was the study of a zero-g deployment of a gravity gradient stabilised, inflatable lenticular structure, with the crew tasked to initiate and observe the deployment and to study and record the dynamics of stabilisation. (Similar types of experiment were performed from STS-77 in 1996.) The *Large Aperture Erectable Antenna (1306)* was in some respects the forerunner of the deployment of the KRT-10 telescope from the rear port of Salyut 6 in 1979. In this AES proposal, a 30-foot parabolic RF antenna was to be deployed from the side of the Apollo Service Module. The assembly featured precisely-shaped panels, with a high strength-to-weight ratio, low thermal distortion, high current conductivity and a high structural integrity. The antenna was to be compactly packaged and deployed automatically on command, unfurling without the assistance of a crew-member. However, contingency EVA deployment would probably have been developed had this experiment moved anywhere near the flight manifest.

14 Sub-system development and testing

The first investigation in this section concentrated on developing techniques and flight equipment for all aspects of crew hygiene and food management. The *Personal Hygiene and Food Techniques (1401)* programme envisaged the evaluation of food storage, preparation and eating, waste disposal, personal cleaning, shaving, the management of human waste, cleaning of clothing, and the development and evaluation of equipment and techniques for working and living in the zero g environment. A forerunner of the Space Habitation experiments and objectives flown on Skylab eight years later, this package of evaluations continues with each new manned spacecraft and programme. In Experiment 1401, the need for a shower for the crew was being investigated, eight years before Skylab flew.

Monitoring and controlling the cabin atmosphere, its circulation and purity, was

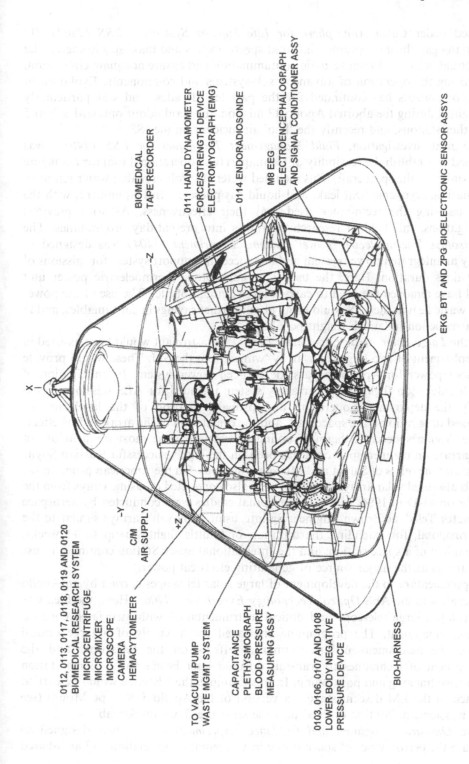

Biomedical experiments onboard an AES Block II CM.

grouped under *Cabin Atmosphere for Life Support Systems – LSS (1402)*. By employing gas chromatography, infrared spectrometers and mass spectrometry, the crew could detect and remove toxic contamination and ensure adequate circulation, monitoring the operation of advanced sub-systems and components. Evolution of these components has continued for the past few decades, and was particularly highlighted during the aborted Apollo 13 mission, the acrid odour onboard Salyut 5 and other stations, and recently the 'bad' air incident on the ISS.

The next investigation, *Fluid Management Techniques for LSS (1403)*, was designed to establish the feasibility of techniques (and operational use) for managing fluids onboard the spacecraft. This included water electrolysis, cabin water removal, sublimation, cryogenic heat leak, and liquid/gas phase control equipment, with the crew assessing the techniques used and their effectiveness. As with previous investigations, this field of research continues into present-day programmes. The *Radioisotope Thermoelectric Power Systems Experiment (1404)* was designed to qualify an integrated power system and advanced life support system for missions of 45–90 days duration. Here, the use of a radioisotope thermoelectric power unit would have removed the need to carry a large amount of fuel. The use of the power unit's waste heat in the LSS could bring about further savings in consumables, and in turn a reduction in launch weight penalties.

In the *Large Solar Array Experiment (1405)*, an astronaut would have assisted in the deployment of two large solar array 'wings' from the SM. These would provide sufficient power for Apollo to pass the fuel-cell power system. In anticipation of extended-duration flight, the solar array system intended for AES missions would qualify the deployment, operation and probably jettison of the hardware, its increased drag ratio on the spacecraft, and the orientation of the array and its effects on the Apollo spacecraft thermal cooling parameters. (The most frequent use of solar arrays on manned spacecraft has been on the highly successful Russian Soyuz family and the series of Salyut and Mir space stations. On the American programme, Skylab also used solar arrays, and tests were also conducted on using arrays from the Shuttle on 41-D in 1984. In 1992, conceptual studies were conducted by aerospace contractor Teledyne Brown Engineering into using deployable arrays similar to the AES proposal, for extending the duration of Shuttle flights to up to 11 weeks, independent of a space station, and the International Space Station continues to use solar arrays as the major source of generating electrical power.)

Supplementary to the development of large solar telescopes carried by an Apollo spacecraft was the AES *Optical Technology Experiment (1406)*. Here, the idea was to develop a large telescope and optical instrumentation with associated communication equipment. The programme envisaged the assembly of a large optical device, the assessment of environmental effects on the telescope, and the development of techniques to transmit diffused laser beams to an Earth station for precise tracking and positioning. In this example, the telescope would be carried in place of the LM descent stage – a version of the Apollo Telescope Mount (see AAP missions, p. 70) that evolved into the structure flown on Skylab.

The *Onboard Navigation and Guidance Experiment (1407)* was designed to evaluate the performance of space-borne instrumentation in relation to Earth-based instrumentation. From inside an experiment 'can', the astronauts would have evaluated the navigation, guidance, stabilisation and control sub-systems that would be employed to identify further technologies required to produce these primary onboard sub-systems on further programmes.

15 Extravehicular operations

When these presentations were issued in May 1965, no American had walked in space, and only cosmonaut Alexei Leonov had made an excursion outside a spacecraft in orbit. The art of spacewalking – Extravehicular Activity (EVA) – had been proven to be feasible by Leonov, but was only in the planning stages at NASA when the plans for AES were being evaluated. The first difficulties in performing activities outside the spacecraft were discovered and addressed during the Gemini EVAs of 1966, but in 1965 it was assumed that EVA would become an integral part of AES missions, building upon experience gained on Gemini and Apollo lunar flights.

In the *Advanced Space Suits Investigation (1501)*, the objective was to validate the integrity, protection and performance of new spacesuits and the technology involved in constructing them, in order to establish advanced design criteria for future applications. In this investigation, EVA engineers and suit technicians would have tasked the crew and the suits to gather data and crew experience on the thermal performance, emergency procedures, leakage changes, micrometeoroid protection, lubrication requirements and degradation of EVA equipment. In a way, this was once again the genesis for similar experiments conducted during Shuttle missions in the mid-1990s, paving the way for extended EVA operations during the construction of the International Space Station.

For the *Manned Locomotion and Manoeuvring Investigation (1502)*, the EVA astronauts were to develop techniques and procedures and to evaluate equipment, for use with or without propulsions systems. They were to evaluate body language movements (probably by means of film documentation), safety tethers and 'cargo nets', hand-holds, astronaut manoeuvring units, the use of tools, centre-of-gravity adjustments, and training techniques (probably pre-flight as well as in flight). Work in this field was aided by valuable in-flight demonstrations of the difficulties encountered during the last four Gemini flights (Gemini 9–12) in 1966. These lessons were well learned, and today EVA is a well-orchestrated and effective tool for the crew in space.

Danger and risk in spaceflight is ever present, and this was clearly recognised early in manned spaceflight and EVA planning. Early manned spacecraft in both the Russian and American programmes did not lend themselves to easy space rescue capabilities, but AES offered the opportunity to explore this option in greater depth. The *Emergency Rescue Operations Experiment (1503)* had the stated objective of evaluating and developing 'failure warning devices and procedures for rescuing injured personnel'. Key techniques and equipment to achieve this was expected to include failure detection and warning systems, development of rendezvous and approach by an EVA astronaut to a passive astronaut, opening spacecraft hatches from the outside or by remote procedures, development of portable airlocks, transfer of injured personnel, and provision of auxiliary life support systems. A wide variety of emergency crew rescue systems and procedures was proposed. (These studies continued into the 1970s, and included the provision of a Skylab Rescue CSM, ready on the pad to bring back a stranded crew from the station – which almost became the case during the second flight. With the advent of the Shuttle, a crew rescue system for orbital transfer was developed, although it was never implemented or even tested on a mission. For the ISS, the development of crew rescue vehicles continues, but for the foreseeable future the Russian Soyuz spacecraft will continue in its primary role of crew escape for resident crew-members, as it has done since 1971 and the first Salyut space station.

Personnel and Cargo Transfer (1504) experiments were developed to evaluate equipment operations and training procedures for the movement of astronauts and cargo by means of EVA. Evaluations were to have included egress and ingress (both with and without airlock tethers), the development of AMUs, transfers via a tether line or conveyor system or by astronaut using only the AMU, and the developments of the training procedures for crews to practise these techniques on the ground. (The Shuttle Astronaut Manoeuvring Unit was evaluated early in the Shuttle programme as a crew rescue mobility system, and several Shuttle flights have included mobility and mass transfer experiments (including obliging fellow astronauts) in a series of Detailed Test Objectives (DTOs) and Detailed Secondary Objectives (DSOs), refining the skills needed to move masses around the outside of the ISS.)

The development of techniques, hardware and training methods suitable for all kinds of space mission was the objective of the *Maintenance and Repair Techniques Investigation (1505)*. Here, the astronauts were to have investigated the assembly, disassembly, replacement and adjustment of various types of hardware, progressing from the simple to the highly sophisticated. (Thirty years later, this has developed into the skills used on several Shuttle flights to rescue, repair and service satellites such as Solar Max, Intelsat and, of course, the Hubble Space Telescope. On Salyut 7 and Mir, cosmonauts also developed their own techniques for prolonging the life of orbital stations beyond their design life or after major malfunctions. On the ISS, both of these skills have been brought together, and the legacy of AES Experiment 1505 is being practised regularly.)

Prolonging the duration that a spacecraft can remain in space depended in part upon an adequate supply of propellants to fuel manoeuvring engines. The *Propellant Handling Techniques Experiment (1506)* on AES was developed to determine the optimum methods for liquid propellant storage, handling and transfer in the vacuum of space. The investigations were originally in sealed models, progressing to more sophisticated systems, in which thermal and insulation characteristics, ullage control, venting, liquid orientation and interface, bladders and capillaries and, of course, crew safety, were all addressed. (A demonstration of the technology of transferring fuel was included on the EVA tasks for Shuttle mission 41-G in October 1984, when plans for the refuelling of satellites by the Shuttle were under review prior to the *Challenger* accident. The art of refuelling spacecraft has been mastered by the Russians in their Progress series of robotic freighters since the mid-1970s, and continues on the ISS. But the dangers of potential leaks from the cryogenic/glycol potentially poisonous fuel have been experienced during incidents inside Mir and on

EVA, including on the ISS, where a strict decontamination system is enforced. During the first EVA of STS-98 early in 2001, during the connection of coolant lines between the newly installed *Destiny* laboratory and the rest of the ISS, a small amount of frozen ammonia crystals seeped out into the vacuum of space. Though quickly stopped, the decontamination rules came into force to ensure that the returning EVA crew brought no particles of ammonia into the Shuttle airlock. They brushed off their suits and equipment by hand, and 'sun-bathed' for thirty minutes to allow solar heat to vaporise any residual crystals attached to their suits. In the airlock, a partial pressurisation was completed before venting again to flush out any missed particles of ammonia. The crew inside the Shuttle worked in oxygen masks for twenty minutes while the orbiter life support system purged the cabin atmosphere, after the two EVA astronauts had re-entered the compartment. No traces of ammonia were recorded, nor any traces smelled or detected by the crew. So, over 35 years after the 1506 investigation was proposed, the techniques and dangers have been fully experienced many times on the Russian stations, the Shuttle and now the ISS.)

The final EVA investigation listed for AES was the *EV Assembly Operation (1507)*. This was a demonstration of techniques and procedures for later more advanced missions. The astronauts would initially use simple tools and breadboards (mock-ups), and then evolve the investigations to include space assembly of structural elements and tankage in support of future mission planning for large scale EVA operations – space construction. Using a dedicated tool-kit, mock-up tanks, work stations (similar to that used on Gemini 12 in 1966), and 'structures' measuring 0.5 feet square and up to 8 feet long or more, astronauts were to hone their skills for later missions to construct space stations and interplanetary spacecraft. (At least, that was the plan in 1965. The first clear demonstration of these techniques came twenty years later in December 1985, as Shuttle mission 61-B astronauts performed space assembly techniques using the EASE/ACCESS equipment, demonstrating procedures designed for Space Station Freedom. On Salyut 7, and more extensively on Mir during the mid-1980s–1990s, cosmonauts evaluated techniques for adding solar arrays, girders, cranes and other equipment to the outside of a space station. From 1998, EVA operations at the ISS have focused on space construction, in conjunction with the use of robotic arms controlled by crew-members inside the spacecraft. With its origins in the science fiction artwork of the 1950s, experiment 1507 is currently space reality on the ISS.)

16 Manoeuvring and docking

The *Orbital Manoeuvring and Docking (1601)* research study was to be divided into two phases, with the common objective of developing capabilities beyond that of Gemini or Apollo in rendezvous and docking with passive, non-cooperative satellites. Phase 1 included the determination of spin, de-spinning, the attachment of grappling devices, 'tow lines' (tethers?) or magnetic coupling; Phase II developed non coplanar rendezvous and docking techniques using the Apollo CSM and LM spacecraft.

The *Observation of Echo Experiment (1602)* was to determine the characteristics of the Echo I and/or Echo II balloon satellites by performing visual (and verbal) observation and photography from close range, using the manoeuvring and proximity skills developed as part of experiment 1601 with a passive target. By using radar measurement uplinked from the ground, the crew could 'find' Echo in orbit and fly in formation with it. They could then provide visual and photographic evidence of any wrinkling and general shape deterioration of the balloon satellite, evaluating its potential for future use and application. The skills learned from the first part of this AES research field would lend themselves to preparing for the *Recapture of Syncom III (1603)* task, the idea being to bring back a satellite to analyse its components, which had been exposed to the space environment.

The flight plan for this seemed straightforward. The launch of a two-man AES (Apollo CSM) into a synchronous, equilateral orbit would be followed by manoeuvring towards the known location of the satellite. Ground stations would aid the rendezvous, and once achieved, the crew were to 'snare' Syncom III and retrieve it onboard the Apollo for examination and storage, pending return to Earth for post-flight analysis. At this time, similar plans were being formed to send Gemini or Apollo spacecraft to Pegasus unmanned satellites – the forerunner of further plans to service space platforms (such as the Hubble Space Telescope) in orbit. (The idea of 'snaring' a rogue satellite, or one that was not expected to be manhandled in space, was re-evaluated during the early Shuttle programme, during which astronauts on mission 51-A in 1984 used the MMU and a special 'stinger' capture device to retrieve the stranded Westar and Palapa Comsats. On 51-D and 51-I in 1985, they also captured, repaired and released the Leasat satellites by hand. Then, in 1992, the first three-man EVA operations were undertaken for the capture and reboost of an Intelsat vehicle. Other plans have been put forward for similar operations but have yet to be manifested. It was twenty years before the ideas proposed by experiment 1603 became part of space history.)

AES experiments and mission assignments

Series 200 flight assignments: Saturn 1B launched missions
Series 500 flight assignments: Saturn V launched missions

Medical – 0100 series
Twenty-one medical experiments, to be assigned to all AES manifested flights

Behavioural – 0200 series
Three behavioural experiments, to be assigned to all AES manifested flights

Artificial g – 0300 series

215	221		0301	Rotating station
219			0302	On-board centrifuge

Living Organisms – 0400 series

218	516	223	0401	Generic effects – micro-organism
218	516	523	0402	Effects on unicellular organisms
218	516	523	0403	'O' g effects on primates

218	516	523	0404	Limb regeneration and wound healing
218	516	523	0406	Space-borne micro-organisms

Space Environment – 0500 series
(All but resupply missions)

			0501	Radiation environment monitoring
513	521		0502	Magnetic field lines
513	521		0503	Comet-like particulate clouds
	521	523	0504	Micrometeoroid collection

Liquid/Gas/Solid Behaviour – 0600 series

209	218	523	0601	Capilliary studies
	218	523	0602A	Kinetics/dynamics if gas bubbles
	218	523	0602B	Liquid drop dynamics
	218	523	0603A	Pool boiling in 'O' g
	218	523	0603B	Nucleate condensate of fluids
	218	523	0604	Fluid density gradient
	218	523	0605	Crystallisation studies
	218	523	0606	Cosmic ray emission

Astronomical Observations – 0700 series

516	523		0701	Emission line radiometry
516	523		0702	Intermediate size telescope
516	523		0703	Manned coronagraph
523			0704	Nearby stars in X-rays
523			0705A	Radio astronomy 1–5 Mcps
	509	516	0705B	Radio astronomy 1–5 Mcps with 'v' antenna

Remote Sensing of Atmosphere – 0800 series

521	509	516	0801	Conjugate aurora and airglow
513	518		0802A	H–R IR radiometer detectors
513	518		0802B	Tests of IR scanning spectrometer
513	518		0802C	Evaluation and calibration of microwave spectrometer
513	518		0802D	Prototype star-tracker

Remote Sensing of Earth – 0900 series

507	215	518	0901	Multi-spectral target characteristics
507	215	518	0902	Synoptic Earth mapping
		518	0903	Multi-frequency radar imagery

Manoeuvrable Sub-Satellite – 1000 series

516	1001	Small manoeuvrable satellite (SMS)

Launch of Unmanned Satellite – 1100 series

521	1101	Launch of OGO

Electromagnetic Propagation – 1200 series

507	1201	Measurement of RF radiation
507	1202	Wide-band width transmission

Space Structures – 1300 series

	522	1301	Effects of environment
229	509	1302	Deploy RF reflective structures
516	509	1303	Extendable rod tests
516	509	1304	Solar sailing passive comsat
	229	1305	Deployment of gravity gradient structure
509	229	1306	Large aperture erectable antenna

Sub-System Development and Test – 1400 series

523	211	1401	Personal hygiene and food technology
230	211	1402	Cabin atmosphere for LSS
	521	1403	Fluid management and technology for LSS
	521	1404	Radioisotope thermoelectric power system
518	507	1405	Large solar cell array
	521	1406	Optical technology
	219	1407	Onboard navigation–guidance

Extravehicular Operations – 1500 series

219	209	1501	Advanced space suit assemblies
219	209	1502	Manned locomotion & manoeuvring
219	209	1503	Emergency rescue operations
219	229	1504	Personnel and cargo transfer
219		1505	Maintenance and repair techniques
	230	1506	Propellant handling techniques
	513	1507	EVA assembly operations

Manoeuvring and Docking – 1600 series

508	1601	Orbital manoeuvring and docking (Phase I and Phase II)
513	1602	Observation of Echo I/II
509	1603	Recapture of Syncom III
211		Lunar survey camera check-out

(This final investigation was part of the planned Apollo lunar programme, and was expected to be tested in Earth orbit first as an add-on to AES series prior to lunar flights, and in application for a proposed Earth Survey (Mapping) camera under development. See Apollo Applications p. 87)

AES SUMMARY

Between 1961 and 1965 NASA completed a wide-ranging study of various Earth orbital laboratory concepts (see the illustration p. 19). The earliest of these featured a rack facility designated Extended Apollo System Utilisation Study (EASUS), carried in the LM adapter. Other early studies proposed the use of the LM ascent stage or one or more newly designed Laboratory Modules, while others used the basic dimensions of the Spacecraft Lunar Adapter as the basis for a large-volume laboratory area. The final approach was to consider the development of a completely new and separate Laboratory Module. It was from this configuration at the beginning of Fiscal Year 1967 that NASA Contract NAS-9-5017 was subject to a revision (Addendum IV) to continue studies under the Apollo Applications Program that evolved into the Spent Stage Experimental Support Module (SSESM) concept which featured in the ensuing phase of the adaption of Apollo hardware to other objectives.

1 *Gemini: Steps to the Moon,* David J. Shayler, Springer–Praxis, August 2001.
2 *Stages to Saturn,* NASA SP-4206; *NASA Historical Data Book*, Volume II, Programs and Projects 1958–1968, NASA SP-4012.
3 NASA Historical Data Book, Volume II, Programs and Projects 1958–1968, Linda Neuman Ezell, NASA SP-4012, 1988, 53–61.
4 *Moonport: a History of Apollo Launch Facilities and Operations,* Charles D. Benson, NASA SP-4204, 1978.
5 *Evaluation of Transport Systems ... C-5 Space Vehicles,* (NASA, M-LOD-F) prepared by J. H. Deese, Chief Facilities Design Section, NASA Launch Operations Directorate, 11 June 1962. From the KSC History Office Archives.
6 *Skylab: America's Space Station,* David J. Shayler, Springer–Praxis, May 2001.
7 AES Experiment Program Discussions, Presentation Charts, 12 May 1965, Advanced Spacecraft Technology Division, Space Station Study Office, NASA, MSC. From the Curt Michel Collection, Rice University.

New applications

When NASA established the Saturn/Apollo Applications Directorate within the Office of Manned Spaceflight on 6 August 1965, there was growing interest at the Marshall Space Flight Center in Huntsville, Alabama, to further develop a concept for utilising the Saturn S-IVB stage as a shelter for habitation and utilisation by astronauts, both in the Earth orbital phase of the programme and possibly on the Moon. On 20 August, MSFC initiated a four-month conceptual design study to further explore this concept, with the assistance of the Manned Spacecraft Center in Houston and Douglas Aircraft, the prime contractor for the stage. Meanwhile, at North American Aviation, prime contractor of the CSM, in-house studies were conducted under contract from NASA into the feasibility of a Block III version of the CSM to support extended-duration missions both in Earth orbit and out to the Moon.

BLOCK III CSM STUDIES

In December 1966 a wide-ranging Congressional staff study was completed by the Congressional Sub-Committee on NASA Oversight into the pace and progress of the Apollo programme. This had been initiated in 1965 to report on the state of the lunar programme approximately half-way into President Kennedy's mandate to reach the Moon by the end of the decade. Most of this report focused on a number of engineering problems that still had to be solved with the S-II second stage of the Saturn V, the Lunar Module, and the Block II CSM. In addition, areas of the budget allocation, long-range lunar plans and post-Apollo studies were covered in the report (these will be described later in this book), although the report also contained discussions of a further upgrade to the Apollo CSM, aimed at supporting the Apollo Application Program and designated Block III.[1]

For some time NAA had been studying the option of a third-generation Apollo spacecraft. In September 1966, members of the Staff Study visited NAA facilities in California to discuss with company representatives the status of NAA's role in the Apollo programme. It was during these discussions that NAA staff presented the planning schedule for the Block III series. It appears that once the approval for the Block III programme had been given at the end of 1966, the final specifications would be completed nine months later in the latter half of 1967. The delivery of the first spacecraft would be achieved about 2½ years later in the first half of 1970, with the first flight of a Block III CSM occurring not before 1971. At the September 1965 briefings NAA staff took the opportunity to underline the need to begin Block III work immediately to ensure a smooth flow in production after completing Block II spacecraft deliveries in 1969.

NAA reported that the Block II spacecraft would complete qualification during 1967, and the workforce would then have to endure a long wait before the beginning of Block III qualification in early 1970. To help fill this gap in 1968 and 1969, the impression from the discussions is that NAA staff were underlying the need for additional work, still linked to the overall support of lunar operations to provide the technical challenge in order to maintain the workforce skills before work on Block III qualification commenced.

During this period, NAA was experiencing time-scheduling problems with Apollo Block II vehicles, but pointed out that these were being resolved. In fact, NAA commented that the CSM was 'getting to be line-production spacecraft' (a philosophy that the Soviets used on their Vostok and Soyuz spacecraft), but did not offer 'major challenges to higher-lever technical engineers'. The initiation of an aggressive Apollo Applications Program from 1968 would provide the experience for engineering and technical teams and allow the opportunity to shift between programmes as required, reduce overall costs, and keep the quality of the workforce at a higher level through the main Apollo landing objective. NAA also pointed out that it was far more expensive to close down the production line down after Block II completion and then reopen it again for Block III rather than keep it going without a break.

NAA had also recently completed the preliminary definition phase, and made a form Block III proposal to NASA, whereupon the Apollo NAS 9-5017 contract was extended, and other alternative approaches to improving the CSM were considered. These included a two-phase programme of short-duration (nominally 14-day) missions using the essentially unmodified Block II spacecraft (Phase I, which evolved into the J series of lunar missions), and then extended-duration missions of around 45 days (in Earth orbit or 35 days in lunar orbit) using Block III spacecraft with new or modified sub-systems (Phase II) .

Block III reference missions
Under Phase II, preliminary definition studies were completed on four 'mission' profile: Earth polar orbit and the synchronous-orbit, and two lunar distance profiles, which included an LEM Taxi escort mission, and the lunar polar orbit mission of 34 days duration allowing about 28 days in lunar orbit. These reference missions allowed NAA to define mission requirements related to spacecraft sub-systems. The Earth orbital missions were:

Earth-polar-orbit mission The study found that maintaining attitude hold using the

RCS would require almost the maximum supply of available consumables and cycling of the communications systems, out of the nominal range of Manned Space Flight Network (MSFN) coverage during retro-fire; and recovery added further complexity to the mission. A further problem concerned the direction from which the Saturns left the Cape to reach orbit. In 1966 the KSC launch azimuth (direction or heading) limited NASA to plan for missions 70°–110°, allowing a maximum orbital inclination out of the Cape at 45°–50°, to avoid flying the vehicle over land masses during powered ascent. If a polar orbit mission was scheduled this way (using a Saturn V) once orbital insertions had been achieved the plane of the orbit would have to be shifted (using the S-IVB) to attain the desired 40°–45°. This method was hugely expensive in propellant values on the launch vehicle, which could normally deliver 100–125 tons into low Earth orbit but only 40,000 lbs in low-altitude polar orbit

Alternative studies by NAA and NASA revealed a method to avoid these problems and send a manned Apollo on a polar orbit mission. Ninety seconds after launch, the Saturn V would begin a yaw steering manoeuvre to take it on a southerly course over the Caribbean islands. Normal jettisoning of the Launch Escape Tower (LET) would probably result in ground impact on the northern shores of Cuba, and to solve this a delay of 100 seconds in LET separation in the ocean south of the island was introduced at the expense of slight payload-to-orbit values. Injection into a 100 × 200-nautical-mile polar orbit would take place as the vehicle flew over the coast of Ecuador, with the orbit circularised at the 200-mile perigee point.

Flying a near-polar orbit over a duration of 45 days would allow surveillance of most of the planet's surface, providing research opportunities in the fields of agriculture and forestry, geology and hydrology, oceanography, geography, and atmosphere sciences, as well as communications, navigation and traffic control. The flight would also provide the opportunity to carry out the first manned observations of both polar regions, as well a provide a set of 45 daily records of data recording global climate changes and environmental systems – a unique resource.

During the mid-1960s the global environmental issue was still in its infancy, and early findings from spacecraft were beginning to identify the value of Earth observations from space. Studies of natural resources, urban growth and land use, pollution and deforestation, and the depletion of the ozone layer, would follow, and it was felt that missions such as the Phase II Apollo CSM polar orbit mission would further advance these studies and assist the promotion of the global benefit of manned space exploration. It was also during this period that the debate into the value of an expensive man-in-space programme over relatively cheaper automated spacecraft was growing. In completing the manned missions it was pointed out that with the crew available to respond to real-time situations this would offer a far broader range of investigation and research capabilities than was possible with fully automated spacecraft, and that both manned and unmanned spacecraft are required to fully explore space in order to gain the most from the investment. This debate still continues four decades later.

At the end of the polar orbit mission, retro-fire would have taken place over the Madrid–Canary ground station, with entry interface beginning at 400,000 ft and a

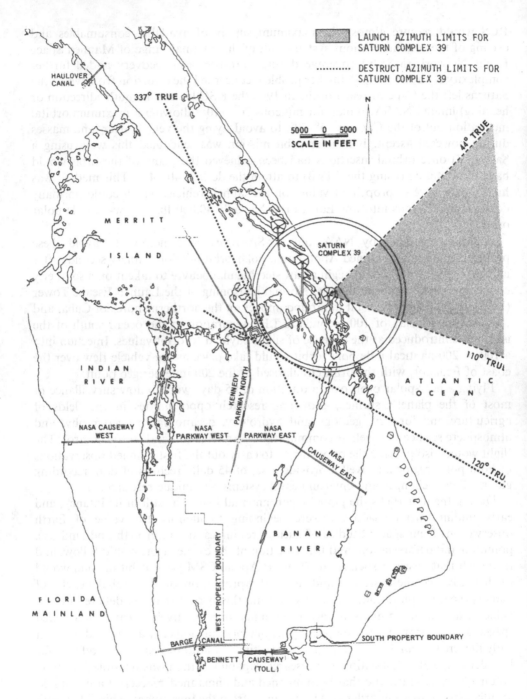

A general view of the Saturn V Launch and Range Safety Azimuth limits out of LC 39. This drawing, produced in 1969, shows a planned but unbuilt third pad to the north.

recovery in the Pacific Ocean east of the Guam tracking station. This profile would require an additional tracking station in Alaska for coverage during retro-fire and the entry phase. NAA recommended a station at Point Barrow, but understood that NASA was initiating a station at Fairbanks to accomplish this type of coverage in the future.

Synchronous-orbit mission This mission was also studied as being of 45 days' duration, but demanded the greatest electrical energy requirement and the worst thrall control problems. The method of reaching the required orbit was also rather complex, beginning with a nominal departure from the Cape out over the Atlantic into a parking orbit. The S-IVB would then be fired as the orbit took the combination near the west coast of Africa over the Gulf of Guinea to take the vehicle into a transfer orbit to an apogee of a synchronous-orbit altitude (stated to be 19,500 nautical miles). A second S-IVB burn would be initiated as the combination passed high over Indonesia, followed 13 hours later by the first phasing orbit burn by the SPS over South America, and then an orbit (several hours) later a second SPS burn to circularise the orbit, nominally at 19,000 nautical miles.

Following this type of profile allowed for maximisation of the planned payload capabilities, and for keeping the spacecraft (described in the report as 'walking into position') within the overlapping coverage of both the Goldstone and Madrid tracking stations. It also allowed for easier preparations for an Indian Ocean recovery at the end of the mission . The requirements for constant communications and ease of recovery therefore indicated which type of profile should be flown, but also required a modification to the S-IVB to perform a second re-light after a long period of coasting.

In the six case studies conducted it was determined that unmodified Block II spacecraft could fly solo mission durations of 10.8–14 days, and with an LM, a durations of 24–30 days. For the Block III spacecraft a programme of major system modifications would be required in order to support the proposed types of missions proposed in Phase II.

However, by February 1966 no further work on solo long-duration Apollo CSM flights in AAP was being conducted by NAA. Only the proposed experiments the flights were to carry received any further development from the AES experiments (listed on p. 42) but were mainly biomedical and did not require much additional stowage in the CM (see the diagram on p. 84). For most of 1966, work at NAA focused on supporting the development of the SSESM configuration using the S-IVB stage, and the use of the Apollo as a ferry craft to and from the SSESM/S-IVB combination.

Although in 1966 the NAA studies of solo Apollo CSM missions were ending, a second company was conducting its own studies into using Apollo spacecraft to support a package of telescopes as part of the Earth orbital AAP programme under development.

MOUNTING TELESCOPES ON APOLLO

Concurrent to the NAA Block III and SSESM studies, Ball Brothers Research Corporation, of Boulder, Colorado, was about to begin a separate conceptual study for NASA (Contract NASW-1318)[2] expanding the proposal for flying large (solar) telescopes on the Apollo Service Module under the Apollo Extension System programme (see AES experiment 0701–0703, p. 29). The final report of this 7½ - month study, submitted on 1 April 1966, focused on the technical requirements for a conceptual Apollo Telescope Mount using the CSM vehicle; but it also had to acknowledge that 'due to the developmental nature of the Apollo spacecraft, much of the data thus derived has fluctuated during the course of the study and now may be obsolete in certain areas.'

With the study going on at Marshall, and similar conceptual designs for utilising the Lunar Module as a telescope mounting being developed at Grumman, the Ball Brothers report focused more on the equipment, types of telescope and sub-systems required for an operational ATM, rather than what type of carrier it would use to gain access to space. However, the report did note a number of guidelines that had been followed in order to conduct the study, which projected an operational concept for using an Apollo CSM as a carrier for a major scientific payload in Earth orbit. The following is an extraction from that Final Report, to demonstrate the direction that the AES was following at its peak as AAP took over in the autumn of 1965.

CSM-MOUNTED TELESCOPES

The study concept by Ball Brothers used the guidelines that the 'mission' would be in near Earth orbit, and be flown by a Block II Apollo. The ATM would be located in Sector I of the Service Module, having minimum interface with the spacecraft, but maximum utilisation of existing technology and hardware. A solar-oriented system was provided as a concept study, with adaptability to a variety of other celestial targets providing the maximum versatility for future Apollo science missions and optimum utilisation of the Apollo astronaut crew, both as operators and observers.

The primary objective of the study was to develop a conceptual ATM system compatible with the requirements of both the scientific instruments it was to carry and the interface constraints imposed by the Block II CSM and a mission in Earth orbit. The study drew upon four solar experiments under development for a hypothetical payload, and upon official documents and technical conferences with NASA and NAA for interface data with the CSM. The Final Report, consisting of thirteen sections, went into some depth as to the design considerations investigated, concept sub-systems developed and their trade-offs, possible extended applications of the system (probably in light of the growing interest in using the S-IVB and LM hardware in the AAP) and a list of conclusions and recommendations resulting from the study. The concept used in the study and discussed here was a key developmental stage in the evolution of the ATM that was launched as part of the Skylab Orbital Workshop (SL-OWS) in May 1973.

CSM/ATM hardware and systems

The ATM was to have been a three-axis orientated solar research platform mounted in Sector I of the SM. It had to accommodate several solar experiments with the capacity to record solar features of approximately 5 arcseconds in diameter, to point the experiments to any desired position on the solar disc, hold the alignment selected to within ±5 arcsec in pitch and yaw, and limit roll about the line-of-sight to approximately 1 arcsec per second during the period of data acquisition. The ATM comprised three main sections: the Sector 1 mounting cradle (1); the extension mechanism (2); and the orientation system (3) (illustrated in the diagram on p. 56). These were the structural elements of one of eight sub-systems of the design. The sub-systems also included thermal control, pointing control, solar monitoring, command, data, power and electrical distribution.

Sector I mounting cradle

This element supported the ATM components, and was the physical attachment to the SM. Mounted to this structure was the extension mechanism and major portions of the data, command and power sub-systems.

The Block II Apollo SM measured 154 inches in diameter and 155 inches long and in cross-section was divided in to six pie-shaped segments (identified as Sectors I–VI) with a central tunnel running the length of the module. The six segments were two each of 50°, 60° and 70°. At this time, Sectors II and V were the location of the oxidiser tanks, while Sectors III and VI housed the SPS fuel tanks and Sector IV was fitted out with the fuel cells and associated tanks. Sector I, a 50-degree segment, was left unoccupied for housing experimental payloads such as the proposed ATM. The dimensions of the Sector therefore determined the maximum envelope of the experiment that it could carry (151 inches long, 50.5 inches high, 17 inches wide at the inboard edge, and 62.5 inches wide at the outboard edge of the sector). This produced a total volume for experiment consideration within Sector I of no more than 308,600 cubic inches.

The ATM would have been stowed inside this Sector for launch until the spacecraft had achieved its desired orbit. The cradle supporting the experiments would be mounted on the central spar via isolation mounts, which minimised the effects of spar temperature changes and distortion on the experiments. It also provided significant structural strength to react to launch loads and to relieve stresses on the gimbals used to extend the unit. The cradle housed the ATM system components, attached at eight points – four in the designed mounting points in Sector I (Sta.290.66), and the remaining four in pairs at the fore and aft outboard ends of the cradle.

Extension system

The extension system and section was a stowed tripod folded in Sector I for launch. It used a pneumatically driven parallel linkage to drive the mechanism to full deployment at 90° once on orbit, and could be operated in 1 g for ground testing of the flight equipment. The orientation section was attached to the top of this structure and would have been extended out from the SM parallel to its stowed position.

When extended, this locked in place along the long axis of the orientation section, parallel to the longitudinal axis of the Apollo spacecraft. At full extension, the centre of mass of the ATM system was approximately located at the cross-section plane of the CSM centre of mass. The extension system was located at the forward end of the cradle (Sta. X = 343.75), and when extended it resembled a parallel-bar linkage stabilised by a strut. On top of the extension structure was a three-axis gimbal platform. The yaw was the inside gimbal, the pitch was the intermediate gimbal, and the roll was the outside gimbal. The roll gimbal shaft was structurally mounted to the extension spar. The platform provided travel of $\pm 170°$ in yaw, $\pm 30°$ in pitch, and $\pm 25°$ in roll, with the system protected from mechanical interface with the orientation section of the CSM in any combination of ATM gimbal angles.

Orientation section

The orientation section included the experiment support structure (spar) that was attached to the 'inside-out' gimbal housing. This supported the experiments, part of the ATM electrical sub-system, solar sensors, solar monitoring telescope, and thermal shield. The experiments were integrated on three sides of the spar, and occupied the majority of the orientation section. Surrounding the entire orientation section structure and experiments was a thermal shield and heaters. These served as the Thermal Control Sub-system, and included aperture covers on the Sun-facing end. It was operated from within the CM for protection of the experiments, hardware and ATM optics. The maximum length of the experiments within this heat shield was restricted to 134 inches. Hatches were mounted on the external surface of the thermal shield for astronaut access to the experiment film magazines during EVA.

This section would have been jettisoned prior to re-entry near the end of the mission, for the reliability of the re-entry manoeuvres and to conserve re-entry fuel. This decision was taken in light of experimental data and calculations about the effects of the mass of the ATM on the CSM centre of gravity, which proved that the SPS thrust line would exceed the operational envelope for a nominal re-entry alignment if the ATM remained attached. There was also the problem of re-stowing the unit, since the SM would be separated and destroyed on re-entry. The added complexity and cost was at that time thought to be unnecessary. The concept *did* include provision for adding a re-stow device, should this have been identified as a future requirement.

The cover of Sector I (an integral element of the SM structure) was deployable with the orientation section, and included reinforced access panels for pre-launch film loading. The study report recommended that due to the structural role of this cover, the contractor (NAA) should a incorporate design to allow for ATM interface requirements by the ATM contractors.

The estimated total mass of the ATM system was quoted at 2,200 lbs, which included 750 lbs for experiments and 600 lbs for batteries – which might be omitted if Apollo spacecraft power could be diverted for use in the ATM system (an option under study at that time).

Thermal control

The design study evaluated a theoretical model of a 14-day mission and its thermal effects on the CSM, with day-night sequencing of about 448 cycles (an average of 16 orbits per day, with one day/night and one night/day transition each orbit for 14 days). Solar radiation and CSM re-radiating from operating the ATM, reposition of the CSM for ATM targeting, and reflective qualities of the Earth and spacecraft hardware were also considered. It was decided to adapt a thermal shield to enclose the ATM comprising two shields in series, the external skin and an inner liner. The outer and inner surfaces of the external skin would have been prepared to ensure that passive thermal properties provided mean orbital external skin temperatures approximately 30° (F) below the desired internal temperatures. The inner radiation shield was attached to the inside of the skin by a series of stand-offs, and had an emissive quality of 0.1.

Apertures were provided for the experiments, fine control sensors and solar monitor sub-system optics, which would be controlled by the crew from inside the CM. An active heating system inside the thermal shield layer was designed to radiate into the orientation section. Designed for a minimum 14-day life, the heaters would be thermostatically controlled to maintain the internal environment to within ±5° (F) of the design temperature. The eight heating zones would be monitored by individual sensors, which would activate when the temperature dropped 5° (F) below desired levels, and switch off at 5° (F) above desired levels.

Several sub-systems and structural components also had to be maintained at a relatively constant temperature – or, at least, within a narrow range – including the soar, the solar monitoring sub-system interference filter and videocon, control Sun sensors, gyros, and critical electronics. This was solved with a range of incorporated thermoelectric devices, including auxiliary filters, passive thermal coatings and conductivity paths out to the outer heat shield (where a low absorption coating allowed excess heat to radiate into space), or aluminised Mylar coatings to shield equipment from other elements.

Pointing control

The design studies of the ATM revealed that there was no need for precise, absolute pointing. The three-axis control system would produce pointing stability of a few arcsec per minute, as long as the ATM did not drift at a greater rate. The rate of CSM drift had to be controlled to prevent image smearing, and the ATM roll rate coupling in pitch and yaw therefore had to be no greater than 4 arcsec per second. There were discussions with members of the Astronaut Office concerning the possibility of manual CSM control over spacecraft G&NC control. As a result, astronauts Roger Chaffee, Dr Joe Kerwin and Dr Curt Michel evaluated this synopsis, and determined that while rates could be manually reduced significantly below the 0°.05 per second (180 arcsec per second) G&NC limit, it would in all probability not be below 0°.01 per second (36 arcsec per second), underlining the need for roll axis rate control on the system.

A DC torque motor was used on each axis, with Sun sensors employed as error sensors in the automatic control mode of pitch and yaw. Automatic mode roll axis

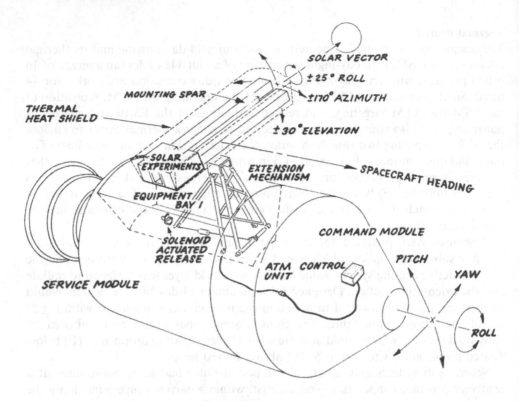

SOLAR VECTOR
± 25° ROLL
± 170° AZIMUTH
MOUNTING SPAR
THERMAL
HEAT SHIELD
± 30° ELEVATION
SOLAR
EXPERIMENTS
EXTENSION
MECHANISM
SPACECRAFT HEADING
EQUIPMENT
BAY 1
SOLENOID
ACTUATED
RELEASE
COMMAND MODULE
ATM CONTROL
UNIT
PITCH
YAW
SERVICE MODULE
ROLL

An impression of the Apollo Telescope Mount/Apollo Command and Service Module configuration *c.*1965, depicting the operational position of the telescope package during the proposed mission. (Courtesy Ball Brothers.)

control would derive from a rate-integrating gyro. The pointing control sub-system would be handled by an astronaut from inside the CM, at a rate limit slewing capability of 6° per second. The operator could then manually position the spar to any desired attitude with respect to the CSM within the allowable gimbal parameters. The Sun sensors in pitch and yaw would be used for course acquisition and for control of the orientation section. The Sun sensors and a vernier positioning capability allowed the operator to position the experiment line of sight to any spot within a 40-arcmin square centred on the solar disc. All rate angles and housekeeping data would be collected by the data sub-system for recording. The retrieved data allowed for post-flight determination of where on the disc the observation was being conducted.

Solar monitor sub-system
With the astronaut in charge of observation via the ATM system, a solar monitor needed to be installed. This would enable the crew-man to examine the solar disc in monochromatic light in order to reveal chromospheric features. He would then align the optical alignment of the given instrument to the feature on the solar disc, within

the precision limitations of the experiment. As he was physically separated from the instrument, the remote sensing system had to duplicate the view that he would have had if he were to align the instrument by means of the axis of the telescope.

In the design study, the solar monitor presented a video image to the observer in the CM, as well as digital inputs from other sensors in the ATM package. This would enable the astronaut-observer to visually track the movement of the experiment while using a 'fly to' control stick to offset the axis of the orientated section out to the edge of the disc. The monitor was designed as a patrol mode of observation, in that the spar was pointed at the Sun while the astronaut could view the whole disc and out beyond the limb. If an important event occurred, the astronaut could offset the orientation section from the solar centre in such a way that the activity target became the new centre of the display screen, and control of the video input switched from his display to the recording instruments to capture relevant data, complementary to the instrument recordings.

Command sub-system

This was the primary interface between the astronaut-observer and the ATM. It enabled the crew-man to deploy the package, control its attitude, control the experiment mode of operation, and monitor the status of the ATM hardware and experiments using controls and indicators on a CM-mounted control unit. There were two types of command: hardware and encoded.

Hardware controls were to be employed when crew safety was involved, or when added complexity would result if the encoded commands were used. These were ATM power control, ordnance control, and ATM attitude control. The 'fly to' control stick handled the ATM pitch and yaw, while a roll control knob handled movement in that axis. A range of selector buttons produced a combination of twenty-five 'discrete commands', resulting in combinations of pitch and yaw on the ATM structure or on the experiments themselves.

There would be two types of encoded command: those operated by discrete switches, and others by means of a nine-button keyboard. This limited the number of conductors between the CM and SM. Commands requiring quick reaction, or frequently used, could be initiated with a single switch. The less frequently used commands required the astronaut to manually input the nine-bit code into the encoder. It was planned to have 512 encoded commands, with 256 available for operating experiments and other ATM systems, and the remainder for direct digital link to the experiment bay and orientation section control panel indicators. These indicators included attitude indicators for constant monitoring of ATM pitch, yaw and roll gimbal angles; programme indicators for monitoring two analogue sources at the same time; and sixteen signal indicators (flags or lights) for conditions such as indication of experiments 'on target', 'experiment power on', and so on.

Data sub-system

This sub-system was responsible for the collection of all data from the ATM and its experiments, and processing it for storage or transmission to Earth via the Apollo telemetry system. As the majority of data to be collected by the ATM were generated by photoelectric instrumentation, two channel capabilities were incorporated in the system: 120 bits/sec and 4,800 bits/sec. The design of this system was influenced by the possibility of using Apollo telemetry systems had this mission flown. The alternative was for total onboard data storage in removable tape magazines. The recommendation of the study report was that a buffer tape recorder be used to gather all data, read out over the Apollo telemetry systems (more commonly known as 'dumping') every two orbits.

Power sub-system
The Ball study evaluated the power requirements of a 14-day Earth orbital mission at 30 kW hours at 28 v DC. Of this, approximately 10 kW hours were available for experiment use from the primary Apollo fuel cell systems. The additional supply could come from silver–zinc batteries or from an autonomous fuel cell as part of the ATM primary power sub-system. The study highlighted the advantages of upgrading the Apollo fuel cells, or even flying a shorter-duration mission to make up the 20 kW difference, stowing additional reactants within the Sector I bay to supply additional fuel reserves.

Electrical distribution sub-system
Wire harnesses were envisaged as primary elements of the ATM electrical distribution system in the bay, the orientation section, and the interfaces with the CM/SM umbilical connections. Major consideration was given in the study to the transmission of electrical signals and power levels across the gimbals, with minimum static and dynamic torque supplied to the pointing control surfaces. The study also recommended flexi-print leads for signals and power, and lightweight co-ax for video transmission.

OPERATIONS OF THE CSM-MOUNTED ATM

In this configuration, the crew would operate the ATM system with the support of scientists on the ground. Several ATM operations would be automatic, requiring no crew effort apart from initiation and periodic checking, but there were still a significant number of operations requiring direct crew participation throughout the mission. This would have included contingency and back-up to automated procedures where possible, and real-time input and decision-making as targets of opportunity occurred during the mission.

Crew selection and training
It was envisaged that in addition to the standard Apollo Block II CSM training, the astronauts would also be given training in solar and stellar astronomy and on ground training units. A period of pre-flight check-out and briefing sessions with experimenters would also be part of pre-mission preparations. The final briefing from Principle Investigators and specialists would be given to the crew a few days prior to launch, providing them with information on current solar activity, updating

specific areas of interest, and including last-minute refinements to the observation programme. Ground and satellite observation of the Sun would be critical in the final detailing of the mission objectives and programme.

One crew-man (the ATM Pilot?) would be the primary ATM operator, but the other two crew-members (Commander and CM Pilot) would also be trained to operate the telescopes to some extent during the mission, as back-up to the primary operator (who was expected to be a member of the science-astronaut cadre). Several prominent astronomers and experimenters proposed that a professionally trained astronomer or solar scientist should be included in the crew for ATM missions. From an early stage, the development of the ATM was monitored by physicists Curt Michel and Ed Gibson and electrical engineer Owen Garriott, from the first scientist astronaut selection in 1965. Had the CSM-mounted ATM progressed to a launch manifest and flight opportunity, then perhaps these astronauts would have been assigned to the flight and/or back-up crew. If other missions were planned, it is probable that members of the second scientist-astronaut selection in 1967 would have been considered as primary candidates for flight assignment. Ideally, those with astronomical or space sciences backgrounds would have been chosen, although (as proved on Skylab) appropriate training made competent observers out of Pilots as well as scientist-astronauts. Once selected to the astronaut programme, assignment to a flight was not necessarily a reflection of academic or career attainment.

Pre-launch operations

The ATM payload would have been prepared and tested separately from the assigned CSM, and then installed in the Sector I bay within the processing facilities at the Cape prior to roll-out to the pad. Installation on the pad was not envisaged. However, two days (or at the last accessible access time) prior to lift-off, one complete set of film magazines would be inserted into the experiments via access covers in the Sector cover panel. A second set (and probably spares) would have been loaded into the CM. It was assumed that a Saturn 1B would be employed as the primary launch vehicle for the mission.

Initial orbital operations

There were no planned ATM operations in the first 24 hours of the mission, during which the ATM would have remained stowed. This provided the crew with time for initial configuration for orbital activities, as well as adequate separation from the S-IVB stage, and dissipation of outgassing from the Apollo. Interestingly, experience on Apollo missions – especially the Apollo–Soyuz Test Project in 1975 – revealed that outgassing and water dumps posed several observation limitations that could have seriously hampered ATM activities from an Apollo.

For the purpose of this study, three years before any manned Apollo flew in space, the recommendation was that, after the initial 24 hours in flight, the Sector 1 cover would be jettisoned and the extension mechanism activated and locked in place. The thermal control system would then be activated. Following this, a second period of inactivity of several hours was recommended, to allow for outgassing of trapped particles within the ATM structure before experiment operations began.

Therefore, it would probably have been late in Flight Day 2 (FD2) that tests of the ATM orientation and control systems would be completed, with full observations beginning on FD3. The ATM yaw angle for proper CSM control would also be monitored as required throughout the mission.

ATM operations

Operation of the ATM would be initiated by manoeuvring the CSM to acquire the Sun in the ATM course target sensor and by setting the ATM gimbal angles to the desired position for maxim data acquisition.[3] When this had been achieved, the astronaut-observer would have activated the ATM manual mode, fine pointing and bore sight systems, followed by reduction of CSM rates to less than $0°.05$ per second in each axis. The ATM would be offset to the general region of experiment interest following either the Crew Activity Plan and detailed flight objectives set for the mission, or real-time data observed by the flight crew or ground observations which would have continued throughout the flight. The thermal control system and the data handling systems would also have been continually monitored.

Experiment operation would have seen the crew-man select the activity region for bore-sight alignment, followed by performing an offset bore-sight manoeuvre. Initiation of experiment data acquisition depended upon which instrument (or suite of instruments) was needed for that research objective. The CSM could be manoeuvred until the astronaut bought the ATM as close to the nominal target as he desired. At any time after the Sun was within the course target indicator field of view (revealed in a square-shaped display 40° each side), the observer could activate the course acquisition system, releasing the gimbal brakes and automatically pointing the orientation section to the centre of the Sun to within $\pm 2°$. If required, the system could be recycled to provide even finer acquisition for even more accurate use of the fine pointing mode.

As soon as the observer was satisfied that CSM rates were reduced as much as possible in the coarse mode, he would then have activated the fine pointing mode to automatically point the ATM to the centre of the solar disc to within approximately ± 1 arcmin. Doing this would also automatically open the aperture cover over the fine pointing sensors, overriding pointing by the astronaut for the final positioning. After a few seconds the sensors would have been switched out of the control loop, but if the observer closed the aperture this would automatically restore pointing control to the coarse mode once again. Fine pointing would have been restored once the covers were opened, to resume the same alignment on the Sun as before.

With ATM pointing in fine mode, the astronaut-observer could manoeuvre the ATM through a small offset angle to any point on the solar disc. He would have observed activity via the visual monitor in the CM, at a rate of 40 arcsec per second, using the 'fly to' control stick to cover a range of ± 20 arcmin from the centre of the solar disc. (It was estimated that it would take 24 seconds for the observer to offset from the centre of the disc to the limb.) The offset would remain in effect until manually removed by the astronaut, and could be held to within ± 5 arcsec for at least one minute, and to within ± 1 arcmin for a period of up to 20 minutes or more.

Throughout experiment data acquisition, the primary astronaut-observer would have controlled management of the experiments in operation mode. He would also have had to control the aperture cover 'open and close' configuration with respect to RCS thruster operation and CM waste-water dumps as required. During periods of data acquisition, the bore-sight alignment would also have to be adjusted as required. The ATM primary astronaut would have been supported by the other two crew-members during data acquisition, as required.

The astronaut-observer would have aligned the bore-sights for each experiment based on pre-flight training and on observations on the monitor in real time during the mission. When the required fine position was reached, the observer would have initiated data acquisition. This varied between individual experiments flown in the science package – ranging from totally automatic, to the observer stopping at a desired point during data acquisition. It also depended on the position in the orbit, during the Sun-side pass, of the CSM when it entered the Earth's terminator, passing out of solar line of sight. The report emphasised that training and in-flight experience would enhance the data recording, and that factors such as water dumps and RCS exhaust gas diffusion rates required further study – presumably during the Earth orbital CSM Block I test flights (Apollo 1 and Apollo 2) prior to initiating this type of mission.

The observer needed to monitor all water dumps so that he could close and re-open the aperture doors. If this occurred during periods of data acquisition with the ground, it could have resulted in the loss of some data. The observer would also have had to ensure that the aperture doors were closed during RCS firings, and for at least two minutes afterwards.

In addition to data-recording by tape, and verbal commentary, the astronauts would have logged events and sketches in the flight log book, and even have photographed the video image to save laborious sketching by the astronaut-observer. Obviously, image copy recording of the video screen was not planned at the time of this study report.

EVA film recovery and stowage
It was expected that nominal mission activity would deplete film magazine stock by approximately the seventh day of the mission. At this point, one of the astronauts (probably the CMP, as in the Apollo J series of missions in 1971–72) would have performed an EVA to retrieve the exposed cassettes and replace them with fresh ones. The ATM would have been repositioned for EVA servicing, with the gimbal locks holding the unit steady relative to the CSM during the EVA. Of course, no data would have been collected during the EVA.

New cassettes stored in the CM would have been taken outside, either attached to the astronaut's spacesuit or in a dedicated storage rack, to prevent inadvertent loss into space. At the ATM, the astronaut would have opened the hatches in the thermal shield and extracted one of the used cassettes, securing it for return to the CSM and then replacing it with a new unit and closing the cover again. A second EVA towards the end of the mission would have retrieved the second set of cassettes, but without installing any further units. Recovered film would have been stored in an experiment rack (according to one reference, in empty food storage boxes), probably beneath the centre couch. Valuable experience in handling crew equipment, waste and storage was gained during the long Gemini missions of 4 days (GT-4), 8 days (GT-5) and 14 days (GT-7) in 1965. It was estimated that about 30 minutes would have been required for each of these EVAs, in addition to suiting, cabin depressurisation and repressurisation and unsuiting activities. Experiences during the Apollo deep-space EVAs (Apollo 15–17) and on Skylab revealed that such operations would normally require approximately 45–90 minutes to perform, providing there were no incidents or any additional EVA activities added to the operation.

Ground control
Although not detailed in this study, it was envisaged that direct communications would be required between the astronaut-observer in space and ground scientists at Mission Control. This would presumably have been from a payload support room, and not directly in MOCR at Houston. Such operations were conducted on Skylab, while separate control operations were carried out for spacecraft systems and procedures. The CapCom role would presumably have been filled by members of the astronaut corps – probably members of the support crew, and probably an astronaut Mission Scientist – with a direct liaison with the Principle Investigator and experimenters in the support room. Data relayed through the CapCom would have been supplemented by ground-observed data, monitoring the air-to-ground transmissions and, it was suggested, a coordinated optical device, which would have transmitted a duplicate of what the astronaut was seeing to a similar display on the ground.

Post-flight activities
Aside from using the second EVA to retrieve the second set of exposed film cassettes and data cartridges for return in the CM, the study suggested that the second EVA astronaut might choose to remain outside the CM (stand-up EVA in the CM hatch?), or 'at a safe distance' (umbilical EVA?) to 'verify that the ATM orientation system jettisons properly.' It is fairly clear that if this mission had progressed to flight operation, the CB office would certainly have objected to a crew-member being in the vicinity of a mass of several hundred pounds being ejected (by explosive bolts) uncontrollably from the end of the extension arm into space. Confirmation of successful ejection could be obtained from remote television, or visually through spacecraft windows or indication lights on the console. If separation failed, then a contingency manual separation EVA could have been performed.

The water recovery of Apollo was designed to ensure that the stowage containers would not be affected by salt corrosion or humidity while in the water or in the CM. It was assumed that the ATM canisters would be stored within the CM and retrieved once the spacecraft was safely onboard the primary recovery vessel. As with all missions, post-flight debriefing of the crew would include medical, technical and scientific issues, and would probably have been initially conducted onboard the recovery vessel, with a more detailed debriefing at MSC in Houston. If further missions were planned, then future flight crews would certainly have participated in a debriefing session with recently returned colleagues (a procedure that was actually followed on the Apollo and Skylab missions).

ATM telescopes

The exact types of instrument to be carried on the ATM were still being evaluated at the time of the report, and several options were listed in the Final Report and in the Briefing Sheets referred to here. During the 1960s, the ATM package constantly evolved, and a description of the instruments finally flown on Skylab has previously been presented by this author.[4] However, four typical experiments discussed in the study included:

Ultraviolet Spectrometer
Principle Investigator: Dr L. Goldberg.
Affiliation: Harvard College Observatory (HCO).

X-Ray Telescope
Principle Investigator: W. White.
Affiliation: NASA Goddard Spaceflight Center (GSFC).

White-Light Coronagraph
Principle Investigator: Dr G. Newkirk.
Affiliation: High Altitude Observatory (HAO).

UltraViolet Spectrograph/Spectroheliograph
Principle Investigator: Dr R. Tousey.
Affiliation: Naval Research Laboratory (NRL).

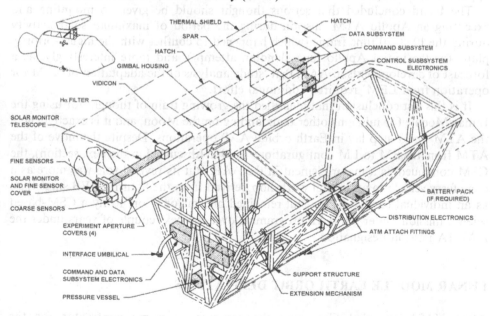

ATM payload hardware. The various sub-systems are shown, but the four experiments are not included. (Courtesy Ball Brothers.)

Contaminants and failures

The report (and later, the actual Apollo missions) helped identify several potential contaminant sources. These included: CSM and ATM sublimation products; poor or faulty ATM film and equipment/system malfunctions; astronaut liquid waste (urine dumps); CM excess water dumps; RCS thrusters gas; RCS Co^{50}; gamma-rays; overheating or freezing due to restricted manoeuvring; crew illness; and sudden solar flare activity requiring corrective action or immediate crew recovery.

Report conclusions and recommendations

The study concluded that a significant amount of space science could be accomplished via an ATM system, and that an astronaut crew could be very effective for the acquisition of scientific data. It also stated that the Apollo spacecraft design presented no foreseen constraints to the successful conclusion of such a mission. Indeed, it was emphasised that the ATM mission could be performed with relatively few changes or modifications to the Apollo spacecraft or Earth orbital mission profile, during pre-launch, launch, in-flight, or recovery. The ATM concept presented in the report was readily adaptable to a variety of other vehicles (including the Grumman LM), and for both celestial and terrestrial targets.

In its recommendations resulting from this design study, the Ball document further stated that investigationa into contamination of the experiment payload in orbit (by residual particles and gas in the vicinity of the Apollo spacecraft) and a further study analysis of ATM system adaptation were required for future applications in support of stellar and geophysical experiments.

The board concluded that serious thought should be given to mounting and executing an Apollo ATM mission during the period of maximum solar activity during the 1968–70 time frame (which happened to conflict with the higher-priority plans for the initial Apollo lunar landing attempt), and more interestingly, as a forecast of developments in AAP, that 'study analysis of the adaptation of ATM for operation from LEM' required additional effort.

It is this last conclusion that identified the growing train of thought for using the Lunar Module for missions other than landing on the Moon, and it is exactly where the ATM's next step lay in Earth orbital AAP operations. Despite the move of the ATM from the SM to LM configurations in 1965 (discussed in the next section), the CSM continued as a major element in the AAP. But its revised role was more as a crew ferry vehicle to an orbital workshop derived from the Saturn S-IVB stage than as an individual carrier. There still remained one major effort to fly a CSM-based science mission under the AAP programme over the next couple of years under the AAP-1A mission designation.

LUNAR MODULE EARTH ORBIT DERIVATIVES

After NASA awarded Grumman the contract as prime contractor for the construction of the manned lunar landing module for the Apollo programme in 1962, studies began into adapting the evolving lander design for other mission applications. These initially formed part of the Apollo Extension System (AES) studies, and were aimed at expanding lunar surface exploration once the initial Apollo missions had accomplished the landing objective. (These lunar versions of the LM are discussed in more detail from p. 197.)

In addition to lunar missions, it was soon realised that the LM design could be adapted for non-landing roles, supporting extended scientific Apollo-type missions in Earth orbit. In September 1965, Dr George E. Mueller, NASA's Associate Administrator for Manned Spaceflight, presented a statement of the status of Manned Space Flight Post-Apollo Planning to the Subcommittee on NASA Oversight. It covered the progress of American manned spaceflight up to and including Gemini 5, the prospects for the initial Apollo lunar landing goal, and applications once that objective had been achieved. The statement was published as a US Congressional document[5] and included details of what Grumman and NASA were planning for the Lunar Module in Earth orbital missions. Grumman subsequently expanded the plans for the Lunar Module during further reports to the NASA Oversight Subcommittee in August 1966 with a plan for adapting the LM to five 'missions' beyond that of its primary role of landing men on the Moon. The 'missions' were: applications, scientific, long-duration, lunar exploration, and space rescue.[6]

In statements, Grumman identified two early plans for adapting the LM to meet the first of these 'missions' centred on Earth orbit operations. Firstly as a worldwide communications resources – specifically for educational television and secondly as a 'continuous orbiting weather station which would permit accurate, long-range predictions and understanding of weather phenomena.' It will be recalled that in the mid-1960s the first weather and communication satellites were just beginning to be placed in orbit, and the benefits from these were still relatively new. At that time there were suggestions from space agencies, weather bureaux, and aerospace and telecommunications companies that manned spacecraft could serve as a source of information complementary to the automatic satellite programme still under development. As Grumman representatives correctly forecast in 1966 in their presentation to the Congressional Subcommittee: 'With our [still to be] developed mass communication systems we do not fully appreciate the enormous impact that worldwide television and broadcast capabilities will have in the underdeveloped and emerging areas of the globe.' Grumman correctly identified that such systems were to become excellent suppliers of weather forecasting and within a generation provide a global network on real-time news coverage as the world moved into a the next era of mobile telecommunications and the Internet.

In the second category it was surmised that flying packages of scientific instruments on the LM could cover such fields as astronomy, solar phenomena, life sciences and space physics. In these applications the LM would have all of its main propulsion and landing sub-systems (such as the descent and ascent engines, landing radar, and landing gear) removed, and the ascent stage would be converted to a laboratory, while the descent stage would be adapted to become the carrier for the scientific experiments.

In the third category, Grumman predicted that man's ability to operate in space for long periods of time would yield significant benefits, one of these being the ability of astronaut to obtain the best data from the experiments and application programmes intended for extended-duration spaceflight. When these plans were presented to the Congressional Subcommittee in August 1966, the question was put to Grumman: 'When you talk about long, how long?... two years?' In reply, S. Ferdman of Grumman stated: 'We are talking about the kind of experience that will enable you to be assured the confidence of going to the planets, I'd say, initially one year in Earth orbit and ultimately beyond that.' (The implication of adapting Apollo hardware to missions far beyond that of the initial lunar goal is discussed from p. 317.)

The fourth area of LM development in the field of lunar exploration is covered on p. 232, while the fifth application – space rescue – recognised that with the increase in manned space mission launches and the desire to further extend these missions, a genuine emergency rescue system would be required. Grumman promoted their LM as a soundly developed spacecraft, developed by an experience work force and benefiting from good operation systems, even though in 1966 no LM had ever flown in space. It was also recognised that the rescue systems under consideration at that time would only work under certain conditions. A disastrous situation would not be solved by the ability of a launch-ready rescue LM, and 'these systems will have to be expanded in the future years to safeguard ... more extensive missions.'

This programme of LM derivatives for future space missions evolved over a period of almost six years from 1963 until 1969. According to Grumman, it was the LM's versatility to fly manned or unmanned, its propulsive capability, and the large payload volume envelope that offered 'attractive options to achieve significant space objectives in the future.' These included increases in scientific knowledge and benefits to man, space technology, and mission time.[7] In the family of LM derivatives, missions of up to two months around the Earth were foreseen, and would last longer with resupply.

LM versatility

According to Grumman's plans, approximately one dozen variants of the basic LM design could be developed, half of which would have direct application to Earth orbital missions: LM Lab I and Lab II configurations, the LM Lab I with a Telescope Mount, and the Spent (S-IVB) Stage experiment. Others had direct application in lunar orbit or on the Moon.

The pressurised volume in the LM Lab I ascent stage was given as 247.5 cubic feet. Removal of the engines and the insertion of canisters could add an additional 85 cubic feet, resulting in a total of 332.5 cubic feet pressurised volume. In addition, the volume between the outer surface of the Lab and Spacecraft Lunar Adapter on the Saturn launch vehicle was quoted as 1,700 cubic feet. If the propellant tanks were not removed (the easiest option that allowed for storage of RCS propellants if required) there was 320 cubic feet of volume available, producing 2,020 cubic feet of total unpressurised volume. Grumman's weight specifications for LM Lab I are listed in the Table on p. 67.

Typical Earth orbit mission weight specifications for LM Lab I

	Low-inclination	Polar	Synchronous
LM Lab I weight (lbs)	11,249	11,249	11,249
Maximum altitude (miles)	200	200	19,350
Maximum orbital weight (lbs)	26,200	32,000	25,623
Saturn	1B	V	V
Maximum payload vehicle weight	14,951	20,751	14,374

Space tug/Rescue LM

In this design, the basic Apollo LM would remain virtually unchanged, with only small modifications to the ascent and descent stages. One application proposed for this was as a Rescue LM, offering the opportunity to utilise the vehicle's highly efficient Main Propulsion System and fully redundant four quad (sixteen thrusters) reaction control manoeuvring system as an efficient orbital manoeuvring vehicle. Grumman suggested that with minor modifications to the vehicle, a 'tug' LM could be used for 'in-orbit interception of, or rendezvous with, other space vehicles, for such purposes as inspection and or repair'. The installation of a Remote Manipulator device was further perceived as an effective space rescue concept. This type of adaptation to the basic LM vehicle was also proposed for a satellite repair and maintenance vehicle, or orbital reboost capabilities similar to those of the Gemini–Agena missions. Grumman also indicated that there were military applications in surveillance and the inspection of satellites that were being investigated, as well as the 'civilian' applications proposed.

LM Laboratory

In the 1965 report, the initial development of Apollo Applications forecast the use of the Experiment Pallet in Sector I of the Apollo Service Module (as proposed in AES, and in the proposal for supporting the ATM from the SM), in conjunction with the development of a Lunar Module Laboratory configuration for early Earth orbital missions. As the missions increased in complexity, so the adjustments to the LM would become apparent. For the LM Lab I configuration the landing gear, descent radar and ascent propulsion would be removed and, changes would be made to the environmental control, electrical power sub-systems – and would have been replaced by CM sub-systems and additional crew provisions to increase the capacity for orbital flight up to about 14 days in a shirtsleeve environment in Earth orbit. This would allow for a variety of mission profiles in varying orbits and with differing experiment packages.

One early application of this type of LM which Grumman foreseen would be in manned Earth resources surveys, in which the package of instruments could map unexplored and uncharted regions of the planet to identify regions of unexploited natural resources. This would be followed by flights of aircraft over those areas to pinpoint ground survey sites derived from the data obtained from space.

The statement that the Grumman LM would be adapted to receive Rockwell CM

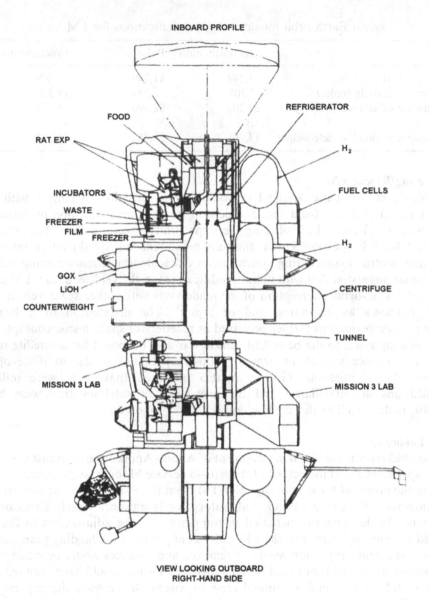

One of several configurations proposed for utilising the Apollo LM as an orbital space operation/bioscience laboratory. In this configuration, two LMs are linked together, with access through a central transfer tunnel provided by the removal of the ascent and descent engines of one of the vehicles. (Courtesy Grumman Aircraft Engineering Corporation.)

systems was interesting. Some systems were easily compatible, while many were not, and the consequence of not following this through to the manned flight phase of Apollo became evident during the flight of Apollo 13 in 1970. The lithium hydroxide canisters (used to keep the levels of carbon dioxide in the LM to safe levels for two men for about two days) were being called upon to deal with three men for four days, and were clearly not coping. When the spare canisters from the CM were moved to the LM it was found that their square design was unable to fit in the round LM receptacle. Some in-flight DIY was required to create a jury-rigged device that allowed the air to be passed through the CM filters attached to the outside of the LM receptacle. Yet this type of system incompatibility had been discussed in 1965, to develop equipment that could be used on either spacecraft.

Other artists' concepts of an LM Lab revealed a variant without a descent stage, with an array of experiments carried in the lower aft of the ascent stage and a modified scientific airlock in the forward access hatch area. Typical experiments envisaged for LM Lab missions included radiometers, spectrometers, a stellar camera, a terrain camera, multispectral camera, X-ray sensors, a day–night camera, and an IR imager. Research was targeted at meteorology, astronomy, Earth resources, surveying and mapping, bioscience, and engineering technology. The LM would have been launched unmanned by a Saturn 1B, or with a manned CSM on a Saturn V.

The LM Lab II configuration was capable of missions in both Earth orbital and lunar orbital profiles. The same modifications incorporated into LM Lab I configuration would be enhanced further to allow missions of up to 45 days, supported by the CSM. These laboratory proposals included the capability of revisit by later crews in a CSM, and in effect could be termed man-tended.

A further LM Lab configuration under study was the Bioscience/Physical Science Laboratory for Earth orbital operations. The illustration on p. 68 demonstrates one version of this concept, with major components and some experiment hardware installed. This concept allowed for the launch of two separate LM modules on different launch vehicles, to rendezvous (unmanned) and dock, and then be joined by an Apollo CSM, providing a three-vehicle 'space station' configuration. This had a projected 90-day orbital life, and was called the Space Operations Laboratory.

These configurations clearly show the former Apollo X and AES experiments discussed earlier being applied to early AAP mission planning, which was the early form of what was to become Skylab. Another early proposal suggested using the LM as an airlock between the CSM and the proposed Spent Stage (S-IVB) experiment. In this design the LM would be connected to the S-IVB without landing gear and ascent or descent engines and associated equipment. Upon entering orbit, the LM would not be separated from the top of the stage once the CSM had docked to it. The LM would contain the environmental control systems as well as information, communication and monitoring consoles, with the ascent stage also providing a large volume for storage, especially for life support expendables and EVA equipment. Both of the triangular windows would provide additional EVA observation viewpoints, and the forward hatch would become the egress and ingress path for spacewalking astronauts, using the ascent stage as a separate airlock module.

In place of the descent engine, an interconnecting hatchway and tunnel assembly into the vented S-IVB hydrogen tank would be installed, and further studies by Grumman indicated that launches of unmanned LMs could be used for resupply missions, thereby extending the operational lifetime of the station.

LM solar ATM
One orbital version of the LM that almost progressed to flight configuration would have seen the LM carry the Apollo Telescope Mount to the Saturn Orbital Workshop Assembly that was under consideration for the major element for Earth orbit AAP missions. The ATM was originally to have been carried in Sector I of the Service Module, but when the S-IVB spent-stage workshop concept began to grow, it was realised that the ATM was more suitable for the LM carrier than the SM. (When the SM lost the ATM package, its role as an independent experiment carrier in the AAP space station programme effectively ended, and it assumed the role of a crew ferry vehicle to and from the OWS. For further details of the evolution of the S-IVB stage in the Skylab programme, see the author's companion volume, *Skylab: America's First Space Station*, listed in the Bibliography.)

In the basic configuration, the ATM rack replaced the descent stage of the LM. It would have been launched unmanned by a Saturn 1B, to perform an automated docking with the spacecraft, and would also have been capable of operating for up to 51 days in an open-ended mode without astronauts onboard the station. Grumman also suggested that in place of the ATM solar telescopes, the racks could be used to deliver supplies or other experiment modules to the OWS.

LM stellar ATM
Grumman also proposed that the basic ATM mission could be used to evaluate component performance essential to the development of more advanced manned solar and stellar observatories. By replacing the solar telescopes with a large-aperture stellar package it could also be used to collect data in the UV spectrum, operating either manned or man-tended. The LM stellar ATM could also be configured for periods of unmanned flight away from the main orbital workshop (and free of emissions from the CSM and station), as a man-tended mission. This was advantageous for prolonged periods of stabilisation and a higher pointing accuracy. Power would have come from solar arrays that could be gimballed to always track the Sun while the instruments were in line of sight of the stellar target. The configuration would have retained the LM guidance and control, and RCS propulsive capability used in the solar ATM version, with probably additional reserves of manoeuvring fuel and direct command and control from the ground or via the crew on the station.

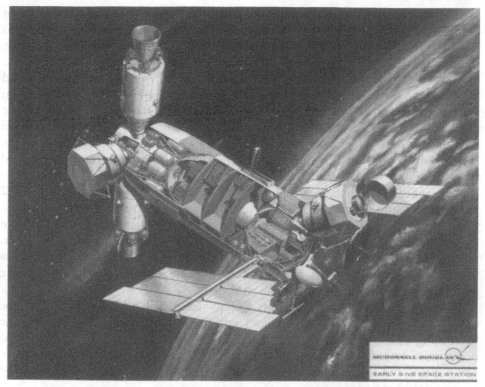

An early configuration proposed for a Saturn Orbital Workshop utilising the S-IVB stage, upgraded Apollo CSMs, and LM-derived experiment carriers. This image from the mid-1960s shows one concept of utilising Apollo lunar hardware for early space station applications in Earth orbit. (Courtesy McDonnell Douglas via the British Interplanetary Society.)

EVOLUTION OF THE LM/ATM CONCEPT

Of all these designs, only the LM/ATM design progressed towards flight hardware, and the remainder were lost in budget restrictions (referred to from p. 267). On 1 January 1966, NASA announced opportunities for study grants for conceptual and preliminary design work, leading to flight instrumentation of an ATM during 1969–75, for inclusion in the AAP Orbital Workshop programme. On 13 July, the management of this project was assigned to the Marshall Spaceflight Center. The design would feature a combination of high resolution, solar orientated telescopes in its initial configuration. These would be attached to the LM ascent stage and docked to the OWS, allowing the crew to manually adjust the instruments to monitor and study specific areas of the solar disc. Nine days later, a recommendation was made that the ATM should be rack mounted, and not part of the manned LM ascent stage configuration. During September 1966, contracts for handling the experiments carried on the ATM were transferred to Marshall and the Goddard Spaceflight Center, to continue the work by American Science and Engineering, Harvard College Observatory, the High Altitude Observatory and the Naval Research Laboratory, that had been progressing since 1964, associated with AES objectives.

The basic concept of using the LM as a carrier for the ATM continued throughout the rest of the year, and on 21 November an LM/ATM Review Team met at the Manned Spacecraft Center in Houston to review the status of the programme and the nature of the proposed design. It was also a chance to establish how or where the MSC could assist MSFC and GSFC in the development of the LM/ATM concept towards flight hardware. The centre at Houston also retained responsibly for overall vehicle safety as a manned spacecraft and its viability as a 'useful' vehicle. The review established that Saturn Apollo missions SA-211–212 were the most probable target missions to fly the LM/ATM to the S-IVB workshop, with the crew launched on a Saturn 1B in an Apollo CSM and the LM/ATM on a second Saturn 1B a day or so later. The LM/ATM could be commanded to an unmanned automatic rendezvous and docking with the S-IVB stage, although a manned docking remained the more favourable option depending on the requirements of the mission finally decided upon.

By early 1967, this LM/ATM mission was designated AAP-4 – the last of the series to the initial OWS. The Apollo 1 pad fire of 27 January 1967 affected the scheduling of the AAP, reflecting the higher priority given to the main Apollo programme to reach the Moon by the end of 1969. However, the development of components and instruments for the AAP programme (and in this case the ATM experiments) continued.[8]

The award of contracts for the production of ATM experiments continued on 15 March 1967, with the NRL subcontracting Ball Brothers Corporation for the production of its intended telescope instrument. This followed a similar contract with Ball for the HAO experiment on 11 January 1965 and the impending Harvard College experiment on 27 December 1967. With the NRL contract, development responsibility was transferred from Goddard to Marshall. Five days after the NRL contract with Ball, MSFC awarded Bendix a $7.4-million contract for the development of the pointing system unit for the ATM, due by August 1965. This unit would allow the crew to target the ATM instruments at selected areas on the solar disc from within the LM ascent stage control cabin.

On 5 May 1967, Grumman presented the costing and statement of work to the NASA AAP programme office at Houston for a final definition of the LM as a carrier for the ATM on the AAP-4 mission. They described the work needed to convert the LM (then planned as LM-6) to the required configuration. During 19–21 June, representatives of the MSC and MSFC Apollo Application Programs Offices (AAPO) held a meeting with ATM Principle Investigators, which constituted a final briefing on the ATM systems, design and experiments. After this, the AAPO would send recommendations to MSFC concerning the suitability of the LM/ATM design and intended experiments, and their compatibility for manned use on the OWS.

A few days later, on 30 June, a series of fact-finding sessions with Grumman was completed. These focused on the definition studies (Phase B) for utilising the LM

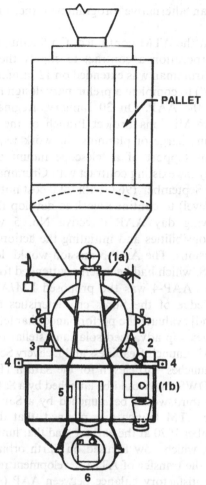

By 1965, mission planners were beginning to identify possible AAP missions for hardware in production. In this illustration, an experiment pallet would supplement the range of research equipment mounted on a modified LM (minus landing gear and landing /propulsion systems), and be assigned to mission 509 (Saturn V launch). 1a) radar antenna; 1b) satellite launch; 2) propellant handling experiment; 3) artificial comets; 4) remote time-lapse cameras; 5) extendable rod experiment; 6) stable platform for a range of astronomical experiments, with a television camera system included.

design, and a statement of work (Phase C) for a final definition of using the LM to carry the first ATM, with cost proposals concerning the manning aspect of compiling this work. No firm dollar assessment was reached at this time. By August 1967 there began to emerge news of a reduction of AAP funding for FY 1968, and projected lower-than-expected follow-on funding for FY 1969, as well as the review of available hardware following the pad fire. Despite continued support for the AAP programme, until the main Apollo lunar landing and return goal had been achieved,

AAP would remain as an 'alternative' programme rather than being promoted as an 'addition' to Apollo.

Meanwhile, work on the ATM continued. On 9 January 1968, MSFC awarded the Perkins Elmer Corporation a contract to supply the Hα telescopes, and the existing contract with Grumman was extended on 12 January (covering the period 15 January to 15 May 1968) to complete a preliminary design review of modifications to the lunar module for the AAP. On 30 January, Reginald M. Machell, Deputy Manager in the Future Missions Project Branch of the AAP Office at MSC in Houston, was placed in charge of planning the work required to modify the LM ascent stage for use in support of a telescope mount on the orbital workshop programme. On 22 May the existing contract with Grumman was again extended for the period 1 June to 30 September 1968 (with total cost not to exceed $3 million, and 380 persons manning level) to continue work to develop the LM/ATM concept for the AAP. The following day AAP directive No. 5 was issued, defining the requirements and responsibilities and initiating the actions required to prepare for the AAP-3/AAP-4 missions. The AAP-3 mission would demonstrate the capability of re-manning the OWS, which had been left unattended for several months after the AAP-1/AAP-2 mission. AAP-4 was the proposed LM/ATM mission that was to 'increase man's knowledge of the Sun's characteristics through solar astronomy conducted in space, [and] evaluate the performance characteristics of a manned solar astronomy system to develop advanced solar and stellar observation systems.'

By 4 June, a new AAP Launch Readiness and Delivery Schedule (ML-14A) revealed a decrease in Saturn launches – to eleven for the Saturn 1B, and to only one for the Saturn V. Of the three OWS, one would be launched by a Saturn 1B, one would remain as a back-up, and the third would be launched by a Saturn V. The schedules also revealed that only one ATM launch was planned, that the programme would not commence until November 1970 at the earliest, and that lunar missions were no longer part of AAP planning, which now focused on Earth orbital activity. In September, NASA HQ authorised the transfer of ATM development responsibility from MSC to MSFC to establish a satisfactory balance between AAP (at MSFC) and Apollo (at MSC). On 9 October the ML-15 schedule was released, revealing a further slip of the first launch to August 1971. Launches were again reduced, to eight Saturn 1Bs and no Saturn V, with only one flight workshop and a back-up and just one ATM mission.

While the contracts and hardware development progressed, several astronauts continued support work in connection with CB input into the LM/ATM design. For several years, technicians, design engineers and professional divers had been conducting underwater simulations of aspects of AAP activation tasks as the programme evolved, using the 1.4-million-gallon water immersion tank at the Marshal Neutral Buoyancy Simulator in Huntsville, Alabama. Full-scale mock-ups of AAP cluster elements lowered into the tank allowed divers to simulate weightlessness for much longer than was possible onboard the parabolic aircraft programme. The idea of using a large water tank for EVA training was vindicated at the end of the Gemini programme in 1966, and the Marshall tank now provided a good simulator for proposed LM/ATM EVA training. On 4 March 1969, three astronauts assigned to the CB AAP Branch Office conducted simulated activities in the tank at Marshall. Ed Gibson, Joe Kerwin and Paul Weitz (who all later flew on Skylab missions) conducted simulated IVA and EVA activities in and around a mock-up LM/ATM and OWS.

On 8 April, the ML-16 schedule revealed a slip of the first workshop launch by three more months, to November 1971. A week later, on 15 April, President Richard M. Nixon recommended a further revision of NASA's FY 1970 budget that included a reduction of $57 million in AAP but added $46 mission for the resumption of Saturn V production.

On 18 July 1969, two days after Apollo 11 left Earth, NASA administrator Dr Thomas O. Paine approved the change from a 'wet workshop' (using a fuelled Saturn S-IVB stage) to a 'dry' concept (in which the laboratory would be fully fitted out on the ground and launched on top of a two-stage Saturn V). This followed a presentation of the dry workshop alternative that included launching the ATM on a rack attached to the station, eliminating the need for a separately launched LM/ATM vehicle. The control panel for ATM would be mounted inside the MDA of the OWS, and not inside the cabin of the LM.

Four days later – as the Apollo 11 astronauts triumphantly headed home after achieving the first manned lunar landing – NASA announced the plan that doomed the LM/ATM. The space station would now be launched on the two-stage Saturn V 'dry' in 1972, with all programme objectives remaining the same as with the Saturn 1B 'wet' workshop of 1971, but now with the ATM attached to the OWS on a single launch. The very next day, 23 July 1969, only three days after Grumman's spacecraft had landed men on the Moon for the first time, and with prospects of at least nine more landings, their celebrations were tinged with disappointment as the contract (NAS-8-25000) to modify an LM ascent stage for use with the ATM was terminated by NASA. It was no longer required to support ATM-A operations, and this ended any chance or hope of flying LM-supported missions in Earth orbit. Over the next few days, other contracts for LM/ATM related sub-systems were either amended to the new configuration or terminated.

It was a frustrating and disappointing time for Grumman, as there were several indications from NASA that to provide the launch vehicle for the workshop at least one Apollo lunar landing might need to be cancelled, and that another LM would not fly. Plans were also being finalised at this point for an ATM EVA using a rack-only configuration, with the telescopes integrated into the OWS design. A review of this – including more than a hundred scientists, engineers and astronauts, ATM Principle Investigators and representatives from NASA HQ, MSC, and KSC – was held at MSFC on 19–21 August 1969, to evaluate methods for ATM film retrieval and replacement on the new design. During this review a full-scale ATM mock-up, without the LM ascent stage, was used.

The revised programme schedule showing the deleted LM-ATM design was released as ML-17 on 13 August 1969. Five months later, on 17 February 1970, the AAP OWS officially became known as Skylab, to be launched by one of the Saturn V vehicles that were to have been used in the lunar programme (see p. 269).

AAP/LM-A MISSION CONFIGURATION

The history books reveal that the story of the LM as a carrier for the ATM officially ended in 1969, and very little has since been known about what exactly the design would have looked like had it proceeded as planned. Study of the design review of the LM/ATM configuration[9] and descriptions of the proposed flight profile of AAP in 1967 (after the Apollo 1 pad fire) reveals one version of what was planned. But this came no closer than planning for the second manned LM to fly in Earth orbit, after the LM-3 test flight finally flew as Apollo 9.[10]

In evaluating the design of ATM consoles in the LM, the review had to take into consideration how easy, or indeed difficult, it was to enter and egress the LM cabin from the CSM or Multiple Docking Adapter. There was also the need to meet both the operational and experimental mission objectives of the AAP-3/AAP-4 mission. The object of the review was to determine the best console configuration to be considered for LM-A (Ascent Stage).

To achieve this, certain ground rules had to be established. There would be two men inside LM-A during ATM operations, but consideration had to be given to operation by one or two men as required, and limited operation with pressure suits. The crew ingress and egress had to be head-first through the overhead docking hatch, with the ability to remove the drogue from the LM side of the docking system for emergency or contingency operations with a two-man EVA (with the astronauts wearing portable life support systems) using the front hatch, considered as an alternative method of gaining entry into the LM ascent stage. Finally, there would be a further alternative capability in the spacecraft life support systems to support a one-man partially suited operation for up to six hours.

In evaluating the designs, a range of criteria was established, including design evaluation in safety, reliability, ingress and egress, crew task capability, the effect on the free cabin volume in the LM of adding the ATM console, crew comfort, restraint and mobility, vulnerability to damage by crew movement; cabin equipment access, control and display proximity for both the ATM and LM systems, one- or two-man operation, available panel area, the effects on cabin stowage volume, rearrangement of LM configuration to accommodate the ATM, structural compatibility, electrical interfaces, thermal compatibility, operation at KSC during pre-launch and on the pad, and the ever-present weight penalties at a time when Grumman was striving to lighten the overall LM design for the lunar missions.

Three types of configuration were used in the definition studies: *panel,* which featured the electrical assembly of components mounted to a common surface, including an electrical interface to a vehicle (LM) harness; *module* was to be a combination of panels on a framework including a common support structure that would interface with an LM structure; and *console* was a combination of one or more structurally independent modules that also satisfied the entire ATM control and display requirements.

LM-A/ATM STRUCTURAL CONSIDERATIONS AND CONFIGURATIONS

Further ground rules applied to the LM construction were that the LM-A should be created from an already completed flight vehicle and not a new spacecraft, and that no major modification should be made to the basic LM ascent stage structure. In the installation of ATM hardware, it was determined that the design had to include components with self-supporting structures, not relying on structural support of the LM. The interfaces with LM components had to be static, and the ATM hard points had to be compatible with current LM load paths and not violate load determinations. As little modification of the structure as possible had to be used to support whichever ATM design was chosen. It was calculated that the inclusion of a 500-lb ATM load in the LM would require revision of primary LM structures and a structural retest programme, which was not advisable due to additional launch weight penalties and the increased costs which this would have incurred. In the review, several design configurations were evaluated, each with an integrated design and location in the LM cabin. In all, six configurations were reviewed, and a series of recommendations was put forward.

The candidate configurations were: (A) full two-man, folding, mid-cabin with a four-bar linkage; (B) full two-man, folding, mid-cabin with a single hinge; (C) two-man, fixed with left-hand side module; (D) full two-man, fixed, mid-section with a 90-degree reorientation; (E) two-man fixed, mid-section, with an upright orientation; and (F) one-man, fixed, with a left-hand side module. These were mostly behind the normal crew station in the mid-cabin location, positioned where the ascent engine cover would normally have been situated in the lunar lander version. This was omitted in the LM-A configuration, although some designs were almost wrap-around configurations in the LM CDR position. One of the major considerations in locating the ATM console in this area was the inward opening top hatch cover, and access to LiOH canisters and other LM systems.

Configuration A: folding four-bar linkage, full two-man (11 square feet). This had the advantages of control and display (C&D) proximity, allowed for either one- or two-man operation with pressure suits, and had good work-station mobility and stowage volume in the mid-cabin. It also required no changes to the LM controls and displays. Major disadvantages included failure potential in flex wire and glycol tubing, and in the mechanisms of deployment; the time it took the crew to set up the ATM station; and difficulty in the ingress, egress and EVA paths. It was found that this configuration also had problems in passively cooling the cathode ray tubes in the displays, and there were restrictions for in-flight access to the mid-cabin equipment, which also left it vulnerable to damage. There was also difficulty in removing the docking drogue, and it was found that access during launch pad operations would be very difficult for maintaining equipment without the addition of more support platforms in the launch vehicle/spacecraft adapter section.

Configuration B: folding single hinge, full two-man (10 square feet). In the second configuration, similar advantages and disadvantages to those in the Configuration A concept were noted.

Configuration C: fixed left-hand side module, two-man (12.8 square feet). This had

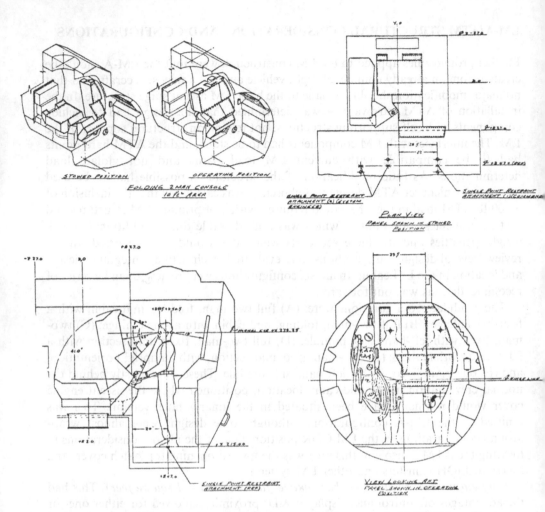

Crew station designs for the proposed two-man folding console LM/ATM configuration. (Courtesy Grumman Aircraft Engineering Corporation.)

the advantage of no moving modules. The proximity of the LM-A C&D access required limited use of pressure garments, but it had good work-station mobility and stowage volume in the mid-cabin. The main disadvantage was the difficulty of ingress and egress and for EVA operations. It inhibited access to the mid-cabin equipment, restricted full two-man operations, required a revision of LM consoles, was vulnerable to damage, presented difficulties in removing the docking drogue, and required complex structural interfaces.

Configuration D: mid-section 90-degree orientation, full two-man (18 square feet). This also had no moving modules, but it also had full two-man operation and minimum effect on free cabin volume and LM ingress or egress. The format featured the rotation of ATM consoles 90° to the forward main LM crew stations and a large

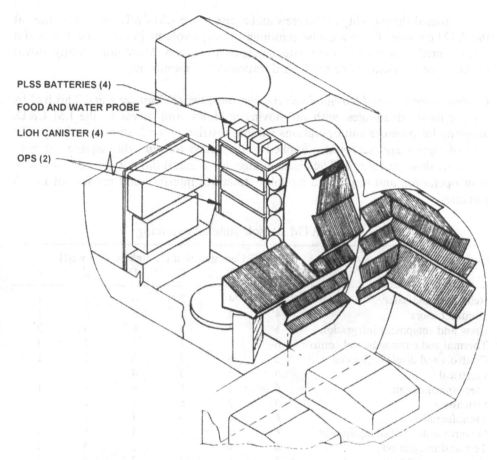

PLSS BATTERIES (4)

FOOD AND WATER PROBE

LiOH CANISTER (4)

OPS (2)

An alternative design for a fixed two-man experiment console in the LM ascent stage. (Courtesy Grumman Aircraft Engineering Corporation.)

panel area with efficient structural interfaces, plus good thermal compatibility and electrical interfaces and good access to cabin equipment. On the negative side, it was difficult to get in and out of the ATM duty station, it required a complex restraint system, and the LM C&D was out of direct reach, restricting crew mobility and task capability. Additionally, there was limited stowage volume in the mid-cabin, the LM panels were vulnerable to damage, and the environmental control system blower and LM waste management system would require relocation.

Configuration E: mid-section upright orientation, two-man (10.2 square feet): This configuration also featured no moving modules, had minimum impact on free cabin volume, and had good equipment access, with little effect on crew movement into and out of the LM. It offered an efficient structural interface, with comparable thermal and electrical advantages to the Configuration D design. However, this design did not allow for full two-man operation, required a complex restraint system, and restricted the mobility of the crew and access to the LM C&D when restrained at the ATM console. There was also minimum stowage volume in the mid-cabin, and it still required relocation of the waste management system. More importantly, it was found to be uncomfortable for long-duration ATM monitoring.

Configuration F: fixed left-hand side module, one-man (9.5 square feet). This had by far the most advantages, with no moving modules, and access to the LM C&D, allowing for pressure suit operations without restricting work-station mobility. The ease of ingress and access also applied to the access to mid-cabin equipment, with adequate stowage volume. The disadvantages were that it did not provide for a two-man operation, and required a complex structural interface and revision of LM-A consoles.

LM-A/ATM console summary ranking

Category	Configuration (1, highest; 6, lowest)					
	A	B	C	D	E	F
Reliability and safety	5	4	2	6	3	1
Human factors	4	4	2	6	2	1
Crew and equipment integration	1	1	3	6	6	2
Thermal and environmental control	6	6	3	1	2	4
Controls and displays (limited)	5	6	3	3	3	1
Electrical	6	5	3	4	1	2
Mechanical design	6	4	1	6	3	1
Structures	3	3	6	1	1	6
Manufacturing	6	5	4	1	2	3
Government-supplied equipment	6	6	3	1	1	1
Test and integration	6	6	3	3	3	1
Operations and test	6	6	4	3	1	1
Totals	60	56	37	41	28	24
Averages (A/B, C/D, E/F)	(58 (+19))		(39 (+13))		(26)	
Ranking	6th	5th	3rd	4th	2nd	1st

In the conclusion to the design review, it was determined that the highest-ranking consoles were the wrap-around one-man (F), which provided the best interface for overall crew operation considerations, and the mid-cabin two-man upright orientation (E) that provided structural and thermal compatibility but had questionable suitability for long-duration crew comfort during ATM watch shifts.

As any of the six configurations could have been installed in the LM-A, the final choice would have been made primarily on the basis of chosen flight experiments and crew operations. This indicated that a one-man wrap around configuration was the most promising and that the development of the overall LM-A/ATM console layout would require integrated design efforts.

The recommendations of the review suggested that working interface groups should be established between Grumman, NASA and the ATM design contractor, to develop the selected configuration in relation to the larger AAP design. Further consideration also had to be given to the application of system design procedures in mission analysis and projected timelines; flight crew function allocation and crew activity timelines; a determination of crew station requirements (with an input of the CB); analysis of communication links between the crew, the LM/ATM and the rest of the OW and/or CSM, and ground control; and the design of equipment in the consoles, restraint systems, mobility aids, waste management and other government-furnished equipment.

In console design, panel layouts and the test and evaluation programme, it was concluded that the design of the consoles would reflect performance and scheduling requirements, restrictions imposed on using the LM-A vehicle, and requirements for crew training and in-flight maintenance, safety, environmental requirements and the size and weight of the flight version.

Panel layouts would be influenced by the performance requirements of each sub-task, simultaneous and/or rapid succession tasks, and the speed and accuracy required to perform these. In test and evaluation considerations, the results of ground simulations, zero-g aircraft flights and use of the neutral buoyancy water tanks, would provide valuable data in the operational decisions made in the final design selection.

LM-A mission profile

The closest the LM-A/ATM design approached to a mission profile was in the constantly revised programme directive issued by NASA's AAP programme office during 1966–69. The LM-A would have been launched under the AA-3 designation, and would have been the fourth and final launch of the short series to the initial S-IVB workshop. (The overall AAP concept is discussed in the author's accompanying volume, *Skylab: America's Space Station* (pp. 19–48), but the mission that the LM-A ATM might have flown is described here.)

Saturn 1B SA-212 would have inserted the 16,675-lb LM-A/ATM payload (which was identified as LM-6) into a 240° circular orbit about 24 hours after the three-man Apollo CSM had been orbited by SA-211 on a resupply mission to the OWS. The OWS would have been visited by the first three-man crew for 24 days, and they would also have delivered the Mapping and Survey System hardware, docked at the opposite port (No. 3) on the Multiple Docking Adapter to that which the LM-A would have occupied. The second flight crew (AAP-3) would have completed a 56-day residency on the station, focusing on ATM operations.

The CSM would have rendezvoused and docked with the LM-A/ATM, extracting it from the spent S-IVB stage in a way similar to retrieving the LM on lunar landing missions. In approaching the OWS, the CDR and LMP would have transferred to the LM, which would then have been undocked from the CSM to approach the OWS and dock to the MDA at the axial port (No. 5). The CMP would have approached the radial port (No. 1) only after the LM had successfully hard-docked to the station. After safing of the vehicle, the OWS would have been activated to provide a shirtsleeve environment throughout the docked vehicle to provide more comfortable crew access to the LM/ATM workstation for the duration of their stay.

The crew would have followed the flight plan including ATM experiment operations. The astronomical/solar instruments for AAP-4 would have included:

S-052   White-Light Coronagraph
S-053   UV Coronal Spectrograph
S-054   X-Ray Spectrographic Telescope
S-055   UV Spectrometers
S-056   Dual X-ray Telescopes
S-027   Galactic X-ray Mapping

The directive indicated that the LM/ATM would also be capable of operating either docked to the OWS in its primary mode, or docked to the CSM in a back-up mode – a secondary demonstration of operating remote (undocked) from the cluster in either tethered mode or free-flight mode (and also perhaps, with an unmanned capability). However, in Curt Michel's hand-written notes he states that there was no mention of tethered operations in either the primary or secondary objectives of AAP-3/AAP-4, which leads to the conclusion that at the time (1967) the idea of undocked operations tethered to the cluster was still being evaluated. At the end of the AAP-3 mission, the LM/ATM would have remained attached to the station while the crew conducted a nominal end-of-mission ocean recovery.

Block III CSM modifications for the AAP
The one element of Apollo hardware that remained constant in the AAP space station studies was the use of the CSM to ferry the crew to and from the station. It has previously been indicated that NAA was considering a third upgrade to the CSM (Block III) to support extended CSM missions under the AAP programme, both in Earth orbit and on lunar distance missions and before the Spent Stage concept was fully adapted. In order to achieve this capability, a number of upgrades to the sub-system needed to be incorporated into the design, some of which were major modifications, while others were considered only minor improvements.[11]

Environmental control system NAA recognised that one of the decisions facing manned spacecraft designers was the development of what type of atmosphere composition should be provided for the crew to breathe. Early weight limitations on US spacecraft indicated a cabin pressure of less than 1 atmosphere, which meant that terrestrial 'air' could not be used, as the lowered pressure would induce the bends. It was decided to use an atmosphere of pure oxygen (100%) at 5 psi, which reduced this risk. However, oxygen is a highly combustible gas, and this introduced a new risk of a potential flash fire in the spacecraft – a disadvantage that would be tragically demonstrated on 27 January 1967. At the time of the Block III studies, the loss of the Apollo 1 astronauts had not occurred, and development of a partial pressure two-gas system (70% oxygen and 30% nitrogen at 5 psi) for the CSM series was therefore not being considered for the Block I series. However, if developed, the Block III series would probably have used the two-gas system that required installation of the nitrogen supply system, a partial-pressure regulating system, and a total-pressure sensing system.

In addition, a change in the method of removing carbon dioxide was being considered: from the lithium hydroxide canisters to a molecular sieve device. The lithium hydroxide canisters absorbed the astronauts' exhaled breath and retained the carbon dioxide by a chemical reaction to the filters. These were suitable for 1.5 man-days, but with a three-man crew they would have to be changed out every twelve hours, thereby adding to the weight penalty of carrying and storing used and replacement cartridges in the spacecraft. As an alternative the molecular sieve does not involve a chemical reaction and instead periodically vents one of the impregnated sieve beds into space and is automatically replenished. Weight trade-offs in adopting this system over the LiOH system balance out at 30 days, but the volume which the equipment would have taken up had an almost immediate effect of saving up to 12 cubic feet, and allowed an extension to the mission duration of up to 45 days. The location of the new hardware would be in the ECS unit in the left-hand equipment bay under the left-hand (CDR) seat. The oxygen surge tank would be relocated to the lower equipment, bay and the glycol system would be installed on the empty shelf immediately above the ECS unit.

Attitude control Increasing the duration of a mission resulted in a growth of propellants required to maintain attitude control to the order of about 2,500–2,600 lbs. At the time of the studies the capacity of Block II RCS propellant in the SM was only 790 lbs, with capacity to reach 1,200 lbs but still well below the AAP requirements. The Block II RCS system was a self-sufficient four-quad arrangement with the tankage on the inside of each quad. The most straightforward option was therefore to add more tanks to supply the thrusters to the required level. The plan was to use already qualified LM RCS oxidiser tanks, which did not require any new development – just a new arrangement of the tanks in the SM.

Electrical power NAA identified the need for a new cryogenic tank design to generate more electricity for extended missions. Modifications could be made to the Block II system, allowing an in-space start capability, which meant that the mission would fly with four fuel cells and operate with initially only three (the fourth serving as a back-up); and once one began to fail, the fourth could be brought on line. This would allow a power range on Apollo of 1,700 W minimum and 2,800 W maximum, to the AAP configuration of a maximum capability of 4,000 W, though not for the full 45 days (which was stated to be 2,730 W). Housekeeping requirements demanded about 1,100 W, which meant that the proposed system could supply 1,500 W continuously, or almost 3,000 W 'very frequently' for periods of up to 45 days, for any sort of experiment grouping 'placed in a lab module'. No experiments proposed by NASA for AAP (by September 1966) would exceed these requirements.

Recovery operations As part of the AAP contract studies, NASA asked NAA to examine the possibility of land-landing capabilities. Apollo had the capability of

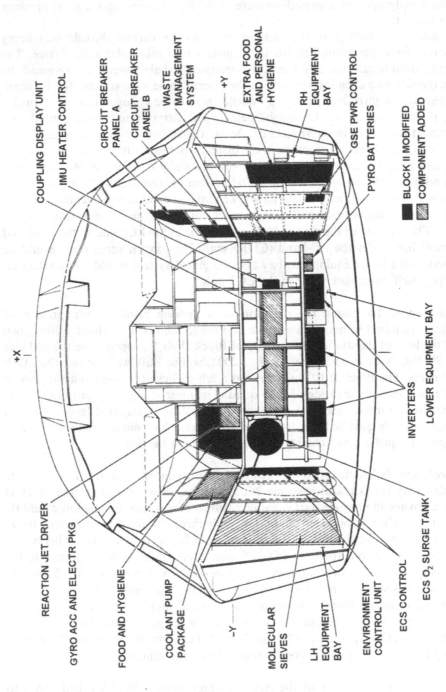

Internal configuration (1965) for the proposed Block III Command Module for use in the Apollo Applications Program. (Courtesy North American Aviation.)

land landing under emergency situations, but retained the water landing by parachute profile similar to that by which each Mercury and Gemini mission was recovered. NAA evaluated four concepts. Configuration I featured the addition of base-mounted retro-rockets (similar to the Soviet Soyuz spacecraft), but retaining the three-ring sail parachutes; but they found that this combination exceeded normal crew impact limits, and it was rejected. Configuration II incorporated base retro-rockets with a parasail canopy instead of the ring-sail parachutes. Configuration III featured a cloverleaf lifting parachute system with a base retro-rockets. Configuration IV had the cloverleaf and retro-rockets in the parachute risers (similar to the Soviet Voskhod). The last three configurations were within crew-impact limits for land landing, but provided a glide range of only 3–4 miles, which was insufficient for manoeuvring towards a landing field (which was an early objective of the Gemini programme).[12] In addition, NAA expressed concern that there was no guarantee that the spacecraft would land in a clear area and thereby avoid obstacles that could overturn the spacecraft and sustain damage to both the vehicle and its occupants.

There were also significant problems in base-mounted retro-rockets in providing blowout holes for retro-firing to reduce impact velocity from about 30 fps to less than 10 fps. Major modifications would be required to the CM heat shield, necessitating an associated test and qualification programme. To resolve this problem a retro-rocket package was deployed in the parachute riser system, providing 20–25 fps reduction in vertical velocity just prior to touch-down. NAA's Columbus Division was already using this concept (on a smaller scale) on the Roadrunner missiles, and the adaption of that system indicated a significant cost reduction over amending the heat shield. NAA considered the use of a land-landing capability as an 'optional extra' for AAP, but recognised it would be an operational (quicker return of experiment materials and the crew to the continental US) as well as financial (no need for the deployment of large naval recovery forces) advantage over water recovery modes.

AAP sub-systems summary
The changes required for the CSM to progress to a Block III design included the following. No changes were to be made to the CSM primary structure, heat shield,

AAP land-landing concepts

	Concept I	Concept II	Concept III	Concept IV
Parachute system	Swivel	Parasail	Cloverleaf	Cloverleaf
Retro system	Base	Base	Base	Riser
Glide range, no wind (miles)	none	3.1	4.2	4.2
Obstacle avoidance	no	yes	yes	yes
Landing stability	marginal	good	better	best
Water landing	better	better	better	good
Crew impact loads	exceeded	OK	OK	OK
System weight (lbs)	782	750	750	695
(AES baseline, 728 lbs)				

recovery systems, or service propulsion systems. Minor modifications were required in the guidance and control (additional redundant components), electrical power (a fourth fuel cell), environmental control (additional nitrogen supply and regulation), and additional secondary structures to relocate items of hardware. There were major modifications in rearranging RCS tank configurations and adding the molecular sieve, which were also newly designed. New development would also be required for larger cryogenic storage tanks and if land landing capabilities were pursued further .

NAA also indicated that a fourth fuel cell would be included in Sector 1 of the Service Module, which was essentially empty of equipment in Block II, along with two larger cryogenic tanks (one hydrogen and one oxygen) and an additional set of four tanks instead of two at each RCS quad. The rest of the SM equipment would be retained for supporting lunar AAP missions. Due to the increased volume built into the SM early in the programme, NAA stated they had not encountered any integration problems in demonstrating these additional features.

Block III CSM production
In September 1966 NAA indicated their forecast for Block III production, including two test vehicles 3TV-1 (for thermal vacuum tests) and 3S-1 (Service Module for static tests), to be followed by six Phase I flights to develop the Block III vehicle and then a further nineteen Block III spacecraft as described above. NAA indicated that these would have been designated Spacecraft 201 through Spacecraft 219. They had also investigated staggered production of Block II – Block III – Block II – Block III, but this interfered with production flows, and it seemed that completion of Block II would be completed before Block III series commenced, although the test articles would be produced after Block II Spacecraft 112.

NAA forecasts, based on 1966 NASA planning schedules, predicted a production rate of eight CSMs (Block II and Block III) per fiscal year, with a maximum capacity of fourteen CSMs per fiscal year. A firm proposal had been made to NASA on 1 January 1966 for the nine-month final definition phase ending in September 1966 (the time of the NAA report to the Congressional Committee on the Block II status), which should have been followed by a preliminary design review that would have seen NASA sign off all specifications and contract documents to begin building the hardware. This would have been followed by a 2-year 7-month production cycle with the first Block III vehicle available for flight in the third quarter of 1969.

NAA estimated the costs of the six development vehicles, nineteen production spacecraft and two test articles, LM shrouds, and flight operations as $.1.3 billion over a seven-fiscal-year period (FY1966 $12.7 million; FY1967, 90.3 million; FY1968, 251.5 million; FY 1969, $360.5 million; FY1970, 347.0 million; FY1971, $192.0 million; and FY1972, $40.4 million). This would be in addition to the funding required to support the Block II Apollo lunar programme and the rising cost of developing the Spent Stage concept – which became the stumbling block for the Block III design, and which NASA did not pursue further in this configuration.

Shortly after the Committee Report was finished, the 27 January 1967 pad fire claimed the lives of three astronauts and grounded the whole US manned space programme for 21 months. While the Apollo programme recovered from the tragic setback in order to still meet present Kennedy's deadline, work continued on preparing the Apollo CSM for the AAP programme in support of the Spent Stage space station concept that evolved into what became known as Skylab. In this guise there was no need for an CSM with a 45-day independent capability, as for most of the mission the CSM was to be docked to the S-IVB cluster and placed in a semi-dormant mode until the end of the mission, when it would again be powered up for the return home. Therefore, although changes would be required on the CSMs intended for the AAP station, they were not of the scale originally intended for the Block III CSM. As a result, further work on the Block III concept was terminated in order to incorporate the changes, resulting from the inquiry into the pad fire, into the Block II spacecraft that would also be used to support the AAP missions.

However, although the AAP Block III CSM would never fly, there remained one independent Block II CSM mission, planned for 1969, that was inserted into the AAP flight schedule and considered as a supplement to the series of missions being planned for the S-IVB Spent Stage concept.

AAP-1A: AN INSERTED MISSION

During 1966, the development of the Apollo Applications Program in Earth orbit centred on the use of the spent S-IVB as the core of an embryonic space station, with the Apollo CSM acting as a ferry vehicle for the crew and the LM as a possible carrier for the Apollo Telescope Mount package.

On 8 November 1966, a new planning schedule showed mission SA-209 launching first, with a three-man crew onboard a Block II CSM on a 28-day mission to the OWS. The OWS would be launched the next day, and would carry the Multiple Docking Adapter and Airlock, with solar arrays as the primary power supply. SA-211 would take a second crew to resupply the station on a flight of 56 days, and SA-212 would deliver the LM-A/ATM hardware to the station.

The SA-209 mission would also deliver a Mapping and Survey Systems (M&SS) module to the station, carried inside the Spacecraft Lunar Adapter on the first launch, to be attached to one of the radial docking ports at the MDA. This was to be an Earth orbit test flight of the proposed Lunar Mapping and Survey System (LM&SS) for later Apollo lunar missions that would map most of the surface of the Moon from orbit. In the programme reorganisation of spring 1967, the heavy flight load for the first crew in activating the station, the employment of the second crew to operate the LM/ATM, and the need to incorporate the safety recommendations following the Apollo 1 pad fire saw the first AAP missions slip to 1969. The M&SS hardware was therefore deleted from the OWS programme.

Although no longer part of the AAP schedule, the mapping system hardware remained an important element of future Apollo planning, and also had a direct application for proposed Earth surface survey by manned vehicles. Therefore, on 8 May 1967 the hardware was re-manifested to a solo Apollo CSM mission

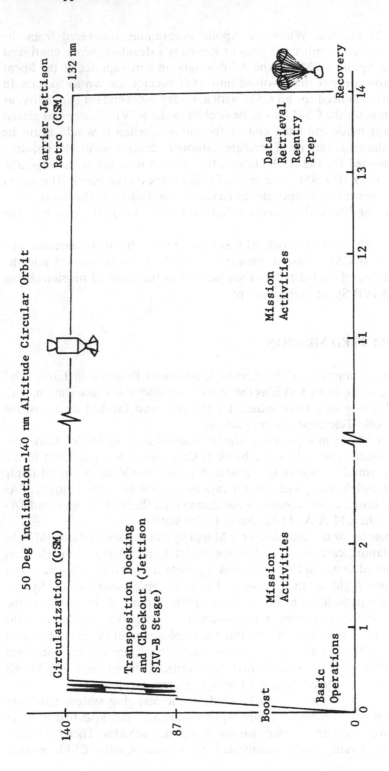

The mission profile proposed in 1967 for the AAP-1A flight in Earth orbit during 1969. (Courtesy Martin Marietta Corporation.)

(designated AAP-1A) planned for a 15 September 1968 launch to provide flight experience with the system in Earth orbit prior to assigning it to a lunar mission. In order to establish the feasibility of mounting such a mission, Martin Marietta, under contract from NASA, instigated a 60-day study between 8 July and 5 September 1967 to evaluate the mission. This was followed by an additional series of studies between 6 September and 30 November of that year, which recommended the type of missions to be flown, the research programme and proposed experiments, and the configuration of the carrier required to support experiment hardware.[13]

NASA's baseline AAP-1A mission profile
In a NASA document dated 19 June 1967,[14] AAP-1A was proposed as a 14-day Earth orbital mission, to be launched from LC 34 by Saturn 1B and flown on a Block II CSM into an 81×120-nautical-mile orbit inclined at $34°$. After orbital insertion, the crew would have separated the CSM from the S-IVB and performed the transposition and docking manoeuvre with the LM&SS stored in the SLA. The CSM would have withdrawn the LM&SS and proceeded to an approximately 141-nautical-mile circular orbit.

For the first five days of the mission, the crew would have conducted the LM&SS mission objective, which was photographic coverage over continental USA (ideally at 120–150 nautical miles) for the maximum time during the period of data acquisition. The plan was to undertake photographic operations with the Sun at its highest altitudes over Phoenix, Arizona, and over parts of the country where it was easiest to establish ground truth. Although problems with the LM&SS would have forced a launch hold, weather conditions affecting photographic operations over the US were not expected to have delayed the mission.

Following the five-day mapping mission, the CSM would have separated from the LM&SS (which would have been abandoned in orbit) and the crew would have conducted their assigned experiments on an open-ended basis, up to the 14-day maximum flight duration. The SPS (with RCS back-up capability) would then have been used to end the mission, with an intended ocean recovery.

In the Martin Marietta report issued three months later, a more detailed account of what was proposed for mission AAP-1A was forthcoming.

AAP-1A mission study report
The AAP-1A baseline mission proposed in the Martin Marietta Report established improved optimum daytime coverage over the United States, for an average of six day-time passes per 24 hours of the mission, at 140 nautical miles altitude and $50°$ inclination. Launch was planned for 10.00 am on 1 April 1969 on a 14-day open-ended mission, with a day-time recovery in the primary Atlantic Ocean recovery area on or about 15 April 1969.

The change to $50°$ inclination enabled an extension to experiment and mission coverage in order to achieve the mission's primary objective – an early evaluation of the operational feasibility of selected Earth resources, and bio-scientific, meteorological and astronomical experiments.

Further objectives of the mission included verification of a human presence in enhancing experiments by monitoring, controlling and interpreting data obtained on orbit, and to gain operational experience of using the hardware in flight.

Considerations established for the study included the opportunity for an early launch, the use of one Saturn 1B booster and a Block II CSM with minimum changes, ensuring crew safety while lowering overall costs, and a proposal for a 14-day open-ended mission.

The study, based on NASA data, found that Saturn 1B performance allowed for 36,500 lbs of payload to be inserted into an 87 × 140-nautical-mile elliptical orbit. This was broken down as 3,947 lbs for the SLA, 25,198 lbs for the fully provisioned three-man CSM and RCS propellants, and an additional 1,472 lbs for SPS propellant. This left 5,883 lbs of performance for both the carrier and all the experiments. The study therefore worked on a 5,000-lb ceiling for the experiments and the carrier, leaving an margin of 883 lbs.

Extensive spacecraft orientation studies were completed, based on the crew visibility, carrier weight, and sensor contamination from RCS firing and CSM outgassing. The most advantageous orientation was found to be in a nose-down attitude. The nominal mission launch time of 10.00 EST, 1 April 1969 was used as a baseline because of optimum lighting conditions over the continental USA throughout the 14-day mission, and the desire for a daylight recovery in the Atlantic Ocean on a descending orbital pass. The optimum solar illumination of the northern hemisphere target areas is between 1 March and 1 October, with solar attitude above 30° at 50° N latitude. By launching at 10.00 am EST, the northern limb of the orbit would be kept in daylight throughout the mission, which allowed for a recovery in the region of 35° N latitude and 60° W longitude between approximately 11.00 am and 12.00 noon EST on 15 April 1969.

The report was used to identify a hardware configuration and develop an approach for integrating twenty-three experiments into the mission. It recommended the development of a separate carrier configuration that would be docked to the nose of the CSM, similar to the LM and earlier designs of the mapping system. The orientation of the combination (Earth vertical), with the nose of the spacecraft pointing 'downwards' to Earth, and with the crew in a head-forward couch position, provided the best balance of crew visibility, carrier weight, orbital decay, sensor contamination, RCS usage and torque disturbance considerations.

Crew operational time lines showed that it was feasible for 100% requested data recovery for fifteen experiments, with a reduced return on the remaining eight investigations, the limiting factors being the availability of crew time, RCS propellant levels and data dump capability. The study established that by using the hardware configuration suggested in the report and with established time lines, ten crew entries into the pressurised carrier would be required to operate the experiments, using the carrier dome as well as side-mounted scientific airlocks and changing film for the internally mounted camera. Expected return stowage weight limits were well within the limits for a Block II CM.

It was also identified that a ground operations sequence making maximum use of existing designs and equipment from factory production to launch would assist in retaining the programme both within budget and on time. As part of the study,

Martin Marietta also presented a programme schedule that followed a flight readiness time line from approval to flight date of between 16 months (minimum) and 20 months (normal). This allowed for lead time for fabrication of the carrier, development and integration of experiments and crew training, and launch preparations. Effectively, the 20 months lead time would have to have been made at the time of the report publication (September 1967), and in the new year for the 16 months lead time, to meet the proposed April 1969 launch date.

AAP-1A experiment carrier
The study also recommended a suitable experiment carrier: a welded 2219 aluminium alloy truncated cone with an 84-inch diameter at the experiment mounting end and 110-inch overall length, with the other end matching the docking system of the Apollo CSM. A truss to support the structure in the SLA would have used the same attachment posts as the LM, and would also have provided structural support for all systems and experiments that did not require in-flight access by the crew.

The pressure chamber featured two major subassemblies: the sidewall and aft closures – also fabricated from 2219 aluminium alloy. The sidewall consisted of four skin panels, four longerons (which also featured attachments for crew IVA hand-holds and restraints), a docking tunnel frame assembly, truss attachment fittings, and the aft closure bolting ring frame. The aft closure would also have been a welded structure, using a spherical segment shell welded to a bolting flange. At the interface between the closure and sidewall sub-assemblies, an O-ring would have been used as the pressure seal.

Two different configurations would have been used for the four truss assemblies supporting the carrier in the SLA. Four primary tubular members attached to the SLA would have supported the experiment platform, fitting on the two Z-axis trusses to balance the structural loading on the Y-axis trusses. Inside the pressure shell, the experiment support frame would have supported several experiments in their operating locations and other experiments in their stowage locations when not being used by the crew. The frame would also have provided direct load paths between the truss attachment fittings at the maximum diameter of the pressure chamber. There was also provision for viewing windows and scientific airlocks, and for wire feed-through channels to pass though the pressure shell with minimum leakage rates, the penetration outlets being machined separately and then welded into cut-outs in the pressure shell.

Meteoroid protection for externally mounted components would have come from the experiment mounting rack skin and experiment platform structures, which would also have formed a shield for two quadrants of the pressure shell. A thin layer of meteoroid shielding would have covered the two exposed pressure shell quadrants. Passive thermal control for the carrier was envisaged as mulitlayered insulation blankets applied directly to the outside of the pressure chamber, experiment platforms and equipment rack.

Power for the experiments would have come from seven silver–zinc batteries, while experiment and housekeeping data would have been handled by a system of three VHF transmitters and one S-band transmitter. An active/passive thermal control system (using Freon 21 coolant and truss mounted radiators) would have provided the required thermal control, and an horizon scanner and gyrocompass attitude reference system mounted in the carrier would have provided a local vertical control back-up system to the primary CSM G&N sub-system.

The internal volume of the carrier was shaped like a truncated cone from the docking tunnel diameter to the pre-flight removable spherical segment aft section. It would have provided sufficient volume for crew IVA activities and stowage. Support sub-systems would have been mounted on shelves in two equipment racks on opposite sides of the carrier cone, which would also have supported the thermal system radiators. The selected design also foresaw the growth of experiment and sub-system components, both inside and outside the pressure chamber.

The preferred carrier design derived from a study of CSM orientation, which concluded that there was superior viewing capability from the CM in the nose-down heads-forward attitude along the ground track. A Carrier Configuration Trade Study evaluated that experiment viewing axially (with a conical central body as the pressurised crew access volume) was superior to all other concepts in terms of crew accessibility, crew ground track viewing, the CSM transposition and docking manoeuvre, design flexibility, growth and practicality. A Carrier Pressurisation Study concluded that a pressurised carrier far outweighed any unpressurised carrier, even though it had the additional disadvantage of structural weight. These conclusions also influenced the selection of proposed experiments for the flight. It

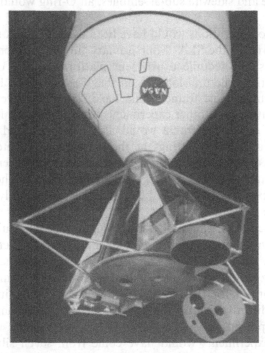

The selected carrier configuration for AAP-1A. (Courtesy Martin Marietta Corporation.)

allowed existing NAA (CM Block I) scientific airlock designs to be incorporated, thereby eliminating the need to develop new hardware, and IVA operations could be conducted in a shirt-sleeve environment (or soft suited), eliminating the need for EVA operations and data retrieval.

AAP-1A proposed experiments
Five different experiment groups were analysed during the period of the study, and following the 9 August 1967 mid-term review, twenty-three were chosen by NASA for further consideration and integration into the AAP-1A mission. This group included eleven early versions of the Applications A and B experiments evolved from Apollo AES and Apollo X studies, and twelve scientific experiments from the Apollo programme. The proposed investigations included:

S-039 Day/Night Camera; S-040 Dielectric Tape Camera; S-043 IR Temperature Sounder These experiments required an S-band transmitter for data handling, which added 40 lbs and $1.6 million to the mission. Additionally, S-039 and S-040 could not play back data simultaneously. A CRT was also required for S039.

S-044A Electrically Scanned Microwave Radiometer The temperature gradients across the antenna face were one of the highest of the package, perhaps the most critical ranging from −10°C to +65°C.

S-048 UHF Sferics Eleven days of continuous operation was required for this experiment.

E-06-1 Metric Camera The temperature gradient across the lens, at 0°C to +32°C, was a critical factor in this experiment.

E-06-4 Multi-spectral camera This featured a self-contained power supply, and required two film cassette changes during the mission.

E-06-7 IR Imager The only experiment that required 400 Hz. The experiment was located outside the carrier, with film feed into the carrier through a film chute to allow retrieval without EVA.

E-06-9A IR Radiometer; E-06-9B Spectrometer Both of these experiments recorded at a high bit rate (10-bit 1,900 SPS), and the experiment data output analogue required encoding.

E-06-11 Multifrequency Microwave Radiometer; At 44,208 cubic inches in volume, this was by far the largest experiment envelope of the payload.

T-002 Manual Navigation Sightings; T-003 Aerosol Particle Analyser Both experiments featured self-contained power systems.

T-004 Frog Otolith Function During crew sleep periods, this experiment would have to be operated via ground control. It used the S-017 data system.

D-008 Radiation monitor This used CM power and two channels of CM data handling.

D-009 Simple Navigation This had self-contained power. At the end of the mission, the crew would return the experiment log-book for post-flight analysis.

D-017 CO_2 Reduction

S-015 Zero-G Human Cell This also used CM power.

S-016 Trapped Particle Asymmetry This utilised the airlock and required accurate spacecraft orientation. It also required a non-concurrent sleep cycle.

S-017 X-Ray Astronomy This required 70 lbs for the data system and 26 lbs for the display and control hardware.

S-018 Micrometeorite Collection This required the use of the spacecraft airlock, and no RCS or waste (water and/or urine) dumps during data gathering.

S-019 UV Stellar Astronomy This also had self-contained power, and required the use of the spacecraft airlock and operator station in the carrier.

S-020 UV X-Ray Solar This required the use of the spacecraft airlock and operator station within the carrier.

The study group conducted a review of requirements for each of these experiments and established a detailed on-orbit time line for experiment operation. This time line evolved from a summary of other industry studies designed to lead to final hardware configuration, an analysis of the requirements of the hardware against the solutions that the experiments were designed to achieve, and a review of experiment availability to meet the proposed time line for launch.

It was established that most of these experiment would have been generated from NASA centres already working on them, and would thus be Government Furnished Equipment (GFE), although there were some experiments that might have originated from outside as Contractor Furnished Equipment (CFE). Following briefings held by the staff of MSC in Houston, it was decided that the majority of the suggested experiments would have been located either inside the pressurised experiment carrier or in the CM, where data retrieval could be achieved without the need for a space walk (EVA). The experiments from which data could be transmitted directly to Earth would have been located outside the carrier, and would have required no retrieval operations by the crew.

Experiment operations

By analysing each experiment requirement, proposed target areas, the ground data network, other crew activities, and the capabilities of the systems and hardware, the study arrived at a proposed experiment time line for the AAP-1A mission. Since most of the data-gathering would have taken place over the United States, the proposed launch time chosen maximised the number of daytime passes in each of the eleven experiment mission dates available to the crew. By verifying the availability of ground stations during every pass on each day, the study developed a sequence for which experiment could or should be operated as the mission progressed.

The *Frog Otolith* (T-004) would have been programmed in the first three days, with some of the shorter experiments conducted simultaneously. The A and B experiments would have been operated repetitively in groups over the five Standard Applications Days, while the astronomy and navigation experiments would have been conducted towards the end of the mission. A review of the most suitable time to fly the mission, based on what the experiments were intending to record, was also conducted. Seasonal requirements were checked, and it was found that the optimum launch time ran from mid-April to mid-October, with the first practical opportunity occurring in the spring of 1969 to allow for adequate hardware preparation and crew experiment and mission training time.

Once the experiments had been identified and their inclusion in the flight confirmed, it became necessary to establish just where on the spacecraft to place them, what support hardware was required and where that would be located, and what modifications to the CM would be required. Two scientific airlocks would have been housed in the carrier, but none in the CM. The final locations for the experiments were decided upon by minimising CM modifications, CM weight, and power and stowage requirements. During this evaluation, two (unidentified) experiments were further analysed for on-orbit disassembly and film extraction, but these were found not to be feasible in their current configurations in time for the mission.

Modifications to the CM were kept to a minimum by the providing the experiment carrier to support experiment operations that were in excess of the basic CSM capabilities. The modifications that would have been incorporated into the chosen flight CM would have included installation of two portable display and control panels (with associated wiring extensions to the existing pressure suit umbilical connections) and additional stowage provisions. New software for the computerised guidance and navigation system to support the mission profile would also have had to be developed, tested and installed.

Of the twenty-three experiments chosen, nineteen exceeded 90% of the operating time required. From eighteen requests for simultaneous operations from the principle investigators, fifteen were completely catered for, and for the remaining three partial simultaneous operations were accommodated. In an evaluation of the availability of experiments for a test and evolution programme, it was determined that three experiments were behind schedule for any systems testing at the time of the report's publication. Of the remainder, eighteen were available to support a full 20-month test programme, but three of these (the Hasselblad multispectral camera, day and night camera, and dielectric tape camera) would not be ready to support the 16-month test programme should launch be targeted for the latter part of 1968. However, an alternative multispectral camera configuration was reportedly an option for the earlier launch opportunity.

Crew activities
The report also presented baseline activities for the flight crew which, if the mission had proceeded, would have been further refined and minutely detailed at MSC Crew Systems Division in Houston, in consultation with the training officers, simulation engineers, flight controllers, and the astronauts themselves.

The time line evaluated in the study focused on a nominal 14-day duration, with eleven days allocated to experiment operation. It featured overall crew activity and individual experiment operations, with the guidelines of sixty-seven daylight continental US passes, flights through the South Atlantic Anomaly radiation area, the minimum possible number of entries of crew-men into the carrier, plus the limitations of each experiment's life-cycle and other operational requirements and restraints. Considerations included simultaneous sleep periods by the three crew-members, allowing for concentrated experiment data collection over the US, plus necessary housekeeping activities scheduled around an 8–10-hour working day during experiment days. Rotating the daily operations would see only two astronauts actively involved in experiment activities at any one time, with the third crew-member assigned to nominal CSM monitoring and control functions but being available to support the other two when necessary. During the study it was discovered that the following experiments would involve a large amount of crew time:

- S-016: in CSM manoeuvres and attitude hold, extending the duty cycle requirements.
- D-009 and T-002: a larger percentage of total mission duration would be required to accomplish the experiment objectives.
- S-039 and S-040: in addition to the normal Manned Spaceflight Network (MSFN) communication loading, unique S-band down-link would be required for these experiments.
- T-003, S-015, D-017 and T-004: where there was potential for complications to crew/mission operations.

Studies also revealed that the time line could be adjusted if a shorter, minimum mission were to be flown or contingency activities were called for, to allow for useful data collection from the meteorological, Earth resources and astronomical experiments, if integrated as part of the total payload rather than as separate elements.

The study encompassed the role of the crew in experiment operations, identified crew-imposed system requirements, the design and integration of crew equipment and carrier interface, crew training and its impact on the limited Apollo type training facilities, and the management of stowage onboard the CM. The results were incorporated in the proposed mission time line.

Primary crew stations for Earth resources and meteorology experiments would have been in shirt-sleeves, with direct window viewing of Earth targets. Several scientific experiments were already in fabrication at the time of the study report (including the NAA Block I scientific airlock), and so the detailed configuration of the layout of crew stations in the carrier could be evaluated to eliminate the need for EVA operations.

Crew station design
The primary spacecraft, flight control and experiment operating stations were located at the CM left and right couch positions, with a portable display and control (D&C) panel located at the central couch position below the main D&C. From these couches, the crew would have had a direct Earth view out of the forward docking windows, looking forward on the flight path and downwards towards Earth. It was evaluated that it would take about 35 seconds to visually acquire a target before overflight (although the Skylab astronauts found that it took a number of attempts to master this skill, which depended upon the ability to reset any instruments in time, and of course required a cloud free field of view). A secondary work-station in the CM (in support of research objectives and the operation of S-017 and S-019) was planned for the G&N station of the lower equipment bay (LEB).

Work-stations in the carrier were designed for a suited crew-man performing experiment operations either at the right couch in the CM or in the carrier itself, with attitude control handled by the second crew-member from the left CM couch position or (in the case of S-017 and S-019) at the CM navigation station in the LEB. Only one crew-man could enter the carrier for experiment operation or data retrieval. Shirt-sleeved entry was desired and expected, but the study provided baseline data in favour of a soft-suited crew-men relying on CM supplied oxygen through an extended closed-loop suit umbilical, which also featured communication and biomedical linkage. The sizing of the interior work-stations of the carrier was determined by using test models of a crew-man in a fully pressurised suit, with a portable life support system independent of umbilical connections. Unmodified Apollo equipment would have been utilised in the mission – notably the International Latex Corporation A7L pressure garment. The tests on crew station facilities also included evaluation of tethers and restraints, stowage, illumination, storage containers for the retrieved data, and voice communications within a mock-up carrier configuration.

Past evaluations of CM stowage had been supplemented by recent drawings from NAA of proposed Block II stowage requirements for launch, in-flight and entry phases of the lunar missions (including the return of lunar samples). For the AAP-1A study, CM-101 – with recent modifications – was used for baseline data. From this, specific stowage locations were identified for each experiment and data cassette, and this revealed that by removing selected transferable items from the CM to the carrier at the end of the mission, a volume of 26 cubic feet and an allowable weight of 560 lbs was available for re-entry stowage. The study further evaluated that AAP-1A re-entry stowage requirements were 6.6 cubic feet and 219.6 lbs. These figures included an additional 50% in volume and 25% in weight for protection to withstand the rigours of entry and ocean landing.

Crew training

A preliminary review of crew training and training equipment was included as part of the Martin Marietta study for the AAP-1A mission. Of the twenty-three planned experiments, the September 1967 report indicated that nine flight experiments had been evaluated for the study, along with four others – such as multi-spectral cameras and Earth resources experiments that were closely related to the 1A mission. It was expected that the first crew training documentation provided by MSC would be updated and supplemented as more experiments came on line.

Primary training equipment identified for IVA activities included a display and controls package installed in an Apollo Mission Simulator at MSC. In addition, an IVA carrier trainer would be required for the Neutral Buoyancy Facility (NBF – more commonly known today as the Water Emersion Training Facility (WETF)) and for a KC-135 aircraft (more commonly known as the 'vomit Comet') for zero-g simulations. Mock-ups of the airlocks, experiments, retrieval of data cassettes and stowage management training facilities for the CM and carrier were also required, as well as normal Saturn 1B/ CM launch, in-flight and recovery training facilities – all to be slotted into the already crowded Apollo lunar training programme.

Budget estimates

From established project costing guidelines and sources, the study estimated that a minimum lead-time launch date from awarding of the contracts to launch (16 months, known as Phase D) would total about $3.5 million. A 20-month lead-time would be about $1 million more, primarily due to the inclusion of two further experiments (S-0389 and S-040). However, as there would have been only one flight in the programme, all costs would be non-recurring and the flight CM and its launch vehicle would have been drawn from already fabricated vehicles. Only the carrier was a relatively new hardware development.

Cancellation of AAP-1A

While the AAP-1A study continued through 1967, there was already pressure on the AAP programme office from a reduced budget appropriation in FY 1967 and FY 1968. The lunar landing remained the primary goal of Apollo, and there was very little left for anything after the first landings to be paid for, budgeted or even planned. With the growth of the Spent Stage concept of the Orbital Workshop using the former stage of an S-IVB, the AAP-1A mission seemed an unnecessary and costly duplication of the objectives of the OWS programme. Therefore, on 27 December 1967 – only a few weeks after the publication of the Martin Marietta report – AAP Director Charles W. Mathews issued a directive to managers at the three centres for manned spaceflight (Houston, Marshall and Kennedy) to halt all activities relating to the AAP-1A mission.

AAP-1A was a doomed mission from the outset. Trying to extend the use of Apollo independent of other facilities such as the OWS stage was a constant uphill struggle for funding and support. However, the dedication of the contractors and engineering team in supporting all proposals was clearly shown by the Martin Marietta study. It was a detailed and in-depth study that almost proceeded to the final definition and flight stages, but it was hindered by budget restraints and overtaken by developments at Marshall in the larger OWS concept that encompassed most of the primary objectives and fields of investigation that APP-1, the Apollo Extension System and Apollo X had proposed over the previous six years or so.

From 1968, the role of the Apollo CSM in the space station programme became simply a crew ferry and logistics vehicle. The solo Apollo flights of the early AAP era were replaced by the development of the wet and then dry workshop that became known as Skylab. Meanwhile, the main-line Apollo programme had continued to evolve to the point where the first astronauts were preparing to man-rate the Apollo spacecraft in Earth orbit – the first small step on the way to the Moon.

1 *Apollo Program Pace and Progress*, Staff Study for the Subcommittee on NASA Oversight of the Committee on Science and Astronautics, US House of Representatives, 90th Congress, 1st Session, Serial F, published in March 1967. Courtesy BIS archives , London.

2 *Apollo Telescope Mount Study Program Final Report* (Copy No. 302), 1 April 1966, Ball Brother Research Corporation, from the Curt Michel Collection, Rice University, Houston, Texas; also an undated ATM/Apollo presentation briefing (probably summer 1965) to the CB office, also from the Michel Collection at Rice University.

3 *Summary of ATM Crew Participation*; unidentified briefing sheet dated 24 February 1966, in the JSC History Office Skylab collection, formerly at Rice University.

4 *Skylab: America's Space Station*, David J. Shayler, Springer–Praxis, May 2001.

5 *Future National Space Objectives*. Staff Study for the Subcommittee on NASA Oversight of the Committee on Science and Astronautics, US House of Representatives, 89th Congress, 2nd Session, Serial O, July 1966. From the library and archives of the British Interplanetary Society, London.

6 *Apollo Program Pace and Progress*. Staff Study for the Subcommittee on NASA Oversight of the Committee on Science and Astronautics, US House of Representatives, 90th Congress, 1st Session, Serial F , March 1967. From the library and archives of the British Interplanetary Society, London.

7 *Apollo Lunar Module News Reference*, prepared by Grumman Public Affairs, Bethpage, Long Island, NY, June 1969. Supplied (1988) via the Grumman History Centre, Grumman Corporation.

8 *Skylab Preliminary Chronology*, NASA History Office, Washington, compiled by Roland Newkirk, May 1973, NNH-130.

9 AAP/LM-A ATM Controls and Displays Consol Configuration Review, MSFC, 15 November 1967, Grumman Aircraft Engineering Corporation presentation sheets. From the Curt Michel Collection, AAP/ATM Files, Rice University.

10 *Program Directive No. 5, Flight Mission Directive for AAP-3/AAP-4*, Apollo Applications Programme, Office of Manned Space Flight, NASA HQ Washington DC, 27 March 1967. From the Curt Michel Collection, Rice University; *Preliminary Data Specification Document for Missions AAP-1a, AAP-1/AAP-2, and AAP-3/AAP-4*, compiled by the Apollo Trajectory Support Office, Mission Planning and Analysis Division, NASA, MSC, Houston, Texas.

11 *Apollo Program Pace and Progress*, Staff Study for the Subcommittee on NASA Oversight of the Committee on Science and Astronautics, US House of Representatives, 90th congress, 1st Session, Serial F. pp. 785–798. From the library and archives of the BIS, London.

12 *Gemini: Steps to the Moon*, David J. Shayler, Springer–Praxis, August 2001, pp. 303–317.

13 AAP-1A Mission Study Final Report, September 1967, Martin Marietta Corporation (NAS8-21004) PR29-81, located in the JSC Skylab Collection, (formally located at) Rice University, Houston, Texas

14 *Preliminary Data Specification Document for Missions AAP-1A, AAP-1/AAP-2, and AAP-3/AAP-4*, by the Apollo Trajectory Support Office (FM13), Mission Planning and Analysis Division, NASA MSC, Houston Texas, 19 June 1967. From the JSC Skylab collection (formerly located at) Rice University.

Block I missions

The plans for adapting the lunar hardware to other missions offered additional advantages to the huge investment of developing the Apollo system. However, the primary goal of Apollo remained the landing of American astronauts on the Moon, and bringing them home safely, by 1969. In order to achieve this, several test flights of the hardware had to be completed before any astronaut could fly on an Apollo spacecraft. Once this had been achieved, then teams of astronauts could evaluate the Command and Service Modules and Lunar Modules, initially in Earth orbit and then out towards the Moon, before attempting a landing. The future of the Apollo system depended on the success of these early missions on which the Apollo Applications Program also hinged.

TESTING APOLLO

During March 1962 a series of ground rules was developed by a small group within the Apollo Spacecraft Project Office at MSC, Houston. This resulted in the basis of a programme plan and schedules once the method of reaching the Moon had been selected from the three main options (EOR, Direct Ascent and LOR). In addition to providing for the establishment of a schedule that was 'realistic', and allowing for second guessing failure which could also exploit early success, the team suggested that planning for circumlunar, lunar orbital and lunar landing missions should commence at the earliest opportunity. Furthermore, development of flight hardware and operational techniques should proceed with the Saturn 1 and Saturn 1B prior to the man-rating of the Saturn V in a series of 'build-up missions', which began with the simplest tasks and gradually progressed towards the more complex objectives. This also included the use of astronaut crews at the earliest opportunity and as often as possible, with safety considerations, to increase the development of the launch vehicles, the spacecraft and sub-systems. This, of course, would also provide the astronaut corps with a pool of flight-experienced astronauts who could progress to more complex missions (the first landing attempts) after completing early test flights in Earth orbit.

The building block concept

The basic development principle behind the Saturn series was the so-called 'building block' approach, in which the hardware was tested a stage at a time, after which the next stage was added and flight-tested, and so on. This also applied to more advanced vehicles in the series, and the Saturn 1 would therefore be flight-tested before the Saturn 1B took over. Only then would the Saturn V enter the flight programme. This cautious approach, favoured at Marshall Space Flight Center, the lead centre for the development of the Saturn vehicles, required several launches during each phase. This was a reliable and cautious approach to flight-testing, and offered a significant amount of work to MSFC and launch teams at the Cape; but it also slowed down progress and increased the costs of production and operations. The Apollo programme could not afford the time delays, and NASA certainly would not be able to sustain protracted development costs.

The Saturn 1 development programme had begun in October 1961, and in the long-range planning a decision had been made to fly manned Apollo missions on at least four Saturn 1 launches using the S-IV upper stage, although it was becoming economically restrictive to fly astronauts on Saturn 1, Saturn 1B and Saturn V. In September 1963, George Mueller had joined NASA as Deputy Associate Administrator for Spaceflight, with the brief to develop the human spaceflight programmes and improve efficiency in the Gemini and Apollo programmes. He worked in close liaison with the three NASA field centres responsible for manned spaceflight – MSC in Houston, the Cape in Florida, and Marshall in Alabama – and brought with him previous experience of the concept of 'all-up testing' that he helped introduce at the Space Technology Laboratory, which had been devised by the USAF.

This concept focused on putting the requirement of contractors to complete systems testing prior to shipment of flight components, which minimised pre-flight testing, and eliminating the need for stage-by-stage flight-testing, as every launch would be treated as the final mission, thus eliminating the need for many of the proposed missions (including manned flights) on the Saturn 1 which would now end with the tenth unmanned launch. Mueller published his all-up concept on 31 October 1963, and with it the decision to go fully into 'all-up testing' released some of the already fabricated Saturn 1Bs for other programmes, as they would not all be required for mainline Apollo – which in turn gave hope of flying them in the developing Apollo Applications Program. In the official NASA history of Apollo,[1] an interesting point was made that had this decision not been made before the contracts had been signed and hardware begun to be delivered, would there have been a Saturn 1 and a Saturn 1B, or Block I and Block II versions of the CSM? It its also worth remembering that it was this decision that would help in freeing up hardware that NASA later assigned to the only elements of the Earth orbital AAP programme that survived: Skylab, and the Apollo–Soyuz Test Project, flown between 1973 and 1975.

Defining the major phase of Apollo development

On 28 October 1964 the Operations Planning Division of the Apollo Spacecraft Program Office (ASPO) defined an example of this multi-phased approach:

Phase I: *Little Joe II flights* A series of unmanned Little Joe II and launch escape vehicle developments carried out at the White Sands Missile Range in New Mexico.

Phase II: *Saturn 1 flights* Block I unmanned Saturn 1 development and Block II unmanned Saturn 1 and boilerplate CSM development.

Phase III: *Saturn 1B flights* Unmanned Saturn 1B and Block I CSM development, unmanned Block I CSM Earth orbital operations, unmanned LM development and manned Block I CSM development, and manned Block II CSM/ LM Earth orbital operations.

Phase IV: *Saturn V flights* Unmanned Saturn V and Block II CSM development; manned Block II CSM/LM Earth orbital operations and manned lunar missions.

Apollo abort modes
In conjunction with the development of the Saturn vehicles and qualification of the launch escape system, four abort profiles were devised for the Apollo missions.[2]

Mode I This utilised the Launch Escape System (LES) tower on the apex of the Command Module, which would separate it from the SM and the rest of the booster to initially a safe distance before separation, allowing initiation of the CM primary parachute recovery system. Mode I was available for use from the final stages of the countdown to shortly after the 500-nautical-mile mark along the ground track (about three minutes of powered flight). Following a sub-orbital-type flight profile, recovery would be made in the Atlantic Continuous Recovery Area (ACRA) just a few miles from the Cape.

Mode II This was available in the 500–2,500-nautical-mile range across the Atlantic (3–9 minutes GET), began once the LES was jettisoned, and continued until Contingency Orbit Insertion capability begins with the threat of a land landing (in Europe or Africa, depending on the direction of flight). This mode required a manual separation of the CSM with a burn to separate it from the S-IVB, followed by CM/SM separation and CM entry orientation. This type of emergency recovery depended on the non-propulsive full lift capability of the CM on a sub-orbital-type trajectory, also landing in the ACRA but much further down-range, and not necessarily within a predetermined point

Mode III This covered the region 2,500–3,000 nautical miles (nine minutes of flight), was initiated once the threat of a land landing affected the Mode II operation, and continued until normal insertion. An abort during this phase consisted of a manual CSM separation and an SPS retrograde (against direction of flight) burn at a fixed altitude, CM/SM separation, and CM entry orientation, and would result in a sub-orbital trajectory that ended with parachute recovery in the Atlantic Discrete Recovery Area (ADRA) in the Eastern Atlantic – this time at a predetermined point.

Mode IV This was also termed the Contingency Orbit Insertion capability, and commenced once the SPS had the capability of placing the CM in a safe low Earth orbit (LEO) by a posigrade (with the direction of flight) burn resulting in at least a 75-nautical-mile perigee, and continued until the launch vehicle attained a safe orbit. COI abort was also known as Abort-to-Orbit mode, and allowed a safe orbit to be attained from which an immediate deorbit could be targeted towards to a designated recovery area or an alternative mission profile.

These modes were evaluated many times on computer simulations, with the programmers on the one hand wanting the profiles to be used to prove their predictions, and on the other hand hoping that they would never be called upon to rescue a crew (which, in reality, they never did during fifteen manned Apollo–Saturn launches between 1968 and 1975). During the development flights of the Saturns, the programmers used data from ground-based radar and from onboard computers to help produce the impact or splash-down points of the returning CM.

Saturn 1 development launches
There were ten unmanned flights of the Saturn 1 booster. The first four were designated Block I, and were distinguished by the absence of aerodynamic fins on the first stage and dummy upper stages that looked like the latter live versions and featured the same centre of gravity and launch weight. This series also did not feature stage separation. The remaining six launches in this programme – the Block II series – featured the 'building block' concept by using live upper stages and by validating the approach of step-by-step qualification of the cluster-design launch vehicle.

SA-1 Launched from LC 34 on 27 October 1961, on a research and development flight of the booster and a test of the S-1 stage propulsion system to verify both the structure and aerodynamics of the vehicle during powered flight. Only the first stage was a flight model. The S-IV stage was a dummy, as was the simulated S-V third-stage structure. The vehicle attained an altitude of 84.8 miles, travelled 214.7 miles down the Atlantic Missile Range, and impacted in the Atlantic Ocean. SA-1 was instrumented with more than 500 gauges for the transmission of data to the ground. The flight of the first Saturn surprisingly revealed a higher than expected degree of propellants moving around (sloshing), which would be resolved by the installation of anti-slosh baffles from SA-3.

SA-2 Launched from LC 34 on 25 April 1962, on a sub-orbital trajectory, again with a dummy upper stage but with the first 'payload' carried on a Saturn vehicle. Onboard the dummy stage was 95 tons (22,900 gallons) of ballast water to be ejected into the upper atmosphere as part of Project High Water, which was designed to study the effects of radio transmissions and changing local weather conditions. The water was released by explosive devices rupturing the dummy tank at 65 miles altitude at GET 2.5 minutes, and just five seconds later ground observers at the Cape saw the formation of a massive 'ice' cloud, estimated to be several miles in diameter, swirling around the ruptured stage high in the Floridian skies.

The launch of the first Saturn 1 – 27 October 1961.

SA-3 Launched from LC 34 on 16 November 1962. This was a repeat of the SA-2 sub-orbital mission and completed the 'Project High Water' experiment and improvements to reduce the propellant sloshing revealed on SA-1. The objective of the experiment in releasing such a vast amount of water in a near-space environment was in studying the consequences of an in-flight explosion of a Saturn stage, the need of range safety deliberately destroying a vehicle straying off course, and the effects of the resulting debris and resulting liquid propellant cloud on radio transmissions (especially if the ground was trying to communicate with a crew riding an abort profile) and local weather conditions.

SA-4 Launched on 28 March 1963 on the final Saturn launch from LC 34 and sub-orbital test flight of the Saturn 1 first stage, reaching an altitude of 80 miles. At GET 100 seconds No. 5 engine (of the boosters eight H-1 engines) was deliberately cut off by an onboard preset timer in a demonstration of an engine-out capability. The redundant systems on the Saturn rerouted the propellant to the remaining seven engines which continued to burn longer than planned in compensating for the lost engine as a study of potential abort scenarios or alternative missions.

SA-5 Launched on 29 January 1964. This was the first Block II vehicle of the series, representing live upper stages and improved engines. It was also the first to be launched from the second Saturn, LC 37B, as LC 34 was to be prepared for the first Saturn 1B launches. This mission scored a number of 'firsts' in flying the initial S-IV upper stage, guidance and control sub-system, and a successful separation of stages. On board were the first of three protected and recoverable camera packages for studying the effects of boosted flight, stage separation and behaviour of propellants inside the tanks, for post-flight evaluation of vehicle and system performance. The other camera systems were used on SA-6 and SA-7. This flight also reached the milestone of placing the first Saturn upper stage (the S-IV) in an orbit of 487.7 × 162.8 miles.

SA-6 Launched from LC 37B on 28 May 1964. This was the first time that an Apollo boilerplate CSM spacecraft (BP-13) had been carried on a Saturn vehicle, and has also been designated Apollo–Saturn 1 flight 1 (AS-101). Launched as part of the payload with the CSM was a production launch escape tower and spacecraft adapter. The escape tower was jettisoned as planned, after which the spacecraft was placed in a 141.0 × 113.0-mile orbit. This was achieved despite the premature and unplanned shut-down of the No. 8 H-1 engine, due to a suspect turbo-pump. However the launch vehicle onboard systems overcame this setback successfully by using the engine-out capability, providing further confidence in the redundancy capabilities of the Saturn design.

SA-7 Launched from LC 37B on 18 September 1964, carrying spacecraft BP-15 on mission designation AS-102. This was a repeat of the SA-6 launch, and this time the spacecraft completed the programme compatibility tests for engineering and design guidelines with the Saturn 1. The tests of the launch escape system were also repeated on this mission.

SA-9 Launched from LC 37B on 16 February 1965 (and designated AS-103). The out-of-sequence numbering resulted from delays in the new industrial contract manufacturing process in which the booster for the designated SA-8, supplied direct from the contractor (Chrysler Corporation), was slipping being the final stage developed and built at NASA's Marshall centre. SA-9 was therefore readied before SA-8, and flew first. SA-9 was also a departure from the previous Block II Saturn 1 missions in the inclusion of the first of three technology satellites, designated Pegasus because of their wing-like meteoroid detection aluminium panels and solar cell arrays for supplying the spacecraft with power. The objective of the Pegasus series was to collect data from meteoroid penetration in aluminium panels of three thicknesses, to determine the requirements for meteoroid protection on future manned spacecraft. On this flight CM BP-16 was also carried, the LES of which was used to separate it from the main vehicle and to place the CM in a separate orbit where it would not interfere with the deployment of the arrays. The Pegasus remained attached to the upper stage of the Saturn throughout its data-gathering mission. Each satellite was to remain on orbit for at least a year, although Pegasus 1 exceeded this by finally re-entering on 17 September 1978, several years after its mission had ended. In the first three months in orbit, Pegasus 1 recorded seventy meteoroid penetrations of the panels. The wings of the Pegasus deployed in about 60 seconds, extended to a 50-foot span, and had a width of 15 feet and a thickness of only 20 inches. The impact sensors were based on a charged capacitor with a thin dielectric, featuring metal foil on the one side and a sheet of aluminium on the reverse. When struck by a meteoroid, a brief 'short' between the metals plates was created, with the discharge burning off the resulting conducting bridges between the metal layers and effectively repairing itself in the process. It was this discharge that was recorded as a 'hit', with onboard sensors relaying information on the frequency and size of the meteoroid and its trajectory, that provided an indication of the power of each recorded hit.

SA-8 Launched from LC 37B on 25 May 1965, on a mission similar to that of SA-9, carrying BP-26 (AS-104) and the second Pegasus satellite, which also remained in orbit far longer than planned, and eventually re-entered on 3 November 1979.

SA-10 Launched on 30 July 1965. This represented the tenth and final launch of a Saturn 1 vehicle. Designated AS-105, it also carried the third and final Pegasus satellite and BP-09 spacecraft. The launch was moved forward from its original date in order to clear the LC 37B area for modifications and preparations to support its use for the first Saturn 1B unmanned missions. This satellite had the added feature of eight large removable detector segments that could theoretically have been 'visited' by later Gemini crews, but this never materialised, and the spacecraft finally re-entered, after only four years in orbit, on 4 August 1969.

All the Saturn 1 launches were completely successful, and set the precedent for the whole Saturn family of launches through 1975. By the time of the second Saturn 1 launch, the Saturn V was under development and the Saturn 1B was being prepared to launch a series of Earth orbital development flights that tested the manned Apollo and the S-IVB stage, which would be used on the Saturn V to take the lunar missions out of Earth orbit towards the Moon. At the time, there was criticism outside NASA that the Saturn 1 series was a programme with no real objective, and that Project High Water and the Pegasus satellites were included merely to demonstrate limited scientific return from the development programme.

Looking back at the programme, it can be argued that the work of the Marshall engineers early in the programme helped in the early success of the series throughout pre-launch testing and check-out prior to the final launch preparations, saved time and money, and directly provided experience in several areas relating to later Saturn 1B and Saturn V missions and in manned spaceflight development in general.

Experience was also gained in preparing the Saturn LH_2 stages, the demonstration of ground processing of cluster configurations, early versions of the guidance and control systems, boilerplate Apollo spacecraft and tests of the launch escape systems, and processing cycles at the Cape and the launch tracking network. In addition to what von Braun called the 'bonus experiments' of Project High Water

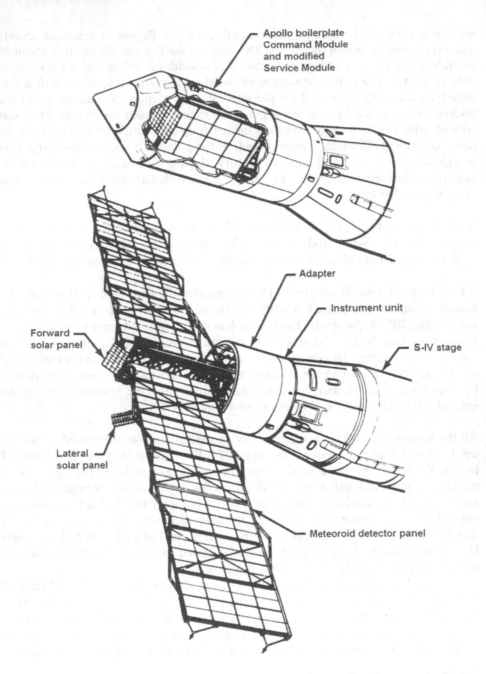

The stowed and deployed Pegasus micrometeoroid detection satellites flown on the final three Saturn 1 operational launches.

and the data from the Pegasus, there was confidence in the impending Saturn 1B launches, and justification of the earlier decision to terminate manned Apollo missions using the smallest of the Saturn boosters.

Saturn 1B development launches
As the Saturn 1 series continued, work was also focusing on a series of unmanned development flights of the next member of the Saturn family: the uprated Saturn 1B. This vehicle used the S-IVB as a second stage, which would be the third stage on the Saturn V, and was planned to be the launcher to carry the first astronauts into space onboard an Apollo spacecraft – both important milestones in reaching the Moon.

SA-201 The first launch of a Saturn 1B, unmanned, occurred on 26 February 1966, and took it to a sub-orbital trajectory reaching a maximum altitude of 303 miles and 5,264 miles downrange, flying a 'short-abort III' profile. The primary objective of this mission was to demonstrate the integrity of the launch vehicle, compatibility of the spacecraft (CSM-009), and verification of the operation of the spacecraft, entry heat shield and recovery systems. At the peak of its trajectory the Service Module was separated and the Service Propulsion System fired for 184 seconds, sending the CM towards a steep descent to simulate re-entry from Earth orbit. Ten seconds after shut-down, the SPS was restarted for another ten seconds to demonstrate its restart capability, which was crucial for the lunar missions. The CM separated from the SM just 37 minutes after leaving LC 34, but it rolled the wrong way and missed its intended target point in the Atlantic. It took several hours to find it, but it was finally recovered two-and-a-half hours after splash-down. The flight encountered several problems which affected the re-entry data. The SPS operated correctly for only 80 seconds, due to a 30% pressure drop due to helium being taken into the oxidiser chamber; steering control was lost when a failure occurred in the electrical system, resulting in an unplanned rolling re-entry for the Command Module; and finally, a short circuit resulted in a distortion of measurements taken during entry. On the whole, however, it was a successful mission, although system testing would be required to solve the persistent problems.

SA-203 Tests on the Block II SPS design were required to ensure that a repeat of the SA-201 malfunction would not reoccur during a lunar mission, especially since there was no back-up to this single engine due to severe weight penalties in providing a second engine. The SPS therefore had to include built-in redundancy as far a possible. The SPS was designed to take the crew in and out of lunar orbit and to adjust the precision of the trajectory to and from the Moon. In the SA-201 situation it was determined that a leaking oxidiser line had caused the problem and would be corrected prior to flying the profile again on SA-202. As a result of these investigations, and allowing time to incorporate changes into follow-on SPS units, NASA decided to delay SA-202 until after SA-203. This was originally the third planned unmanned Saturn 1B development mission, and was not scheduled to carry a payload but to instead place an S-IVB stage in orbit to study the effects of weightlessness on the liquid hydrogen propellant. Therefore, on 5 July 1966 the

Check-out of the Apollo 011 spacecraft in the Manned Spacecraft Operations Building at KSC. The spacecraft was being prepared for the SA-202 unmanned mission launched by a Saturn 1B.

second Saturn 1B left LC 37B, with a nose cone topping off the two-stage vehicle. The resulting orbit had the parameters of 114.9 × 117.4 miles. Ground control monitored operation of the third stage during four orbits, using an onboard television camera to view the 10 tonnes of liquid hydrogen in its tanks. A simulated restart of the J-2 engine was planned, and despite consideration of adding an actual restart to the mission objectives, Marshall engineers opposed such a move as adding unnecessary complexity to the mission. The television camera provided views of the successful demonstration of settling the hydrogen by the venting of gas into the tank, which pushed the remaining propellants towards the rear of the tanks, where valves took the fuel into the engine system. The controllers were so amazed by the television pictures from inside the tank that they thought they were looking at the earlier simulations (which featured a barrel filled with an oil and water mixture). They had to confirm that the simulator was turned off, and that what they were watching was indeed SA-203 hardware in orbit. A deliberate break-up of the stage during the fifth orbit was the official end of the mission, providing in-flight data to verify the data obtained from ground tests on structural test articles representing the S-IVB stage and supplying additional data from break up due to a vacuum. The second Saturn 1B flight also provided an opportunity to evaluate the instrument unit operations in orbital flight – yet another important milestone in flying the unit on each Saturn V launch.

SA-202 Launched from LC 34 on 25 August 1966, carrying CSM 011 on a sub-orbital trajectory that attained a maximum altitude of 710.2 miles in a second test of CM re-entry conditions in a much more complicated flight profile than the earlier SA-201 mission. Four firings of the SPS engine resulted in 200 seconds of burn time, this time without incident. The CM successfully accomplished a re-entry profile (termed a 'long-Abort III') that 'skipped' through the upper layers of the atmosphere to raise heat loads to put maximum loads onto the heat shield. The re-entry was steeper than planned due to a miss-computed centre of gravity of the CM, resulting in a 230-mile overshoot of the planned landing area and a long search to find it. Pick-up of the CM from the Pacific Ocean – two-thirds of the way around the world – was achieved ten hours later.

After thirteen Saturn launches, and despite a few small problems both in the Saturn 1 and Saturn 1B flight programmes, there were high expectations of achieving the first manned flight in a Block I CSM launched by a Saturn 1B in the first quarter of 1967. However, the prospects of an early manned Saturn V launch were hampered by delays in developing the S-II stage, although two or three unmanned Saturn V test flights were planned for 1967, leading to the first manned Saturn V launch hopefully by the end of that year. By September 1966 there were also six three-man teams of astronauts in training for the first three manned Apollo missions in Earth orbit, also manifested for 1967.

PLANNING THE FIRST MANNED MISSIONS

By December 1965 NASA was half-way through the eight-year programme to put Americans on the Moon by the 'end of the decade', as decreed by President Kennedy in 1961, and was preparing to launch the first CSM with a crew onboard within the following twelve months.[3]

From 1966 the launch manifest for Apollo included the three unmanned test flights of the Saturn 1B (described above), two manned tests of the Block 1 CSM on Saturn 1B boosters in Earth orbit, two or three unmanned tests of the Saturn V, and one or two unmanned tests of an unmanned LM on a Saturn 1B booster. These development flights were to be followed by a series of manned Apollo missions launched by Saturn Vs at six-month intervals, designed to evaluate both the launch vehicles, spacecraft, astronauts and procedures, in ever higher Earth orbits that eventually encompassed manned lunar fly-by and orbital flights, but without attempting a landing on the Moon. Only when as many of the remaining questions concerning the safety and procedures of the landing mission had been answered by these flights would a landing be attempted to place astronauts on the lunar surface.

The overall plan was to fly at least two unmanned missions prior to each manned phase of the programme, the only exception being the single unmanned test of the

LM prior to the manned mission in Earth orbit, though a second unmanned flight remained an option if required. Built into the schedule was the additional safeguard that any flown mission could, in theory, be reassigned to repeat a previous flight that was not wholly successful

In brief, these original 'Apollo' missions were:

SA-201, SA-202 and SA-203 The three unmanned Saturn 1B development missions described above, planned for January (SA-201), June (SA-202) and July (SA-203), were designed to tests the basic compatibility of the Block 1 Apollo CSM on the booster on sub-orbital flights, and to evaluate the behaviour of liquid hydrogen inside the S-IVB second stage that was to be used as the third stage on the Saturn V to dispatch the lunar missions from Earth orbit on a trajectory to intercept the path of the Moon three days later. The missions as described above were actually flown in February (SA-201), July (SA-203) and August (SA-202) 1966.

Apollo 1 (CSM-012, AS-204) Planned for October 1966, but slipped first to December 1966 and then February 1967 due to difficulties in preparing the hardware for flight. Apollo 1 was the designation given to the flight by Deke Slayton, Director of Flight Crew Operations at MSC, and the members of the Astronaut Office (CB), while the rest of NASA and the contractors identified the missions by the spacecraft hardware and Saturn flight designations (CSM 012/AS-204). The first manned mission would have flown for up to 14 days, using a Block 1 CSM with a primary mission objective as the full check-out of the spacecraft, its systems and the SPS engine. The crew would also have conducted a programme of several onboard experiments.

Apollo 2 (CSM-014, AS-205) Had Apollo 1 flown as planned, this flight was manifested for the end of December 1966 or in the spring of 1967. This flight was planned to include the first re-rendezvous with the Saturn S-IVB stage to demonstrate the technique that would be used to extract the LM on later missions. There would also be an opportunity to complete any of the experiments not performed on Apollo 1. This would be the second and final manned flight of the Block I, again open-ended in duration for up to 14 days. Should the Apollo 1 mission fall short of completing its major objectives, then this mission would provide a 'back-up' opportunity to achieve the Block I goals.

AS-501 The first of the unmanned test flights of the Saturn V launching a Block 1 CSMs in an evaluation of spacecraft/launcher compatibility as well as evaluating the design of the CM heat shield at lunar re-entry velocities. The first launch was expected in December 1966 but slipped to April and then November 1967, in an initial in-flight demonstration of the first (S-IC) and second stages (S-II) of the three-stage vehicle.

AS-206/LM-1 A Saturn 1B (AS-206) was to have taken the first unmanned LM into Earth orbit in March 1967, in a mission to test-fire both the descent engine and ascent engine of the lander.

AS-502 The second Saturn V was originally expected to fly in June 1967, but slipped to November of that year. It would launch the final Block I spacecraft.

Apollo 3 (CSM 103/LM2) This launch was originally planned for July 1967. It featured a dual launch mission, the first being a three-man Apollo (CSM-101) on a Saturn 1B (AS-207), and the second a Saturn 1B (AS-208), launched 24 hours later, which would carry the unmanned LM-2 to orbit. The flight crew would perform a rendezvous and docking with the LM, and extract the lander from the top of the S-IVB stage. Two of the astronauts would then transfer to the LM to perform repeated undocking and docking exercises. Mission duration was planned for up to five days, and featured the first manned use of a Block II CSM. The missions were jointly designated of Apollo 3 by the CB, and either AS-207/208 or AS-278 otherwise. If this mission was not successful, then two other 'missions' were scheduled as back-up flights.

Apollo 4 In August 1967 there was a full dress rehearsal using all hardware planned to send to the Moon on one launch: the Saturn V, the CSM, and the LM. This was planned as the first manned launch of an Apollo Saturn V, but in high Earth orbit to further develop CSM and LM rendezvous and docking techniques over a mission duration of about three days.

Apollo 5 Launched in January 1968, this would be a repeat of Apollo 4 or an early attempt at the complete lunar mission profile, but in high Earth orbit or perhaps a lunar orbit mission

Apollo 6 Planned for the spring of 1968 at the earliest, and the first in a series of manned missions (through to at least Apollo 9) flown every three months to rehearse the lunar mission in ever-higher Earth orbits. These missions would be the qualification missions to attempt the first lunar landing. Apollo 6 was also possibly the first lunar orbital tests flight of the LM.

Apollo 7, Apollo 8 and Apollo 9 Initially planned as lunar landing qualification missions or as back-up flights to continue the test phases of the lunar missions. Any one of these missions would probably have been the first landing attempt that was being developed as an eight-day mission, with one short-duration EVA (three hours) during an 18-hour surface stay time, and could have been flown on any of these missions, depending on the pace and progress of the programme. Following Apollo '9', the main Apollo programme was expected to be closed out after December 1969, having achieved the primary goal set by President Kennedy in 1961, and any hardware remaining would be moved to the lunar-AAP phase using more advanced spacecraft.

Apollo Applications – Earth Orbit At the same time as the development of the Apollo lunar missions there would be the series of Apollo Applications missions in Earth orbit from 1968 and continuing through 1970, using spacecraft and launch vehicles, as well as astronauts, not required for the lunar landing missions (see the author's *Skylab: America's Space Station*, listed in the Bibliography).

This was an ambitious schedule that relied on successful tests of the Saturn V, the Block II CSM and the LM, which in turn followed the two planned manned test flights of the Block I spacecraft.

Selecting the crews

While the hardware and facilities were being prepared and tested unmanned, several members of the Astronaut Office were receiving assignments in the Apollo programme. Some of the earliest assignments in the CB were in new technical assignments issued on 23 January 1963, when John Glenn was assigned to Apollo issues, Scott Carpenter was assigned to LM development, and Walter Schirra received a dual assignment to Apollo and Gemini training issues. It was still too early in the programme's development for any crews to be formed, but several astronauts from the first three groups participated in numerous meetings and simulations representing the CB as the spacecraft, hardware and systems were developed over the following two years. Many of these assignments were concurrent to assignments in the Gemini programme (see the author's *Gemini: Steps to the Moon*, listed in the Bibliography).

From early 1964, classes in geology increasingly became a major focus of astronaut training and reflected the need of acquiring such a skill for the lunar landing programme as well as more advanced Earth orbital missions under the Apollo Extension System that became AAP. Initially, all current members of the CB commenced Series I Geology instruction (twenty-five sessions over fifty hours) both in the classroom and during field trips from March to June 1964. A second series, for twenty-two astronauts not directly assigned to Gemini, was begun in September 1964 and completed in March 1965. Geology Series III training was conducted between June and December 1965, and featured eleven astronauts who focused mostly on developing the skill and training required for the forthcoming Apollo landing missions.

The third group of astronauts was the first to focus on Apollo, as well as Gemini, at the begininng of their training, reflecting the growing important and imminent series of manned flights on which they were expected to fly. Deke Slayton had determined that the members of the CB office from the first two groups would fly the early Gemini missions, with the former Mercury astronauts flying as Commanders, and with the rookies from Group 2 as Pilots. Flown Group 2 Pilots would rotate to command later Gemini missions before moving over to Apollo in senior roles for the more demanding missions that would lead to the first lunar landing. The Mercury astronauts would be reassigned from Gemini after their first mission, head up the early Apollo crews to test the Block I CSMs, and probably rotate to take command of early lunar missions.

On 8 July 1964, Slayton issued new technical assignments to the current astronauts, featuring the Apollo Branch Office headed by Gordon Cooper from Group 1, with Jim McDivitt, Ed White, Pete Conrad and Frank Borman from

Group 2. The members of this group were also in line for assignment to long-duration Gemini missions by the end of the month. Gus Grissom was head of the Gemini branch, and was in training with John Young for the first manned flight of the series (Gemini 3), while their backs-ups, former Mercury astronaut Walter Schirra and Tom Stafford from the Group 2, would fly the first rendezvous and docking mission (Gemini 6). The Operations and Training Branch, headed up by Neil Armstrong, assisted by Elliot See and Jim Lovell, all from Group 2, included all fourteen members of Group 3 on various technical assignments supporting Gemini and Apollo issues.

The third group of astronaut candidates had been selected in October 1963, and Slayton had tentatively calculated that the eight members with previous test pilot experiences (Bassett, Bean, Collins, Eisele, Freeman, Gordon, Scott and Williams) would be good Pilots for Gemini, and would move on to flying as Senior Pilots on Apollo for the more demanding role of flying the Command Module on solo phases of the Apollo missions. The remaining six had previous experience in operational naval flying (Cernan and Chaffee), Air Force engineering positions (Aldrin and Anders), and civilian scientific research after former military flying careers (Cunningham and Schweickart). To Slayton, these six astronauts would be more suited to development issues on Apollo.[4]

The technical assignments issued on 8 July reflected this thinking. For his PhD, Aldrin had worked on orbital rendezvous techniques, and was an obvious selection for supporting Gemini rendezvous and docking techniques that were to develop skills necessary for the Apollo programme, and he was therefore assigned technical duties under mission planning. Anders had a degree in nuclear engineering, and was assigned to environmental controls systems, radiation protection and thermal protection. Cernan was assigned to spacecraft propulsion and the Agena docking target, working operational rendezvous and docking issues for Gemini and Apollo. Chaffee worked on communications and the Deep Space Tracking Network being established for Apollo. Cunningham was assigned to electrical and sequential systems, and used his experience in monitoring experiments on unmanned satellites that were related to current and future manned programmes. Finally, Schweickart was assigned to future manned programmes such as Apollo and AAP that included space station studies as well as in-flight experiments for Gemini and Apollo.

These six astronauts, along with Eisele, Freeman and Gordon, were expected to receive early assignment in the Apollo programme, though some also filled out assignments on later Gemini missions. By the end of 1965 the first manned Apollo missions on Block I CSMs were expected to occur within the following twelve months, and the time to select and train the crews for those missions was approaching.

The first astronauts for Apollo
In his memoirs, Slayton revealed that he had originally chosen Al Shepard to command the first manned Apollo (from here on termed Apollo 1), but that developing medical programmes forced the first American in space to remain grounded for several more years. In his still confidential planning, Slayton turned to

Gus Grissom to command the mission. Grissom had earlier worked on Apollo issues prior to assignments in the Gemini programme. As there were no plans to fly a lunar module on this mission, the rest of the crew did not require any previous flight experience, and so Slayton assigned Don Eisele as Senior Pilot and Roger Chaffee as Pilot. (The Senior Pilot was an early designation for what became the CM Pilot, while Pilot was the initial designation for the LM Pilot position.)

The back-up crew was intended to be rotated to be the crew to test fly the first manned LM in Earth orbit (Apollo 3), and perhaps even rotate a second time as a complete crew on a lunar landing mission. Jim McDivitt had been working on Apollo since September 1965, and was assigned as back-up Commander; and as the Senior Pilot required previous spaceflight experience to fly the CSM solo in space, Slayton assigned Dave Scott, whose training to fly Gemini 8 in the spring of 1966 would provide him with experience in rendezvous and docking. Rusty Schweickart was assigned as Pilot, and would accompany McDivitt in testing the LM, with the pair also considered as a possible landing team.

For the second Block I mission, Apollo 2, which again would not feature an LM, Slayton chose Walter Schirra as Commander, with Ed White as Senior Pilot and Walter Cunningham as Pilot. Unlike Grissom, Schirra was not expected to rotate to command a later Apollo landing mission, as he already indicated his plans to retire after the flight (his third). In addition, Slayton was planning to move White and Cunningham to Apollo Applications after Apollo 2 (along with Eisele and Chaffee from Apollo 1). The back-up crew for Apollo 2, however, was to rotate as the crew of Apollo 4 – the first manned flight of the Saturn V. Frank Borman was to have been back-up commander of Apollo 2, with Charles Bassett as Senior Pilot and Bill Anders as Pilot. Therefore, in Slayton's planning the Block I assignments were:

Apollo	Position	Prime	Back-up
1 (204)	Commander	Grissom	McDivitt (reassigned as Apollo 3 Commander)
	Senior Pilot	Eisele	Scott (reassigned as Apollo 3 Senior Pilot)
	Pilot	Chaffee	Schweickart (reassigned as Apollo 3 Pilot)
2 (205)	Commander	Schirra	Borman (reassigned as Apollo 4 Commander)
	Senior Pilot	White	Bassett (reassigned as Apollo 4 Senior Pilot)
	Pilot	Cunningham	Anders (reassigned to Apollo 4 Pilot)

Of the twelve astronauts tentatively assigned to the first two flights, all but Eisele, Chaffee, Schweickart and Cunningham were involved with, or just completing, Gemini assignments. In late December, Slayton quietly informed the Apollo 1 crews of their assignments. However, these plans soon received a setback when in December 1965, prior to the official crew announcement, Eisele dislocated his left shoulder during weightless training onboard a NASA KC-135 aircraft, and aggravated the injury further by playing handball, forcing his removal from flight status. Prior to official release of the assignments, Slayton merely swapped Eisele with White from the planned Apollo 2 crew, allowing him time to heal from corrective surgery.

In January 1966 the Apollo 1 assignment was announced to the Astronaut Office, followed a few days later by Schirra informing Eisele and Cunningham that they were to be the prime crew for Apollo 2. On 3 February, Grissom was named Chief of the Apollo Branch Office, and with White and Chaffee he had begun training for the Apollo 1 mission. The crew assignments had still not been made public when on 28 February the planning received a second blow with the deaths of the Gemini 9 prime crew, See and Bassett, in a T-38 jet crash. Slayton needed an experienced Senior Pilot to replace Bassett on Borman's crew, and would assign Tom Stafford as soon as he finished his Gemini assignments. Slayton also moved Anders over to fill in a dead-end back-up role in Gemini, and planned to bring in Mike Collins Pilot on Borman's crew after completing his first spaceflight on Gemini 10.

On 21 March 1966, just five days after Scott returned from Gemini 8, the crews for Apollo 1 were named officially, and manifested for a flight in the first quarter of 1967, but unofficially targeted for a launch in just eight months time (November 1966), with MSC aiming at a possible joint mission with Gemini 12, the last of the series which could rendezvous visually and inspect the Apollo (in a way similar to that of the Gemini 7/6 mission in December 1965). Official approval for this mission was opposed at NASA Headquarters, and the delays in preparing Apollo for the first flight would have prevented this occurring before the planned end of the Gemini programme by December 1966.

While the public and media followed the training and preparations of the Apollo 1 crew, the crew of Apollo 2, also in training, would not be officially announced until 29 September 1966, more than eight months after the formation of the prime crew. By then, another two members of the back-up crew had joined the training group, with Stafford arriving in June (after Gemini 9), and Collins in July (after Gemini 10). In September 1966, the Block I Apollo assignments were:

Apollo	Position	Prime	Back-up
1 (204)	Commander	Grissom	McDivitt
	Senior Pilot	White	Scott
	Pilot	Chaffee	Schweickart
2 (205)	Commander	Schirra	Borman
	Senior Pilot	Eisele	Stafford
	Pilot	Cunningham	Collins

It is interesting to note that the Borman crew was now the only non-rookie Apollo crew who would take a wealth of flight experience to later Block II missions, and perhaps even to the Moon. Borman had logged 14 days on Gemini 7, Stafford had performed rendezvous manoeuvres on Gemini 6 and Gemini 9, and Collins had performed docking and rendezvous activities on Gemini 10 and also had EVA experience – all suitable qualifications for a landing crew.

Apollo 2: a mission to nowhere

At the time of the official announcement of the Apollo 2 crew assignments, the mission had already began to suffer from the lack of enthusiasm of Schirra and the duplication of Apollo 1 objectives.

In addition to completing the lunar mission it was widely expected that the manned Apollo development missions would include a large programme of technical, scientific and medical experiments, incidental to achieving the primary goal of landing on the Moon, provide tangible evidence of real benefit to the American taxpayer who was paying for the programme, and pre-empt those investigations planned for AAP. A similar programme had been incorporated into the Gemini series, and some of these experiments would be re-flown on Apollo.[5]

The problem with incorporating science on Apollo was that it was a new complicated programme that was planned first and foremost as an operational transport system and secondly as a experiment platform, and that any experiments flown would not interfere with qualification of the system to achieve the first landing. To some extent this included that first landing attempt. It has already been stated that the ideas of using Apollo for other missions in addition to the lunar programme (Apollo X, AES and AAP) could satisfy these criteria but would not be at the expense of the primary Apollo goal. In addition, NASA was very careful in trying to incorporate anything that might supplement or even follow Apollo as extensions to the lunar system, and not a new-start programme that was in competition for limited funds and hardware that could threaten funding for the lunar goal.

The Office for Space Science at NASA Headquarters in Washington was responsible for assigning science and applications research on missions, and managers within that office viewed manned spaceflight as one of a number of techniques that could support the so-called pure science over more technical objectives. On the other hand, at the field centres responsible for manned spaceflight – notably at MSC in Houston – the primary objective of early experiments on Apollo flights in Earth orbit was to contribute to the body of data required for the landing missions, and anything that would cause a reassignment to these later more demanding lunar missions would be clearly undesirable. If an experiment did not support the lunar effort or if it could be reassigned to later missions after a landing had been achieve then it would be reassigned.

Mission of the Block I Apollo 2

As its primary mission, Apollo 2 had the objective of repeating and developing the engineering tests planned for Apollo 1, and the secondary task of obtaining scientific data from the spaceflight environment on an open-ended mission of up to 14 days. Both Apollo 1 and Apollo 2 were termed 'open-ended': they would continue to the maximum allowed duration (14 days) or until some technical problem or emergency forced an early return. To achieve this, both spacecraft would use roll-stabilisation during most of the flight, with marginal consumables in the amount of cryogenic propellants stored onboard. These were used in the main SPS engine, the fuel cells and breathing oxygen, and for the SM RCS quads, also used as a back-up to the larger SPS system.

Apollo CM 014 – the planned spacecraft for the Apollo 2 Block I mission (AS-205), and the sister spacecraft of CM 012 assigned to Apollo 1 (AS-204). This spacecraft would not fly, and was used to determine the cause of the tragic pad fire. (Courtesy NASA/ North American Aviation.)

Since they were planning for a long mission, Grissom and his crew were firmly intending to surpass the record set by Borman and Lovell on Gemini 7 in December 1965 (13 days 18 hours 35 minutes) by at least 15 minutes (Apollo 1 – 13 days 18 hours 50 minutes), while the competitive Schirra wanted his Apollo 2 mission to go even further – by a massive 48 minutes (Apollo 2 – 13 days 19 hours 33 minutes).

In addition, Apollo 1 would conduct eight firings of the SPS, while Apollo 2 would complete only four firings although it would carry more experiments – fifteen (eight medical and seven scientific) as opposed to eleven on Apollo 1 (one medical pre- and post-flight, four in-flight, four scientific, and one technological). Other primary objectives for the twin manned Block I missions were:

Apollo 1 Verification of both spacecraft and crew operations, the determination of CSM systems performances, an evaluation of the S-IVB stage and Instrument Unit check-out in orbit, demonstration of the S-IVB attitude control system to determine its acceptability for subsequent re-rendezvous operations on later missions, and demonstration of crew/CSM/launch vehicle/mission support facility performance

Apollo 2 All of the above, and in addition, CSM/S-IVB manoeuvring for IMU alignment, star tracking, and Earth landmark tracking; CSM transposition and simulated docking with the S-IVB; separation, re-rendezvous and simulated docking with S-IVB; and venting of the S-IVB LH_2 tanks six times and the LO_2 tanks twice.

As early as summer 1966, despite its additional objectives and research programme (including a frog experiment), mission planners were voicing concerns that Apollo 2 was not advancing the technical objectives far enough over Apollo 1, and requested the addition of more rendezvous experiments to Apollo 2 in order to provide continuity for the more demanding AS-207/208 mission. It was also found that the CM-014 destined for Apollo 2 was progressing through check-out at NAA far quicker and smoother than the 012 planned for Apollo 1, and that NAA engineers had named the spacecraft 'the great leaping frog', joking that it might even fly before the Apollo 1 spacecraft.

Apollo 2 experiments

Proposals for early Apollo flights first emerged in May 1965, following eight months of discussions between NASA Headquarters, the field centres, and outside scientists. And many followed on from the Gemini programme. Initially, Apollo 1 was to include nine medical, two scientific, and one technological experiment of manual navigation, totalling twelve experiments. Following detailed assessments, four of the medical tests and the technological experiment were found to be incompatible with the plans for Apollo 1, and were dropped; but another medical experiment was added, providing the flight with eight experiments by the end of 1965. Apollo 2 was assigned fourteen experiments – eight medical and six scientific.

Block I experiments, January 1966

	Experiment	Apollo 1	Apollo 2
M-3	In-flight Exerciser	x	
M-4	In-flight Phonocardiogram	x	
M-5	Bioassays Body Fluids	x	x
M-6	Bone Demineralisation	x	x
M-7	Calcium Balance		x
M-9	Human Otolith Function	x	x
M-12	Exercise Ergometer	x	x
M-20	Pulmonary Function		x
S-5	Synoptic Terrain Photography	x	x
S-6	Synoptic Weather Photography	x	x
T-3	In-flight Nephelometer	x	
T-4/S-14	Frog Otolith		x
S-15	Zero-g Single Human Cells		x
S-16	Trapped Particle Asymmetry		x
S-17	X-ray Astronomy		x
S-18	Micrometeoroid Collection		x

Most of the medical experiments flew on the long-duration Gemini flights (GT-4, GT-5 and GT-7), but M-12 and M-20 were new, as were the technical and scientific experiments.

By early 1966 there were concerns voiced in the CB that crew time would be limited for achieving the engineering tasks without the additional medical and scientific assignments – not only during the mission, but also in the training programme. On 20 January, prior to the first official announcement of Apollo crew assignment, Slayton suggested a reduction in the experiment payload to accommodate the increasing time required for new simulations and check-out operations. These facilities were already failing behind schedule, and were delaying the already pressured training programme and constantly forced refurbishment of the simulators so frequently during the year that Grissom eventually hung a lemon in the CM simulator at the Cape.

Slayton felt that re-flying experiments – especially the medical experiments from Gemini – was essentially a waste of time, and would not yield any significantly new data; and if they had not worked on Gemini, then they certainly would not be compatible with Apollo. The Principle Investigators argued that the tests would provide additional correlative data to support the earlier research to validate the findings even further. Ironically, this repetitive flight-validation philosophy for experiments so unpopular to space engineers was the same method used to validate spacecraft hardware and systems.

Some of the astronauts assigned to the Apollo 2 mission have commented on the crew response to the ever-increasing experiment load on the flight. In his 1974 autobiography, *Carrying the Fire*, Mike Collins recalled that some of the planned experiments required the simple flipping on or off of a switch, and were hardly a challenging task for professional test pilots; while Walter Cunningham stated in his 1977 autobiography, *All American Boys*, that Grissom would consider taking anything non-engineering off his flight and dumping it on Apollo 2. (In a recent (2001) detailed analysis of the two Block I missions, author John Charles did not find much evidence of this in the archived documentation.) To make matters worse, Schirra tended to be more anti-science and anti-scientist than was Grissom.

Cunningham also suggested that Schirra's heart was never in Apollo 2, and was puzzled by his uncharacteristic acceptance of the growing experiment load on their flight. Cunningham mused that perhaps Schirra never intended actually flying the mission, as he had already announced his desire to retire from the programme, and could well have been performing a dead-end assignment until Deke Slayton (who was grounded from the Mercury 7 mission in 1962) was medically cleared. Slayton could easily have taken command of Apollo 2, on what would be his first mission, and on which there would be no demanding rendezvous or docking objectives with an LM, or planned EVA operations. Any shortcoming in Slayton's performance would therefore not seriously hinder the progress of the lunar programme with the Block II series.

Apollo 2 research objectives
By September 1966, the experiment assigned to both Block I missions had established somewhat, but Apollo 2 still contained the heavier programme. The difference was that the experiments assigned to Apollo 1 required minimum in-flight

Changes in planned Apollo 1 experiments

Experiment		1965 May	1965 Jun	1965 Aug	1965 Sep	1966 Jun	1966 Sep
M-1	Cardiovascular Deconditioning	x	x	x	x		
M-3	In-Flight Exerciser					x	x
M-4	In-Flight Phonocardiogram			x	x	x	x
M-5	Bioassays Body Fluids	x	x	x	x		
M-6	Bone Demineralisation	x	x	x	x	x	x
M-7	Calcium Balance						
M-9	Human Otolith Function	x	x	x	x	x	x
M-12	Exercise Ergometer	x	x	x	x		
M-17	Thoracic Blood Flow	x					
M-18	Vectorcardiogram	x					
M-19	Metabolic Rate Measurement	x					
M-20	Pulmonary Function	x					
M-48	Cardiovascular Reflex Conditioning					x	x
D-9	Simple Navigation	x					
T-3	In-Flight Nephelometer				x	x	x
S-5	Synoptic Terrain Photography	x	x	x	x	x	x
S-6	Synoptic Weather Photography	x	x	x	x	x	x
S-28	Dim Light Photography						x
S-51	Sodium Vapour Cloud						x
Totals		12	7	8	9	8	10

Changes in planned Apollo 2 experiments

Experiment		1965 May	1965 Aug	1966 Apr	1966 Aug
M-4	In-Flight Phonocardiogram	x	x	x	x
M-5A	Bioassays Body Fluids		x		
M-6	Bone Demineralisation	x	x	x	x
M-7	Calcium Balance	x	x	x	x
M-9	Human Otolith Function	x	x	x	
M-11	Cytogenetic Blood Studies	x	x	x	x
M-12	Exercise Ergometer	x	x	x	x
M-19	Metabolic Rate Measurement	x	x	x	x
M-20	Pulmonary Function	x	x	x	x
M-23	Lower Body Negative Pressure			x	x
S-5	Synoptic Terrain Photography				x
S-6	Synoptic Weather Photography	x	x	x	x
T-4/S-14	Frog Otolith	x	x	x	x
S-15	Zero g Single Human Cells	x	x	x	x
S-16	Trapped Particle Assymetry	x	x	x	x
S-17	X-Ray Astronomy	x	x	x	x
S-18	Micrometeorite Collection	x	x	x	x
Totals		14	15	15	15

effort, while those assigned to Apollo 2 were more complex and required substantial crew involvement.

The medical experiments assigned to the early Apollos were designed to identify and explore physiological problems that might develop in flight and which could interfere with the landing attempt or otherwise pose a health threat to the crew. Many could be done before or after the flight but some did require the participation of the crew during the mission to record or gather samples and data. The following are those experiments specific to Apollo 2 during the latter part of 1966. (Those for Gemini are discussed in the author's *Gemini: Steps to the Moon*, pp. 331–359; and those assigned to fly on Apollo 1 are discussed later in this chapter.)

M-012 Exercise Ergometer A collapsible version of an exercise bicycle, to be used every four days in the flight and designed to stimulate the heart, lungs, blood vessels and muscles. By completing this exercise, the aerobic deconditioning of the astronauts could be measured and later compared with the loss of tolerance when standing upright following recovery. Inside the normally cramped CM it would have been quite a task to uses this device effectively.

M-019 Metabolic Rate Measurement Used in conjunction with the M-12 experiment, this was a respiratory gas analyser.

M-020 Pulmonary Function Measurement Also used when 'riding' the ergometer, this experiment would measure lung volumes and airflow during inhaling and exhaling. None of these three experiments required anything more than minor modifications to the CM.

M-023 Lower Body Negative Pressure Developed to simulate the gravity loads on the cardiovascular system by decompression of the lower body (abdomen and legs) to pool the blood to simulate standing up on Earth. Evaluation into the changes to the astronaut's lower body was monitored in the response of the heart and blood vessels to the gravity load placed on them. The design could also incorporate a bicycle ergometer, and evolved from the ideas being put forward for experiments flown on the AES series of missions. In 1966 NASA decided to limit the use of LBNP devices to just pre- and post-flight studies with a US LBNP device finally flown onboard Skylab seven years later.

S-015 Zero g Single Human Cell Designed to study what influence weightlessness had, if any, on development of human cells and tissue cultures. The experiment featured time-lapse photography, through a phase contrast microscope, of single human cells.

S-016 Trapped Particles Asymmetry One of two experiments located in a scientific airlock mounted in the side hatch of the CM. During passes through the South Atlantic Anomaly, studies of the proton spectrum, differential energy spectrum, pitch-angle distribution and trapped heavy particles located in the Van

Allen radiation belts would be conducted by directly exposing a nuclear emulsion stack.

S-017 X-Ray Astronomy One of two experiments located on the Service Module. This experiment was placed above Sector 3, behind an access door in the forward bulkhead fairing below the CM aft heat shield. A set of sensors would study the quantity of X-ray emission sources emanating from outside the Solar System. The crew-member would have used the optical telescope to align the sensors to acquire the target star of interest, and then orientate the whole spacecraft towards the target.

S-018 Micrometeorite Collection The second experiment in the airlock, and requiring an eight-hour break in the use of the S-016 experiment to collect small micrometeorites for post flight analysis and for measuring the flux of large particles.

T-004/S-014 Frog Otolith The second experiment carried in the SM Sector 1 bay. Two immobilised bullfrogs were laced in a small self-contained life support environment located within a small centrifuge. Each hour a crew-member would activate the centrifuge to initiate the experiment. Over ten seconds it would spin up to 58 rpm, then maintain ½ g for a further ten seconds, and then spin down over the final ten seconds. The cycle was then repeated an hour later, and so on, with the main emphasis on the experiment taking place during the first three days of flight. Attached to the frogs' vestibule nerves were surgically inserted microelectrodes that would register nerve impulses generated from the stimulation of the animal's otolith organs and recorded onboard for post-flight study. The objective of the experiment was to provide an indication of the frogs' vestibule systems to adapt to prolonged weightlessness, from indications by their response to the periodic exposure to a g force.

The inclusion of a scientific airlock on Apollo 2 was a first for a manned spacecraft, and the hatch was suitably modified to accept the hardware. It was installed in place of the side hatch window, and allowed direct exposure of experiments and samples to the space environment without depressurising the whole spacecraft and requiring the astronauts to wear pressure suits (as was the case in the Gemini programme), and allowing for a far greater use and expansion of the system on later missions.

Three of the experiments manifested for Apollo 2 (S-016, S-017, T-004/S-014) had evolved from experiments short-listed during 1965 for integration into the proposed experiment pallet to be inserted into the vacant SM sectors under the AES programme. The idea of flying 5,000 lbs of scientific payload on missions of up to 14 days' duration on the SM from 1968 was abandoned in late 1966 as the Spent Stage concept evolved under the AAP programme. However, the idea was reactivated as the Scientific Instrument Module (SIM-bay) carried on the Apollo 15–17 lunar missions in 1971–1972 (see p. 242)

Overall, the experiments assigned to Apollo 2 represented a repetition of Gemini and Apollo 1 investigations – which the test pilot astronauts probably did not fully appreciate as breaking new ground. However, by providing additional points of data, especially for the medical experiments, John Charles observed in his analysis of the Apollo 2 saga that 'such flight experiments would require more than just a few Gemini crew-members as subjects if they were to be meaningful; repeating them on Apollo 1 could provide up to three more subjects. Further repetition on Apollo 2 could provide up to ten subjects in total for some investigations, perhaps finally reaching the level of statistical validity expected by the medical science community.'

Charles also noted that the provision of an airlock on Apollo 2 eliminated the Gemini requirement for EVA to retrieve sample particles and provide samples without the contamination of launch for retrieval during flight, and that some of the experiments assigned to Apollo 1 and Apollo 2 later found their way to the AAP programme, these included the scientific airlock in which two were incorporated in Skylab in 1973, providing not only a location for scientific studies but also a deployment point for the parasol sunshield structure.

In 1967, the Frog Otolith experiment also moved to AAP as an experiment for the first manned flight, and was then reassigned to the AAP-1A solo flight until it was cancelled in 1968. By then the experiment was authorised for modification as an experiment on an unmanned flight. After many delays, the Orbiting Frog Otolith (OFO-A) was launched by a four-stage Scout rocket on 9 November 1970.

Cancellation of the Block I Apollo 2
During the middle of September 1966, prior to the official crew announcement, the Apollo 2 prime and back-up crews discussed the difficulties being encountered during their preparation for the mission. The result was a two-page 'crew list' of sixteen ultimata designed to streamline the preparation towards their mission, and certain things would be done during the flight. This list did not appear to affect the announcements of the crews to the mission at the end of the month, and possibly gave the crew added confidence in that their demands were at least being considered.

However, a more serious event would influence the fate of Apollo 2. On 25 October 1966, SM-017 was undergoing a ground test when the propellant tanks exploded, damaging the Service Module. The cause was soon identified and corrected, but this event affected mission planning. SM-017 was being prepared to fly with CM-017 on AS-501 – the first unmanned flight of the Saturn V with the primary objective being to test the CM heat shield at lunar re-entry speeds. To keep the mission on schedule, NASA decided to transfer the SM (020) from the second unmanned Saturn launch (AS-502) to the first mission, but in doing this a Block I SM was required for CM-020 on the second launch – the last use of a Block I CM. As the Block I CM was not compatible with the Block II Service Modules further down the production line, the SM-017 had to be rebuilt to fly the mission or an alternative Block I SM needed to be found.

Apparently without warning, Schirra and his crew were told on 15 November 1966 that their mission had been cancelled and that they would be reassigned to back up Grissom's crew on Apollo 1, the sole remaining manned Block I mission. The public announcement was made two days later in a press release that stated that the mission was cancelled in a change in the launch schedule, 'because of launch vehicle and spacecraft development problems'. The elimination of the Apollo 2 mission –

partly recognising the repetitive nature of the planned mission to that of Apollo 1, and the requirement to proceed to man-rate the Block II series – also solved the problem of a Block I SM for the second unmanned Saturn launch, as SM-014 was shipped to the Cape to fly (without the science experiments) with CM-020 on AS-502. Meanwhile, CM-014 remained at NAA to be used in altitude chamber tests. Five days after the fire it was shipped to the Cape as an intact Block I CM, to be used in the investigation as CM-012 was carefully taken apart. Following the investigation, CM-014 was used for supplying spares for mission simulators and in a variety of tests and simulations before being scrapped in May 1977, a decade after it should have flown in space.

Crew changes

On 17 November, the reassignment of the Apollo 2 crew became official with the news that the new mission of Apollo 2 would be the first manned lunar module check-out flight for which the original Apollo 1 back-up crew, lead by McDivitt, had been groomed, but which was not officially announced on that date.

Schirra must have been irritated at the 'demotion ' of his crew from prime to back-up, and with no confirmation of a new mission to which to rotate. Slayton thought it more sensible to use both prime Block I crews for preparing to fly the solo Apollo 1 mission, and to allow the McDivitt and Borman crews to concentrate on the more demanding Block II missions to qualify the CSM and LM on the first manned Saturn V missions.

On 29 November, Slayton also issued a memo[6] on new designation for the crew position on Block II missions. The Block I designations of Commander, Senior Pilot and Pilot were ideal for Apollo 1, but more descriptive identifications would be desirable for mainline Block II missions:

Commander (CDR) Responsible for the overall command of the mission, launch to orbit booster monitoring, LM descent and landing, lunar surface operations, LM ascent and rendezvous, and LM guidance and navigation

Command Module Pilot (CMP) Primary responsibilities included being second-in-command of the mission, detailed knowledge of the CSM systems, CSM G&N, transposition and docking, translunar injection and mid-course corrections, the command of the CSM in lunar orbit, CSM pick-up of a disabled LM in lunar orbit, trans-Earth injection and mid-course corrections, and Earth re-entry and landing

Lunar Module Pilot (LMP) Responsibilities were an operating knowledge of CSM systems, detailed knowledge of LM systems, LM G&N, back-up to LM descent and landing, lunar surface operations, lunar experiments, and back-up for ascent and rendezvous. Slayton stated that on AAP missions the LMP would receive the designation Mission Module Pilot (MMP), whose responsibilities would be designated as mission plans evolved (which in reality became the Science Pilot (SPLT) on Skylab missions and on the ASTP the Docking Module Pilot (DMP) – which, ironically, would be filled by Deke Slayton).

In the reshuffle, Slayton wanted to keep the McDivitt crew intact due to their training experience in preparation for the first manned LM flight, and he therefore assigned them as prime crew on the new Apollo 2 (AS 205/208) mission. Slayton also wanted Borman to command the first manned Saturn V flight, which now become Apollo 3, and took Stafford off Borman's crew to command his own back-up crew for Apollo 2, with Young as CMP and Cernan as LMP. Collins therefore had to move up to CMP, and, as he later recalled, this 'promotion' cost him the chance of flying the Lunar Module and joining an early lunar landing crew. Anders returned to the Borman crew as LMP, and back-up crew positions were filled by Conrad (CDR), Gordon (CMP), and Williams (LMP), all recently off Gemini assignments.

It was not until 22 December 1966 that the new crew assignments for Apollo 1, Apollo 2 and Apollo 3 were officially named in a NASA News Release (66–326), revealing the first assignments of a third, support, crew that would feature in all crew announcements through 1975. These were filled by members of the most recent astronaut intake – the class of April 1966 (Group 5) – who were training as either CM or LM specialists – a skill that would have an influence on their future careers as astronauts. Support crews were provided partly in response to concerns in the CB that with Apollo contractors spread all over the country and with the added complexity of each mission crew, representation was being lost in important discussions and developments. They also alleviated the prime and back-up crews from some of the more mundane crew participation requirements, and provided some of the newer astronauts with mission preparation experience that could result in a rotation to a back-up crew and, eventually, a flight crew position on a later Apollo mission.

Mission	Block	Position	Prime	Back-up	Support
Apollo 1	I	Commander	Grissom	Schirra	Ron Evans (CM specialist)
		Senior Pilot	White	Eisele	Ed Givens (CM specialist)
		Pilot	Chaffee	Cunningham	Jack Swigert (CM specialist)
Apollo 2	II	Commander	McDivitt	Stafford	Al Worden (CM specialist)
		CM Pilot	Scott	Young	Ed Mitchell (LM specialist)
		LM Pilot	Schweickart	Cernan	Fred Haise (LM specialist)
Apollo 3	II	Commander	Borman	Conrad	Ken Mattingly (CSM specialist)
		CM Pilot	Collins	Gordon	John Bull (LM specialist)
		LM Pilot	Anders	Williams	Jerry Carr (LM specialist)

Slayton's system of 'back-up one flight, miss two, and fly the next' meant that the three back-up crews could have flown Apollo 4–6, and in turn their back-up crews would be in line for assignment to Apollo 7–9. However, in the reshuffle after the cancellation of Apollo 2, it was clear in the CB that Schirra would not fly again. Slayton also later revealed that his intention was to move White, Chaffee, Eisele and Cunningham over to the AAP programme under its CB Branch Chief Al Bean after Apollo 1. Slayton was also thinking of moving Schweickart and Anders to AAP after flying their missions.[7]

Grissom was the lead contender for the command seat on the first landing crew. Based on their experience, the crews headed by McDivitt and Borman were also suitable for landing crew assignments. McDivitt and Scott were thought by Slayton to be a good crew with much experience with the LM, which would be very useful on the first landing attempt. With Grissom losing the rest of his crew to the AAP, and Schweickart expected to go, one option would be a Grissom–Scott–McDivitt crew (certainly sufficiently experienced) to receive the first landing assignment on Apollo 7, after serving as Apollo 4 back-up crew.

With the new crew announcements, the preparations continued for the first manned Apollo flight schedule for a February 1967 launch. Training with the new back-up crew had begun in the third week of November, and resumed a few days after Christmas 1966. The next major milestone was the plug-out test on the pad by the Grissom crew on 27 January 1967, clearing the vehicle for launch the following month.

THE UNFLOWN MISSION OF APOLLO 1

The events of 27 January 1967, and the tragic fire that claimed the lives of the prime crew of Apollo 1 during a pre-flight training exercise on LC 34, remains one of the lowest points in American manned spaceflight history. It resulted in emotional turmoil not only for the families and colleagues of the crew, but also within NASA, the aerospace industry and the American general public. The events of that tragic day continue to be deeply remembered and debated almost four decades later. The fate of Apollo 1 and her crew has been well documented over the years,[8] but not so the mission the astronauts should have completed had they flown as planned in February 1967. What then, was the planned mission of Apollo 1?

Objectives
In November 1966, the *Indianapolis Star* published an article by Gus Grissom on his interpretation of his impending mission.[9] Grissom wrote: 'My fellow crew-members and I are finding that our Apollo spacecraft is infinitely more complex than were Gemini and Mercury. And so is the flight plan, even for our own Earth orbital mission ... We'll be using a [navigation] system designed for far-out space where reference points are the stars and the Earth is a small globe a long way behind. Scientists [at the Massachusetts Institute of Technology] tell us that it can be used for space flights as far into the Universe as man cares to go. In the Earth orbital mission

[we] hope to be flying soon [and] helping to write the book by which future crews will fly spacecraft. Our job will be to operate and observe and evaluate all of the spacecraft systems. Where necessary we must come up with suggestions for solutions to any problems we encounter . And this we can do only in actual space flight. We may spend anywhere from three to fourteen days in orbit ... learning as much as we can about the spacecraft performance. Even as we fly the mission, people on the ground will be working to make the Apollo lunar spacecraft an even more sophisticated vehicle than ours.' Grissom also commented: 'The Apollo program, like Mercury and Gemini, is one of research and development [and] we are finding things we want to add and/or change as we go along. Certainly I can't say what lies beyond Apollo, but I can't see man calling it quits after we've reached the Moon.'

The primary objective of Apollo mission 204 (Apollo 1) was to evaluate all the systems of the Apollo parent spacecraft – the Command and Service Modules. Apollo 1 would use the Block I configuration, which would enable future missions to proceed using Block II to qualify the lunar spacecraft for operational use, initially in Earth orbit and then on subsequent lunar missions prior to the first landing attempt. Essentially, Apollo 1 was to be a shakedown test flight of basic Apollo CSM systems to ensure that they worked before flying the more sophisticated spacecraft designed to support the lunar missions.

The first manned Apollo flight was planned as an open-ended mission – that is, the flight was expected to reach a maximum duration of 14 days unless some reason was found to bring it down prior to that goal. The mission was described as a 'six-orbit to two-week' mission, with the capacity to be recalled at different times during that period. The crew believed that it would be advantageous (as long as the systems were functioning and it was safe to do so) to keep the spacecraft flying as long as possible. There was no defining moment as to when to terminate the mission or to extend it – not even to surpass the 14-day target – but only to maximise the supply of information. The crew wanted was to fly as long as was practical but that certainly did not mean a flight in excess of two weeks, which was well beyond the capabilities of a solo Block I CSM. It was considered that CSM-012 could only marginally have remained in orbit to 13 days and a few hours, which would challenge the record set by the Gemini 7 crew in December 1965.

Responsibility for mission success and crew safety lay with Grissom, while the primary assignment for White was on the computer software and crew training programme. Chaffee took care of the flight plan, the back-up Commander handled the spacecraft testing, the back-up Senior Pilot was responsible for the onboard flight data book and the tracking and communications network, and the back-up Pilot worked on the flight operations procedures. Although each man held primary responsibilities for different aspects of the preparations, the hardware and the mission itself, each was cross-trained for mutual support and to provide a human redundancy in all aspects of the flight.[10]

Launch

In the AS-204 basic flight plan published on 9 December 1966, and used as a baseline reference for this account, a 1 December 1966 launch date was used for planning

The Apollo 1 prime crew enter their spacecraft – here in an altitude chamber at the Cape, but in semblence of a scene in the White Room on LC 34 had the mission taken place.

purposes. During the Apollo News Media Symposium of 15–16 December 1966, Chris Kraft, lead Flight Director for the mission, predicted the launch 'somewhere around February or March', and showed a chart (NASA S-66-68010) that indicated that the launch window of 3 hours 30 minutes opened at 1700 GMT (12 noon EST) and was constrained by lighting conditions in the primary recovery area. The window closed at 2030 GMT (15.30 pm EST), was constrained by horizon visibility for Mode III aborts. Kraft added that depending on sunrise at the Cape the launch could be moved back to 11.30 or even 11.00, but that the planning was for 12 noon to ensure adequate lighting conditions in the primary landing area following a full-duration mission. At the time of the accident, Apollo 1 was scheduled for launch on Tuesday, 21 February 1967.

At 12.00 noon on Tuesday, 21 February 1967, astronauts Grissom, White and Chaffee were to have begun their fourteen-day flight onboard Apollo 1 with a launch by Saturn 1B from LC 34 at KSC. Here, the launch of Apollo 7 on 11 October 1968 depicts what the beginning of the Apollo 1 mission might have looked like.

On launch day, the three astronauts would be occupying the couches in the CM – Grissom on the left, White in the centre, and Chaffee on the right. They would be assisted into the spacecraft by the Cape Launch Support Team and probably one of the back-up or support crew astronauts, who would have spent the previous few hours configuring the spacecraft switches in readiness for the launch. In the blockhouse was astronaut Stuart Roosa,[11] who was on duty as CapCom (known as 'Stoney '– from the original 'stone' blockhouses occupied by the early launch control teams), and who communicated with the astronauts in the spacecraft on the pad as the CM was prepared for flight, and during lift-off, until MCC in Houston took over the control of the flight following clearance of the launch tower.

At T–3 seconds, the engines of the Saturn first stage would have ignited, with

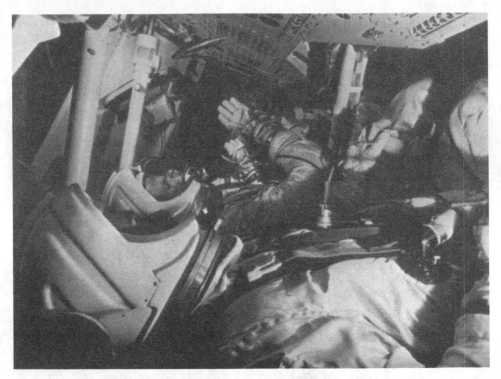

Inside the Apollo 1 spacecraft, the three astronauts monitor their instruments during ascent. In the foreground is Pilot Roger Chaffee, in the centre couch is Senior Pilot Ed White, and at the rear is Commander Gus Grissom.

launch control reporting their ignition as the blockhouse CapCom (Roosa) counted down the final seconds to launch. At T–0 the vehicle would have left the pad to the call of 'Lift-off!' by ground control, and Grissom reporting that 'the clock has started'! The onboard mission events timer would be running as Apollo 1 left the pad and headed out over the Atlantic Ocean. As the launch tower was cleared, controllers at Mission Control in Houston would have assumed responsibility for the flight until splash-down at the end of the mission.

Ten seconds into the mission, the roll programme would have begun in order to aim the spacecraft to the correct launch direction, completed 28 seconds later. Throughout the powered ascent, the crew would be updated on the changing abort modes as the velocity and altitude increased and the range arced over the ocean.

At a GET 2 minutes 10 seconds, with all systems functioning normally, MCC would have advised Grissom: 'Apollo 1, you are GO for staging,' to which Grissom would reply 'Roger, we're GO for staging'.[12] Approximately eleven seconds later, the command to cut off the inboard engines of the Saturn first stage would be given, followed three seconds later by the outer engines. At engine shut-down the crew would be thrown forward against their restraining harnesses.

The physical separation of the spent stage at 2 minutes 25 seconds into the flight would have been followed two seconds later by the ignition of the single J-2 engine of the S-IVB second stage. The crew would have reported the event to Mission Control, being pressed back into their couches as the velocity once again increased. Grissom and White had experienced staging (more commonly known as 'the great train wreck') on previous occasions, but the Saturn was a larger, more powerful vehicle. For Chaffee, this would have been his first rocket launch experience. They would now have awaited the separation of the launch escape tower and boost protective cover that had covered the CM during the early minutes of the ascent through the atmosphere.

At 2 minutes 45 seconds, the crew would have reported jettisoning the tower, acknowledged by MCC. As the vehicle climbed, the status of the trajectory would have been reviewed at 60-second intervals in a series of 'Go/no go' reports. Nine minutes 49 seconds after leaving the pad, the S-IVB would have shut down, and once again the crew would be flung against their harnesses; but this time they would remain there as they prepared for post-orbital insertion check, confirmed by the ground at GET 11 minutes 30 seconds into the flight. Apollo 1 would now have been in orbit of 85 × 130 miles.

Initial activity
About twenty-four minutes into the flight, White would have unstrapped himself from his couch and headed to the lower equipment bay to unstow some of the scientific equipment and photographic hardware. Chaffee would assist him. One of the first tasks would have been to photograph the spent S-IVB stage, setting up the 16-mm sequence camera with an 18-mm lens mounted on the left-hand docking window. A second 16-mm sequence camera, with a 5-mm lens, would also have been prepared, and an SWA (wide-angle) Hasselblad and a standard Hasselblad were also available.

White would also have set up the G&N optics and hand-holds, and removed the flight data files and the LEB work shelf. As the spacecraft continued to orbit, attached to the S-IVB, a 'Go/No Go' decision would have been made on an orbit-by-orbit basis until the sixth orbit. With the first hour of the mission approaching, the crew would have commented on the conditions of the windows, mobility inside the spacecraft, and visibility through the spacecraft optics.

All three astronauts would have remained in their suits, although they could have opened their faceplates as required until the end of the first SPS burn. The main activity of these first orbits (apart from setting up the cameras) would have been the establishment of communications with each of the ground tracking stations, and in performing navigation sighting and IMU alignment checks.

At 2 hours 24 minutes into the mission, White would have secured the LEB and returned to his couch to prepare for S-IVB separation manoeuvres. Grissom would have initiated separation from the S-IVB at 2 hours 54 minutes into the mission, near the end of the second orbit. He would have fired the aft thrusters on the SM for three seconds, resulting in a 1.15-fps velocity increase as the explosive bolts separated the SM from the spacecraft. Using the spacecraft's aft-facing RCS thrusters, Grissom would have gently moved his spacecraft for 70 seconds until it was about 80 feet in

One of the first tasks for the crew upon entering orbit was to perform a station-keeping exercise with the spent S-IVB stage. No rendezvous manoeuvre was planned for Apollo 1, as this would be attempted on subsequent flights.

front of the stages, and then halted his manoeuvre by firing the forward thrusters. Then, firing one set of forward and one set of aft thrusters, he would have turned Apollo 180° to face the top of the stage they have just separated from, to allow White and Chaffee to film the stage with sequence, Hasselblad and television cameras.

During the third orbit, Grissom would have maintained a formation flight with the S-IVB, while LOX vent and LH_2 venting operations were recorded visually and on film. This whole manoeuvre would have duplicated the repositioning of the CSM to retrieve the Lunar Module that would be housed on the top of the S-IVB during docking missions. After twenty minutes the formation flight would be terminated, as the RCS quantity reached 739 lbs and the spacecraft entered darkness. There were no plans for Grissom to manoeuvre the spacecraft to approach the S-IVB/SLA on Apollo 1, and he would maintain only a station-keeping exercise. The rendezvous experiments with the stage would be assigned to later missions.

After the completion of the S-IVB rendezvous sequence, Chaffee would have made his way to the LEB to stow photographic equipment and set up the television camera at the station, with a wide-angle lens to record crew activities as required. He

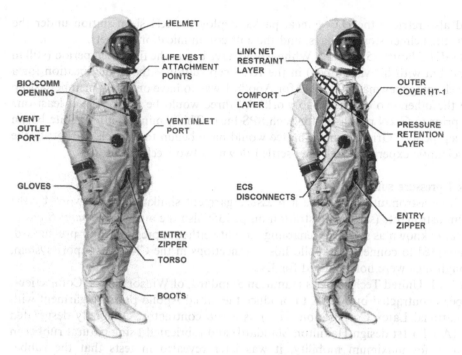

The Block I A-1C pressure suit as worn by the Apollo 1 crew. (*Left*) overall view; (right) cut-away view.

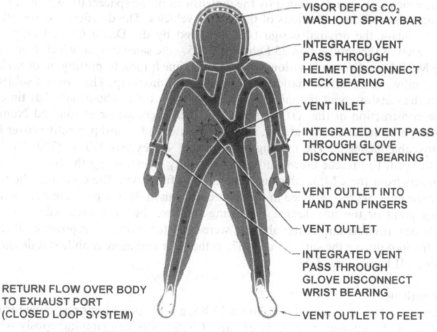

Apollo Block I pressure suit assembly ventilation distribution system.

would also retrieve the Day 1 meal packs, deploy the RH sleep station under the couch, the helmet stowage bags, and the soft communication helmets.

At GET 5 hours 25 minutes, White would have begun the first sleep period (still in his suit but with his visor open) in the RH sleep station – the prime location for a single sleeping crewman. The plan for Apollo 1 was to have one crew-member asleep while the other two continued to work. All three would be awake for at least one hour prior to G&N power-up for each SPS burn. Thirty minutes after White began his sleep period, Grissom and Chaffee would have begun to unpack and set up the first scientific experiments as they settled down to two weeks in orbit.[13]

Block I pressure suits

The three astronauts each wore a pressure garment similar to those worn by the Gemini astronauts (see the illustration on p. 135; also the author's *Gemini: Steps to the Moon*) known as soft suits, enabling mobility either unpressurised or pressurised, and adapted to connect to Apollo hose connections in the CM life support system, although they were not designed for EVA.

In 1961, United Technologies Hamilton Standard, of Winsor Locks, Connecticut, had been contracted by NASA to produce the initial Apollo pressure garment with International Latex Corporation (ILC) as a sub contractor. Originally designated A1H (Apollo 1st design Hamilton Standard) and fabricated using natural rubber in its joints for maximum mobility, it was later revealed in tests that the rubber degraded when it was exposed to the space environment. The incorporation of additional thermal barriers resulted in the suit being too broad for the Apollo couch design, which was already limited by the dimensions of the spacecraft, which in turn was controlled by the limitations of the Saturn vehicles. The decision was therefore made to adapt the original design (A1C) offered by the David Clark Company, which received the contract on 25 February 1965 – the same day on which members of the MSC Crew System Division evaluated the time it took to putting on or taking off the suit while at various stations in a Block I CM mock-up. The two test subjects put on the suits in about ten minutes, and took them off in about half that time.

The construction of the A1C suit was of either uncoated or aluminised Nomex high-temperature-resistant material, while the helmet had a hard protective cover for the moveable faceplate. The melting point of the Nomex was 370° C (700° F), but after the pad, fire rescue crews had great difficulty in extracting the bodies of the astronauts from the CM, as the suit material had fused with the couches. The fire, fed by the pure oxygen supplied in the spacecraft, raised the temperature beyond the melting point of the suit design. Following the fire, the suits were redesigned to include new fire-resistant materials, and were used for training purposes until they were dropped during the summer of 1967, as the safer and more mobile suit designed for Block II became available.

Daily activities

The crew for Apollo 1 trained to follow a 16-hour work and 8-hour rest cycle for the duration of the mission, with Grissom and Chaffee sleeping simultaneously while White remained awake to monitor the spacecraft systems. However, in reality this would probably have been adjusted in flight, following the difficulties experienced during long missions in the Gemini programme. The sleeping crew-man was often disturbed by the other crew-man going about his routine chores. Gemini was much more confined than was Apollo, but with three men inside the new vehicle, there would still be problems encountered.

Here we can also draw upon the actual flight experiences of a similar mission to that planned for Apollo 1. During Apollo 7 in October 1968, the CMP (Eisele) occasionally drifted off to sleep while on duty during the CDR (Schirra) and LMP (Cunningham) sleep shift. According to Cunningham this resulted from Eisele's sleep period being scheduled too early for his normal circadian rhythm, and this experience lead to the practise being abandoned by the time that Apollo 8 flew in December 1968, during which all three astronauts slept simultaneously. This experience would probably also been revealed during Apollo 1.

For the first day, partial suiting would be allowed for an evaluation of the effects of this mode on spacecraft environmental operations. Up to 5 hours 30 minutes into the mission, all three astronauts could have had their faceplate open as required. After that time, both duty crew-men could have removed their helmet and gloves, while White would have remained in his suit, but with his faceplate open, and would try to catch some sleep. It would not have been until some 27 hours into the flight (and after the first burn of the Service Propulsion System) that they could finally take off their pressure garments and remain in their constant wear flight coveralls, until they put on their suits again prior to de-orbit near the end of the mission. However, as part of the full suit removal, the crew would have conducted a suit don and doff exercise to practise and evaluate the procedure in zero g.

Apparently there would be no naming of the spacecraft in the tradition of 'Molly Brown', which Grissom had called Gemini 3. Throughout the mission the spacecraft would have been referred to by name ('Apollo'), with the ground referred to by the tracking station name in range (for example 'Texas'). During launch the ground would have called 'Command Pilot' when addressing the left seat (Grissom), and throughout the mission the CapCom should have identified the crew-member by his position CDR (Commander, Grissom), SP (Senior Pilot, White), and PLT (Pilot, Chaffee), but in reality this would have evolved into simply Gus, Ed or Rog, as was the case with the astronauts on later Apollo missions.

All three crew-members would have eaten together for two of the three daily main meals. Forty-five minutes were allowed for each meal break, and no other crew activities were scheduled for this time.

Housekeeping chores

One of the daily chores that the astronauts would have carried out was the changeout of the LiOH canister every twelve hours, which began with the removal of No. 1 canister ten hours into the mission. If carbon dioxide partial pressure had exceeded 7.6 mm of mercury at any time, the crew would have removed the oldest canister and changed it for a new one without waiting for the end of that twelve-hour interval. Another daily housekeeping task featured a check of the ECS redundant components designed to monitor the main oxygen regulator, although the crew could modify or delete this task as required.

Daily recording of gas chromatograph operation required the crew to activate the chromatograph in warm-up mode about 80 minutes prior to the intended recording time. After the data had been recorded, an optimum 60-minute cool-down period was required prior to deactivation of the instrument.

Other periodic systems checks on the spacecraft by the crew would have been conducted real-time, and acted upon as necessary. These included fuel cell purges, battery charges, status checks on ECS redundant fans, compressors and glycol pump checks, ECS surge tank leakage checks, hydrogen and oxygen purges, SPS heater checks, CM/RCS heater checks, de-orbit programme (P32) pre-thrusting checks and communication station checks. To record mission events, the spacecraft mechanical clocks were scheduled to rewind every 96 hours to retain proper tension on the main spring.

Throughout the mission the crew would also have had to complete a number of humidity surveys of the new spacecraft. These were to be completed in their partial-suited mode (prior to removing the pressure suits completely for the first time) at one-hour intervals during the first day in orbit. For the next three days after taking off the pressure suits, Grissom and his crew were required to conduct humidity surveys every eight hours, and, from the fourth day, every twelve hours during the remainder of their mission.

The surveys would be recorded during both daylight and night-time passes (to establish the effect of direct sunlight on the levels of humidity inside the spacecraft) with the crew partially suited and unsuited, and focused on specific areas of the cabin. These included the general area of the right-hand equipment bay, focusing on the G&N station, the general area of the left-hand equipment bay, the ECS outlet hose, the cabin air return valve, the outlet of the cabin heat exchanger, and the nearby left sleep station.

One of the engineering tests of the mission would be made on the thermal conditions of the spacecraft during an extended period of flight (fourteen days) with approximately seventeen day/night cycles each 24-hour period. If the spacecraft was not rolled along its longitudinal axis occasionally, but allowed to free drift (no RCS firing) in the same orientation, surface areas would receive prolonged exposure to sunlight or shadow – termed 'hot soak'. On the other hand, allowing the spacecraft to conduct a programmed roll (using the RCS thrusters) ensured surface temperatures to be controlled at constant levels across the spacecraft – 'cold soak'. On the later lunar flights this cold soak was conducted during the translunar and trans-Earth trajectories, and became known as the 'bar-b-que' mode.

In addition, a cold soak survey (with RCS thrusters firings) conducted on the spacecraft windows would determine their clarity for visual observations and photography. During the Gemini programme, it had been found that deposits from thruster firings degraded the optical clarity of the windows. During this flight, the contamination would be recorded, in order to devise countermeasures for Block II spacecraft. The photographs would have been taken through each window, with the astronaut holding the camera 2–3 feet from the inner pane, using a variety of Sun angles through the window for illumination.

Ground support

Flight rules established that, unless contacted by the crew, ground stations would not attempt voice contact until after communications or telemetry lock-on and inspection of the data, and that each ground station would not interrupt the onboard activities of the crew except during any contingency or emergency condition of the spacecraft.

Once a day, following the activities over continental United States, the Crew Report would have been relayed to the ground by Grissom, and would include the particular flight plan items which the crew had accomplished, including the revolution and the approximate time that they accomplished it. Changes from normal operational or experiment procedure due to any experiment equipment malfunctions or spacecraft anomalies would also be reported, as would the quantities and types of photographic film used, an estimation of the quality of voice-recording tape used, and the number of LiOH canisters changed during the previous 24 hours.

Updates to the flight plan would include star chart and orbital map updates (node updates), any changes or verification of flight plan items, impending experiments, mission objectives, and biomedical operations. Reporting would use a standard format, including identification or mode number of the experiment or objective, the GET of the initiation time, and any other applicable remarks.

Flight control teams

Preparations for the flight control of the first manned Apollo mission, from Mission Control in Houston, had gained pace in early 1966 when Chris Kraft handed over Flight Director duties to his colleagues, veteran Flight Directors Gene Kranz and John Hodge, who, along with new Flight Directors Glynn Lunney and Cliff Charlesworth, would handle the remaining five Gemini missions. Kraft would be lead Flight Director for Apollo 1, with Hodge joining him after Gemini 8 in March 1966, and Kranz after Gemini 9 in June, operating a three-shift system developed during the Gemini series.[14]

As lead Flight Director, Kraft and his Red Team of controllers were to handle the launch phase (after the tower had been cleared) and support the first three orbits before handing over to Kranz's White Team at GET 5 hours. The White Team had what was termed the 'systems shift' with the responsibility for solving any problems with the various systems onboard the spacecraft. After the crew had gone to sleep, the White Team would hand over to Hodge's Blue Team, who would work on developing the flight plan updates for the next 'day', and after waking up the crew would update them with these changes to the flight plan and then hand over to Kraft's Red Team. The Blue Team shift was also known as the Planning Shift, but was more commonly called 'the graveyard shift', as it was normally very quiet.

The Apollo 1 (AS-204) CSM during mating operations with the Saturn LM Adapter No.5 in the Manned Spacecraft Operations Building at the Cape in early January 1967. The SPS engine was to have been test-fired up to eight times during the planned mission.

SPS burns

One of the primary objectives of Apollo 1 was the operation of the Service Propulsion System by a flight crew in Earth orbit. A total of eight SPS burns were planned during the fourteen days in space. During the News Media Symposium (15–16 December 1966) Grissom stated that he would do the flying on three of the burns, navigating on two and engineering on three. Ed White would do the same, noting that one of his burns was actually only half a second long. As the new boy on the crew, Roger Chaffee would only have been assigned to burn the SPS twice.

The crew-man responsible for 'flying' the burn would occupy the left seat, for 'navigating' he would have occupied the centre couch, and for 'engineering' he would have positioned himself in the right couch. These planned crew positions would probably have changed during the mission, depending on real-time events and crew decisions, but the planning documents for the eight burns of the SPS list them as in the following Table.

Apollo 1 SPS burn crew positions

SPS burn	Left couch	Centre couch	Right couch
1	Grissom	White	Chaffee
2	Chaffee	White	Grissom
3	White	Grissom	Chaffee
4	Grissom	Chaffee	White
5	White	Chaffee	Grissom
6	White	Chaffee	Grissom
7	Chaffee	Grissom	White
8	Grissom	White	Chaffee

There was 9,705 lbs of propellant loaded in the tanks of the SM, with 9,376 lbs available for the planned burns. A summary of the Apollo 1 SPS burn history is presented in the following Table.

Apollo 1 SPS propellant usage

Burn number	Time of burn GET (hrs:min)	Duration (sec)	Propellant used (lbs)	Purpose
1	25:12	10.2	758.7	Orbit, 86 × 234 miles
2	46:57	10.3	759.9	Crew performance in darkness
3	94:14	22.3	1,600.2	SPS performance test
4	142:21	43.1	3,045.4	SPS performance and propellant gauge test
5	188:48	0.56	43.5	Minimum impulse
6	236:29	21.0	1,444.3	SPS performance and PU valve test
7	281:39	3.8	268.7	Altitude correction
8	330:20	8.0	1,037.1	De-orbit burn
Totals		119.26	8,957.8	

The Guidance and Navigation (G&N) scheduling sequence supported the SPS burns and navigational sightings. All G&N optical sights for IMU orientation and alignment were scheduled for night-pass operation, except for one scheduled for a day pass at GET 44:10, prior to the second SPS burn. This was for evaluation of optical sighting during daylight portions of the orbit.

The first two SPS burns at the beginning and end of the second day were to be followed by six further burns separated by 50-hour intervals. The first burn would use a four-RCS-jet ullage burn, while all subsequent burns would use two jets. In total, the Apollo 1 SM housed 328.4 lbs for a nominal mission, less de-orbit requirements, which committed reserves to redline (absolute minimum) levels (87.4 lbs). The redline de-orbit burn requirement at the end of mission apogee was 364 lbs,

Apollo 1 RCS propellant usage

Objective	Propellant (lbs)	
Separation and transposition	22.5	
Inspection of SLA	29.8	
Translation control	0.0	This objective would have been satisfied during the inspection of the SLA. The estimated propellant requirement to perform this test as a separate item was 15 lbs.
Retro orientation	5.8	
Landmark tracking	3.3	
Optical tracking of target vehicle	13.7	
AGCU/FDAI back-up alignment	3.6	
Burn No.1	28.1	
Heat rejection test	1.5	
Burn No.2	21.9	
Thermal soak (cold)	17.8	
Window visibility test	3.4	
Burn No.3	22.6	
Photograph sodium cloud (one observation)	11.4	
Lunar landmark/star sighting	1.9	
Thermal soak (hot)	17.8	
Burn No.4	22.0	
Burn No.5	25.0	
Burn No.6	24.1	
Burn No.7	27.7	
ECS radiator test	9.9	
Daytime align, IMU align	6.2	
De-orbit burn	87.7	

providing a total of 692.4 lbs total supply. There was 790 lbs in the tanks, providing a safety margin of 97.6 lbs. The RCS reserves available to Grissom during the Apollo 1 mission are shown in the above Table.

Navigation and star sightings
Located in the lower equipment bay, at right angles to the main control and display panels (and by the astronauts feet when in the couches) was the Guidance and Navigation station, which could be made more accessible by folding the central couch. This was comprised of three sub-systems: an inertial guidance unit, an optical unit and a computer. Gyros in the guidance unit sensed movement in the spacecraft around the three axis (pitch, yaw and roll), and how far it had moved and in which direction. The astronauts would have used the optical equipment (a sextant and a telescope) to take sightings of the Moon, landmarks on Earth, and known stars. Both sets of data from the guidance unit and astronauts sightings were fed into the computer which, also updated from data supplied by tracking stations on Earth,

The CSM-012 (Apollo 1) flight instrumentation panels (with couches removed) during preparations in September 1966.

constantly updated the accurate position in space and relayed this data to the onboard navigation system.

During course corrections, the G&N determined the steering angle (direction) of the engine bell and the amount of thrust required, combined with precise calculation of the length of time that the engine should be fired for. During Apollo 1 the crew would have had the opportunity to try this system several times both with and without engine burns.

A cold soak count of stars by the crew would be recorded at GET 52 hours, and at GET 100 hours during a hot soak period, the task would be repeated. Data received from the Manned Spaceflight Network (MSFN) would be entered manually or electrically into the computer. In order to observe as many stars as possible, the window shades would have been installed on necessary windows, and cabin lights would have been lowered to a minimum level for one complete revolution.

During each revolution of the two tests, one astronaut would count the number of stars seen through the spacecraft telescope six times: mid-way through the period of darkness; approximately four minutes after sunrise; approximately eight minutes after sunrise; about thirty minutes after sunrise; eight minutes before sunset; and

Astronauts Jim McDivitt (left) and Rusty Schweickart perform systems checks on the guidance and navigation display station in Apollo 012 (Apollo 1). Some of the panels not required for this test were removed for configuration on later tests. The cables at right are part of the ACE cable system.

four minutes before sunset. The astronaut would inform the ground of the number of stars observed and the conditions he observed after each sighting, allowing modification or cancellation of subsequent sightings as required during the orbit. If no stars could be seen through the SCT, or the number observed (more than 50) became impractical to record, then the test would be discontinued.

Experiment activities

The mission of Apollo 1 (although essentially an engineering test flight of the Apollo CSM with the crew on board) also featured a small programme of nine experiments. To accomplish these, at least 60 mission hours were incorporated into the flight plan – except for S-005 and S-006, which were dependent upon real-time events. Grissom and his crew would have had to follow a defined set of guidelines to achieve the best results from each experiment.

T-003 In-Flight Nephelometer Designed to measure the aerosol particle concentration and size distribution in the atmosphere of the CM. All three astronauts would have participated in this experiment by collecting measured samples for post-flight analysis. The first readings were to be taken as close to orbital insertion as possible, and then Grissom and Chaffee were required to take their readings before and after each sleep period. Generally, the crew would have recorded data on this experiment after each third LiOH canister change, and towards the end of the mission before and after they put on their pressure suits. There were seventy filters stored in Apollo 1 for use by the crew, who were instructed not to take readings if the temperature exceeded 90° F, if there was a relative humidity of more than 85%, or if visible moisture 'clouds' were present inside the spacecraft.

To operate it, the astronaut unpacked the instrument from its enclosure bag and located the analyser on a food shelf. The cycle was initiated by pressing the cycle button, followed by extracting an air sample from the cabin. The GET was recorded in the log, along with location, digital read-out, and filter number. At the end of the sampling, the next filter was advanced until a new sequence number appeared, whereupon the unit was ready for the next use. The astronaut returned the device to the stowage bag, and either repacked it in a locker or velcroed it to the bulkhead until the next use.

If an eye irritation was experienced during the flight, one sample would be taken after advancing the filter prior to sampling. Other locations used for sampling included the ECS suit outlet and inlet.

M-048 Cardiovascular Reflex Conditioning This was the primary responsibility only for Ed White, prior to final suit donning towards the end of the flight.

S-005 Synoptic Terrain Photography Designed to improve and extend the techniques of synoptic geographic and topographic photography up to orbital altitudes, using a hand-held camera with no sunlight or stray light (for exposure control). To minimise distortion and fading the astronaut would align the line of sight as near perpendicular as possible to the window surface. Ideally, no roll manoeuvres would be performed, to allow a continuous view of the intended ground target. Conducted by all three crew-members, this was aimed at stressing new photodocumentation of the southern hemisphere, as well as areas pre-marked on flight maps of previous targets of interest. In addition, the crew were instructed to take photographs during consecutive orbits and during periods of hot soak when, during the sunlight portions of the orbit, the spacecraft was not rolled, to maintain even temperatures on its surface.

Completed during a hot soak, window shades were installed as necessary on windows not to be used, and interior lights were turned off. The crew would prepare the Hasselblad camera, choose the appropriate film, and attach the appropriate filter. Control of the spacecraft to record the photographs of selected areas would have depended upon available fuel supplies, but nadir photography was essential for the experiment, with windows parallel to the Earth's surface, controlling the roll to maintain the observation windows in shade.

Data recorded in the flight log included the GET of the exposures, exposure number, film back number, and area photographed. The targets were chosen from the marked-up map on board, or by real-time updates from the ground. Strip

photography consisted of one photograph every five seconds, while stereo
photography featured a set of two exposures five seconds apart. The forty-five
targets for Apollo 1 were:

01	South-West Coast of Africa
02	South-East Coast of South America (IR film, strip)
03	North-West Australia (strip)
04	Central West South America (strip)
05	Eastern and Southern Africa (strip)
06	West Pakistan (IR film, strip)
07	Southern India (strip)
08	South-West United States
09	Mexico (Southern Mexico and Central Africa) (strip)
10	Mississippi River Mouth to Blue Water
11	Amazon River Mouth
12	River Mouths off South-East United States
13	Florida Straits
14	Tongue of the Ocean
15	Echo Bank (no visible island)
16	Currents off West Coast of South America
17	Johnson Island
18	Colorado River (mouth)
19	San Diego
20	Oyster Bay, Jamaica
21	Argue Island (Texas Tower)
22	Hawaii
23	Naguria Island, Solomon Islands
24	Ontong Java Island, Solomon Islands
25	Niutau Island (Ellice Islands)
26	Danger Island, Nassau Islands
27	St Andrew's and Old Providence Islands
28	Seychelles Group
29	Yucatan Coastal Areas
30	Dangerous Ground, South China Sea
31	Christmas Island
32	Area North-East of Phoenix Island
33	Wake Island
34	Pointe De Ca Nau, Vietnam
35	Liberia
36	Lake Chad
37	Bolson De Mapima, Mexico (dry lake)
38	Laguna De Myran, Mexico (dry lake)
39	Guadalupe, Mexico (dry lake)
40	San Luis Potosi, Mexico (dry lake)
41	Laguna de Sayula, Mexico (dry lake)

42   Diego Garcia Island Group
43   North-East Coast of South America
44   Congo River Mouth
45   Kilinailua Island, Solomon Islands

S-006 Synoptic Weather Photography The objective of this experiment was to improve techniques of weather interpretation from photographs taken at orbital altitudes, using constraints similar to those on experiment S-005. Again, this was a task for all three astronauts during the mission. It would also have been attempted during hot soak periods and for interesting weather areas that developed during the flight, and would have been accomplished on consecutive orbits.

Using the 70-mm Hasselblad camera with a 30-mm SWA lens, the astronauts would have operated this experiment in hot soak or per ground update, depending upon real-time weather developments. In the log, they would have recorded the GET, exposure number, film magazine number, and area description. Spacecraft control was dependent upon fuel reserves on board at the time (as with S-005), allowing a controlled roll to keep the observation windows in shade. There were seventeen sequences to be used on Apollo 1, depending on ground updates:

01   Cloud eddies
02   Diurnal cloud variations
03   Sun-glint showing wave action
04   Cirrus clouds
05   Severe thunderstorms
06   Thunderstorms showing wind sheer (stereo pair, 5-sec delay)
07   Dust storms
08   Tropical storms
09   Cirrus bands from tropical convergence areas
10   Cities (showing smog)
11   Oceanic meteorological sounding stations
12   Anomalous cloud lines
13   Persian Gulf
14   Lombok Straits
15   Bahamas Bank
16   Area east of Bermuda
17   Ocean currents

M-003 In-Flight Exerciser The objective of this experiment was to define an in-flight isometric-isotonic exercise and to monitor the heart rate response during isotonic exercise periods in order to document the loss of work (aerobic) tolerance during extended weightlessness. The crew would have used this before meal periods in nine exercise sessions per 24 hours of flight, throughout the mission. The crew had to be careful not to interfere with the experiment M-004.

The exerciser was to be unpacked and attached to the LiOH container near the right-hand equipment bay. With the right hand couch at 172° or the centre couch at

96°, the test subject would have floated into the couch and put on the harness, recording the experiment data and GET in the log or on voice tape. The astronaut then had to pull the exerciser forty-five times with the arms, and then, holding the exerciser handle with his stomach, push with the right and then the left leg forty-five times each. He would next brace his arms against his thighs, hold his legs fixed in the device, and pull forty-five times, using his back and arms. He then had to repeat the entire sequence. Each session would take at least five minutes, and with as little rest as possible between each step, it was designed to 'place stress on the heart.'

Using the isometric exerciser, a series of sustained contractions (followed by complete relaxation of muscles) would then be conducted over a further five-minute period. Each muscle contraction on each exercise would begin slowly at first, building up to a maximum contraction of 6–8 seconds, but no longer than 10 seconds.

The astronaut would then have to place his hands on his forehead or behind his head, and press his head against his hands. He had to push his head left or right, with his hands resisting that pressure. Alternatively, using interlocked hands, he would have conducted exercises such as placing the wrist in the palm and pressing, pressing hands together overhead, pressing hands against an overhead or 'down' onto the couch, arching the back to contract stomach muscles, stretching the legs with pointed toes, or crossing his legs at the ankle and pushing laterally. Once the data was recorded, a two-minute rest period would precede the eighteen-step isotonic exercise programme, and then a three-minute rest period followed the session.

M-004 Phonocardiogram This experiment was to be used to evaluate cardiac functioning by measuring differences in the times of the various cardiac phases (filling, contraction, emptying, rest) from pre-flight data. This schedule was to be followed by Grissom and White only, being careful not to interfere with M-003 – so that this experiment had to be scheduled before an M-003 run, or two hours afterwards. M-004 used the onboard Medical Data Acquisition System (MDAS) for data-recording during the sessions to include scheduled rest and duty sessions for each astronaut, separated by an eight-hour period. The MDAS timer was scheduled to reset each 24-hour period.

Data were recorded for five minutes in the log or on voice tape, including experiment number, GET, and initials of test subject.

M-009 Human Otolith Function This experiment was designed to estimate the effects of spaceflight on the human balance (otolith) organs of the inner ear, by measuring the rotation (or torsion) of the eyeballs (controlled by the otolith), which relates to the astronaut's sense of orientation within the CM. Each astronaut completed this once per flight day. When the scheduled run was missed, the crew would try again no later than four hours before the next scheduled run. The runs were planned to include two of the astronauts, with one of them assisting the person completing the run, eliminating the need to reset the bias. A back-up procedure was available for two pilots to record data if no camera procedures were scheduled.

The crew would have conducted this experiment from the centre couch at the 96-

degree position, with its head-rest removed. In addition, window shades would be installed on all windows, and the primary floodlights would be turned on to 'full bright'. After preparing the experiment hardware and recording cameras, the test subject then floated into the couch and tightened the harness to provide a firm degree of tension.

Wearing the experiment apparatus on the head, helmet-like, the astronaut would keep his eyes closed until he was ready to make the visual judgement of the bite board and record his findings on a tape recorder. Grissom would have used a solid black design, while White would have evaluated a board of horizontal white stripes on a black background. Chaffee's pattern consisted of vertical white stripes on a black background. The GET was marked at the beginning of each test.

S-028 Dim Light Photography This would have been attempted during the cold soak periods for exploratory twilight studies as the spacecraft traversed into or out of the night-pass part of the orbit. No propellant or electrical power would have been allocated, and only one of the astronauts was required. The crew would use the 16-mm sequence camera and an 18-mm lens, with three settings for normal (cold soak), one for drifting flight, and two for a 70-mm camera if the 16-mm camera failed. A hand-held 16-mm camera was also available. Positioning the spacecraft so that the Sun was 15° to the side of the field of view, the astronaut would have begun the sequence ten seconds prior to the lower limb of the sunset, and would have stopped after four minutes. Then, four minutes prior to sunrise, he would have begun recording again, stopping as the lower limb of the Sun cleared the horizon.

S-051 Sodium Cloud Photography This required at least two of the astronauts. There were also specific propellants allocated from the reserves for one sighting and tracking. The experiment featured the launch of two rockets – from Hammeguir (15-second turnaround), the original French national launch site at the Colomb Bechar military base in the Sahara Desert, in French Algeria, and from the Thumba (24-hour turnaround) Equatorial Rocket Launching Station, near Trivandrum, Kerala, on the west coast of India. This provided the crew with the opportunity to see clouds on the same orbit, and to allow them to determine lighting conditions and altitudes versus the cloud graph.

Two of the crew would have prepared the 70-mm camera with a 80-mm lens, and the 16-mm camera with a 100-mm lens, while the third crew-member handled the orientation of the Apollo based on updates from ground control in order to record the data on film and on voice tape.

Spacecraft habitability

In addition to the engineering and systems checks, the SPS burns and navigational sightings, and the programme of in-flight experiments, the crew's comments on the habitability of the Apollo for up to two weeks were an important part of their mission, with applications for future lunar distance flights. The onboard food supply and storage systems, waste management system and crew comfort ware clothing would all have been evaluated by the crew. In addition, an evaluation of the stowage capabilities on the Apollo was to have been an important, if not headline-making, crew activity.

Experience on the Gemini series, especially during the long-duration missions, revealed the difficulties in providing stowage of equipment for two men for fourteen days, in how to effectively handle waste products and manage the daily requirements of living and working in zero g. Apollo 1 may have been billed as a test of the spacecraft for fourteen days to evaluate its effectiveness on a flight to the Moon and back, but the mission was also an important test of the three astronauts to determine how well they could live and work in Apollo, both comfortably and safely.

Apollo was described in the press at the time as a luxurious spacecraft compared with the more cramped Gemini and even smaller Mercury. Many astronauts have since revealed that although Apollo was adequate for its role, the CM habitability was more comparable with a two-week camping trip than a hotel suite, and with no shower and a very basic toilet system the onboard environment could rapidly become much worse. The Apollo 1 astronauts would certainly have pioneered this new ground in evaluating everyday life in space.

During the mission the astronauts would have occupied different couches for each of the eight SPS burns, and would have assumed their original launch positions for the re-entry sequence at the end of the two-week mission.

De-orbit and landing

Throughout the flight, data blocks of landing co-ordinates would be transmitted, so that SPS de-orbit data would be available for each orbit throughout the mission should the need arise to bring the crew home early. This data also included reserves of RCS fuel supplies and allowed for an RCS de-orbit burn return, and water was available in case of primary SPS failure early in the mission. As the mission progressed, the data required to achieve a recovery in either the primary or contingency landing area with SPS or RCS engines would be updated for each approaching orbit. This provided the crew with a real-time situation of re-entry consumables should they be called upon.

Towards the end of their mission the crew would have spent many hours over several days stowing gear and equipment in preparation for the last major hurdle: moving from orbit and returning to Earth. Preparation for entry would have officially begun during the orbit 196 pass over the continental US, approximately 312 hours (thirteen days) into the mission. As White would be four hours into his final sleep period in orbit, Grissom and Chaffee would begin stowing all loose items. Then, three orbits later, on pass 199 (317 hours GET), Grissom and Chaffee would have been in their final sleep period, and White would continue to stow the remaining experimental equipment following the deactivation of the gas chromatography experiment. Grissom and Chaffee would end their last sleep period at about GET 324 hours on orbit 203, and then all three men would have taken their final meal together in orbit.

At GET 325 hours, the astronauts would have conducted the final M-048 experiment procedure prior to putting on their pressure garments for landing. With the G&N powered up at GET 327 hours, a pre-thrust systems test would have been conducted on orbit 206 with an IMU alignment on the same orbit

On orbit 207 pre-entry preparation would have begun by filling the ECS entry tank in the CM and configuring the ECS. Then, 26 minutes before the de-orbit burn, the crew would begin CM/RCS engine pre-heating.

At 330.2 hours, the eight-second de-orbit burn would be initiated to take the vehicle out of orbit during orbit 208. Using data from Apollo 7 as a guide, a reconstruction of the Apollo 1 recovery sequence can be determined more accurately than from the flight plan. Separation of the SM (which would burn up on entry) from the CM would have occurred approximately four minutes after the end of the de-orbit burn, followed by rotation of the CM to place the heat shield towards the direction of flight. Ten minutes later, entry into the denser layers of the atmosphere would have begun, with all communications cut off for about four minutes by the ionised sheath surrounding the spacecraft, while the crew experienced about 3 g. Emergency from black-out would be followed three minutes later by deployment of the drogue parachute, and then the three main parachutes, in less than a minute. During this time, communication with the crew would have been established by recovery forces as the CM headed towards the ocean.

Splash-down in the Pacific would have occurred at GET 330.50 hours on the 208th orbit of the mission, giving a (planned) duration of 13 days 18 hours 50 minutes, and landing on Tuesday, 7 March 1967. The recovery forces would have deployed para-rescue divers from helicopters to attach flotation gear to the spacecraft and to assist the crew out of the spacecraft for helicopter pick-up, after which they would have been flown to the prime recovery vessel for preliminary debrief and medical examinations.

After the mission
The astronauts – the first crew to qualify Apollo – would probably have remained onboard the prime recovery vessel until it docked at Hawaii. They would then have been flown back to Houston for a full mission debriefing and a homecoming welcome by family and friends. Following initial medical examinations and crew debriefing, a post-flight crew press conference would have been held at MSC for the crew to take the opportunity to describe their mission and discuss the results with the world's media – the first small step for Apollo, accomplished on its way to the Moon.

As for the crew, it would have been the opportunity to seek out new goals in their careers, which could have meant remaining at NASA or moving on. Grissom would probably have remained in line to command an early – perhaps the first – lunar landing mission, while White and Chaffee (according to Deke Slayton's long-range planning) seemed to have been in line to fill in assignments on the Apollo Applications space station programme, and would not have been rotated to a new back-up crew for assignment as prime crew members on later lunar landing missions.

Following Apollo 1, the programme would have moved towards the launch of the first manned Block II mission and manned test of the LM in Earth orbit, probably designated Apollo 2, under the command of Jim McDivitt. However, before Apollo 1 had a chance to fly, the events of 27 January suddenly terminated all preparations for future manned missions, and placed the lunar programme on temporary hold while the tragedy was investigated. Apollo 1 can never be recorded in the pages of Apollo history as a 'flown' mission. Perhaps in a small way this account compensates, and recognises, at least in spirit, that the 'mission' of Grissom, White and Chaffee was a part of every Apollo mission that *did* fly.

1   *Chariots for Apollo: A History of Manned Lunar Spacecraft,* Courtney Brooks, James Grimwood, Loyd Swenson, NASA SP-4205, 1979 pp.128–131.
2   *Abort/Re-entry Processing and Testing on Saturn 1B launches,* James Summers (IBM retired) posted on the web, Space History Newsgroups 26 March 1998, and 1 July 1999; *Skylab: America's Space Station,* David Shayler, Praxis–Springer, 2001 pp. 98–9; also various NASA Apollo, Saturn 1B and Saturn V press kits.
3   *Seven Minus One: The Story of Astronaut Gus Grissom,* Carl L. Chappell, New Frontier Publications, Mitchell, Indiana, 1969, Chapter XI, The Apollo Project, pp. 129–130.
4   *DEKE: US Manned Space from Mercury to the Shuttle,* Donald K. Slayton with Michael Cassutt, Forge Books, 1994, pp. 132–140.
5   *The First Apollo 2: Science and Operations planned for 'The Leaping Green Frog',* John B. Charles, PhD, Houston, Texas. *Quest,* **9,** No. 2, 2002, pp. 26–42.
6   *Block II Apollo flight crew designations.* Memo from Director of Flight Crew Operations (Deke Slayton), 29 November 1966. From the Apollo History Archives, formerly (1988) at NASA JSC, Houston, Texas.
7   Private correspondence from Michael Cassutt to the author, 18 October 1993.

8   For example, *Disasters and Accidents in Manned Spaceflight*, David J. Shayler, Springer–Praxis, 2000, pp. 99–115 and Bibliography.

9   *Indianapolis Star,* Magazine Section, 13 November 1966, pp. 9–22.

10 *First Manned Apollo Crew Press Conference*, 3 August 1966, Downey, California, with the prime crew of Grissom, White, Chaffee, and the original back-up crew consisting of McDivitt, Scott and Schweickart.

11 Stuart Roosa was on console duty on 27 January 1967, and was one of the group of (19) astronauts selected by NASA in April 1966.

12 *Flight crew/ MCC-H-Working Agreements for Apollo 1*. Memo from Deke Slayton, Director of Flight Crew Operations, 28 November 1966, concerning voice communications and flight procedures.

13 The primary reference for the author's interpretation of planned crew activity during the Apollo 1/AS-204 was extracted from the document, *Apollo AS-204A Flight Plan (Final)*, 9 December 1966, Mission Operations Branch, Flight Crew Support Division, NASA MSC, Houston, Texas.

14 *Failure is not an Option: Mission Control from Mercury to Apollo 13 and Beyond,* Gene Kranz, Simon & Schuster, New York, 2000, pp. 168 and 195.

Small steps

The primary objective of this book is to recall the missions of Apollo that never took place due to budgetary restrictions, hardware shortages or programme redirection. So far the discussion has recognised that Apollo hardware could be used in other applications in addition to the lunar goal, and in testing smaller Saturn launch vehicles and Block I CSM hardware to achieve this.

The end of 1966 had seen the establishment of the Apollo lunar mission profile, several astronauts had begun mission training, and unmanned development flights of the Saturn 1 and Saturn 1B launch vehicles had been completed. The prototype unmanned CSMs had completed sub-orbital and orbital missions, and a wide variety of ground tests had been performed. The series of Mercury (1961–1963) and Gemini (1965–1966) manned missions had provided valuable experience in living and working in space for up to two weeks, and had helped develop many of the basic skills that would be needed for Apollo and the AAP. However, the mighty Saturn V had yet to fly, no Lunar Module had proven itself in space, and no Apollo crew had flown in space, let alone journey a quarter of million miles to the Moon and return over the same distance.

The loss of the Apollo 1 crew was a tragedy that reached far beyond the astronauts' families, NASA or the space community. More than three decades after the events of 27 January 1967, the deaths of the three astronauts remain a painful memory for many Americans – similar to the *Challenger* accident during Shuttle mission 51-L on 28 January 1986. For months following the pad fire, the thousands of workers at NASA, the spacecraft prime contractors and the sub-contractors worked many long hours to ensure that such an accident would not occur again, and that Apollo would fly safely with astronauts on board as soon as was practically possible.

By February 1967 there still remained many hurdles to overcome in order to reach the Moon by December 1969, just 34 months away. The grand plans for what could or should follow Apollo were still being defined and were still unfunded, the recovery from the pad fire had begun, and progress towards the Moon resumed – slowly at first, but soon the momentum would return. The road to the Moon would be taken in distinct stages. First, by unmanned spacecraft continuing a programme, begun several years earlier, to send probes to the Moon in order to scout for potential landing sites. Secondly, a sequence of unmanned tests of the Saturn V and LM was scheduled to complete the testing of the final elements of the hardware. Thirdly, astronauts were selected for a series of ground simulations and as flight crews for a series of Apollo missions designed to develop the techniques and procedures of the Apollo system and flight profile, thus allowing an attempt at the lunar landing at the earliest opportunity. Finally, a plan was drawn up for the initial landing – not only narrowing down the choice of where to land but what to do when the astronauts arrived there.

Each of these events was a small steps towards the larger goal of achieving the Apollo landing by the end of 1969. Some were to become historic milestones in the history of manned spaceflight, while others are not so well remembered, but nevertheless were important developments that contributed to the success of Apollo.

ROBOTIC EXPLORERS

When the decision was made to send Apollo to the Moon, it was clear that it would be some years before the first astronauts could make the trip. However, with the Moon as our nearest neighbour in space it became the target for some of the earliest missions. To support the manned lunar effort, NASA adapted and expanded a programme of unmanned scouting missions that could provide preliminary data on the structure of the lunar surface and environment. These missions would be accomplished with attempts to hit the Moon with a package of survivable instruments, with fly-by or lunar orbiting for photography and environmental studies, and with soft landings to study the structure and surrounding environment of the landing sites, providing important information to the designers of the Lunar Module and those planning manned surface exploration.[1]

Pioneer (1958–1960)

This initial series of missions – originally a USAF programme to send a probe to the Moon – was transferred to NASA in October 1958 to become the agency's first spacecraft family. Pioneer missions were divided into two phases: lunar (1–4) and then solar orbit (5–9) spacecraft. (From Pioneer 10, the series also became interplanetary probes.) The lunar probes were the first American spacecraft designed to reach the vicinity of the Moon, and employed a direct ascent trajectory.

Pioneer 0 (USAF launch, 17 August 1958) Intended to be a lunar orbiter, containing a simple imaging system, but the Thor–Able launch vehicle exploded just 77 seconds after leaving the pad.

Pioneer 1 (First NASA launch 11 October 1958) The first spacecraft launched under the direction of the civilian space agency, and also intended to orbit the Moon. It suffered a launch abort due to an error in the Able second stage cut-off speed. After returning data on the Van Allen radiation belts, the probe burned up in the atmosphere two days later after reaching 71,600 miles into space.

Pioneer 2 (8 November 1958) The third lunar orbiter suffered a failure in the Altair third stage when it failed to ignite, and reached 960 miles altitude and returned data on Earth's flux and energy levels before returning towards Earth and burning up after 412 minutes in flight.

Pioneer 3 (6 December 1958) Planned as a lunar fly-by mission with a payload of radiation detector and a test of a camera-triggering instrument but without an imaging system. The Juno II (Jupiter) launch vehicle failed to attain Earth escape velocity by about 640 mph, but the probe still recorded data on the dual bands of radiation around the Earth. After reaching 63,500 miles altitude it burned up in the atmosphere after a flight of 38 hours 6 minutes.

Pioneer 4 (3 March 1959) The first US probe to escape Earth's gravity, achieving a fly-by of the Moon (4 March) at 37,000 miles (instead of the planned 20,000 miles).

Pioneer P-3 (26 November 1959) Planned as a Pioneer-class lunar orbiter with a television scanner, but the payload shroud failed after just 45 seconds into the mission.

Pioneer 5 (11 March 1960) Launched towards Venus in a test of technology interned for lunar and planetary probes of the future, and entered a 312-day solar orbit.

Atlas–Able 5A (25 September 1960) and *Atlas–Able 5B* (14 December 1960) Both intended to orbit the Moon, but suffered launch failures. On the 5A attempt, the second stage shut down prematurely and resulted in the burn-up of the payload 17 minutes into the mission, while the 5B attempt failed after 68 seconds when the Atlas–Able launcher exploded.

Ranger (1961–1965)
The first two probes (Block I) of this series (Ranger 1 and Ranger 2) were designed to test the systems and hardware of the main spacecraft carrier (termed a 'bus') that supported the scientific payloads. The next three spacecraft (Ranger 3, Ranger 4 and Ranger 5) carried both a television camera system and hard-landing instrumented capsules (Block II), while the remaining spacecraft (Ranger 6–9) carried only television cameras and no other experiments (Block III).

Ranger 1 (23 August 1961) Carried instruments designed to gather data on solar plasma particles, magnetic fields and cosmic rays. Planned as a high-altitude test, the probe was stranded in Earth orbit when the Agena B upper stage failed to restart, and the probe re-entered the atmosphere seven days later.

Ranger 2 (18 November 1961) Planned as the second high-altitude test flight, but again was stranded in low Earth orbit when the attitude control system of the launch vehicle failed, re-entering just two days after launch.

Ranger 3 (26 January 1962) The initial flight of the spacecraft carrying a television camera and a hard-landing instrumented capsule with a seismometer. Unfortunately a fault in the launch vehicle resulting in the probe missing the Moon by 20,000 miles, and a course correction error led to the probe flying by the Moon at 22,800 miles on 28 January. No images were returned due to a failed central computer and sequencer system onboard the probe.

Ranger 4 (23 April 1962) Carried a television camera and hard-landing capsule, but a failure in the computer and sequencer system resulted in no data being returned. The uncontrollable spacecraft impacted the Moon on 26 April (15.5° S, 130.5° W).

Ranger 5 (18 October 1962) Launched successfully, but suffered a power failure which disabled all systems and experiments. Fortunately, four hours of gamma-ray data were received before the onboard batteries died. The spacecraft passed the Moon at 450 miles and entered solar orbit.

Ranger 6 (30 January 1964) The first of the Block III series to carry an array of six television cameras to relay a sequence of images of the approaching lunar surface in the final twenty minutes of the descent, providing high-resolution close-up details of selected *mare* (lowland) target areas in support of the Apollo programme. Ranger 6 impacted on the Moon in the Sea of Tranquillity at 9.39° N 21° E on 2 February, but due to a camera failure no images were returned.

Ranger 7 (28 July 1964) The first successful spacecraft of the series when it struck the Moon on 31 July after transmitting 4,316 pictures and providing the first close-up views of the lunar surface in the northern rim area of the Sea of Clouds (10.7° S 20.7° W). The data returned were used to indicate that the surface would support a manned landing. In April 1972 the lunar orbital cameras of Apollo 16 SIM-bay photographed Ranger 7's resulting 9-foot impact crater.

Ranger 8 (17 February 1965) Transmitted 7,137 images of the equilateral mare regions of the Sea of Tranquillity, which was just one of several leading areas of interest for targeting the first manned landing mission. The 20 February impact was determined at 2.59° N 24.77° E – within fifteen miles of the planed target area.

Ranger 9 (31 March 1965) Targeted for a more scientifically interesting area after the two previous missions had gathered the survey data required for targeting the first Apollo to a mare landing site. The spacecraft was aimed for the 24-mile diameter crater Alphonsus (13.3° S 3.0° W), where it impacted on 24 March after returning 5,814 images – 200 of them live on American commercial television in what was billed as 'the first TV spectacular from the Moon'. Image resolution was refined to less than 1 foot prior to impacting just 4 miles off target, creating a 9-foot impact crater which was imaged from Apollo 16 seven years later.

Surveyor 3 sits on the Ocean of Storms with Apollo 12 LM *Intrepid* in the background. A similar pinpoint landing was planned for a later Apollo mission to the landing site of Surveyor 7 in the early 1970s, but budget cuts terminated that objective.

Surveyor (1966–1968)

This was originally a two-part programme consisting of an orbiter and a soft-lander. Then, following setbacks with the early Ranger probes, new Apollo requirements were issued from the Office of Manned Spaceflight for more detailed images of potential landing sites from orbit, resulting in a refinement to the Surveyor objectives. What emerged were two separate programmes: Surveyor would be a programme of ten soft-landing spacecraft carrying scientific instruments to the lunar surface, while Lunar Orbiter would provide the high-resolution images from orbit. Initially, weight constraints and the growing importance of Apollo changed the role of Surveyor from pure science to engineering test bed. Each Surveyor was fitted with a television camera for imaging the surrounding terrain, a sampling scoop, and data collection devices to record surface magnetic properties, and radar and thermal reflectivity.

Surveyor 1 (30 May 1968) The first US spacecraft to achieve a fully controlled soft landing on the Moon: 2 June 1966, in the Ocean of Storms (2.45°, S 43.22° W), nine miles off target. The achivement was a huge success for what was originally planned as an engineering test of the design. The spacecraft returned 11,350 high-quality images from horizon views of the surrounding mountains to close-ups of its own reflective mirrors, and selenological data on its landing site over a period of six weeks (three periods of lunar daylight).

Surveyor 2 (20 September 1966) Crashed onto the Moon on 22 September after one of its three vernier engines failed to ignite during a mid-course manoeuvre. It impacted (5.5° N 12° W) south-east of the crater Copernicus.

Surveyor 3 (17 April 1967) Landed on the Moon, after bouncing twice, on 19 April, about 230 miles south of Copernicus in the south-eastern part of the Ocean of Storms, within an Apollo landing area. The spacecraft returned 6,326 images, and data on soil gathered by the sampler from four trenches, seven bearing tests and thirteen impact tests. Three years later the astronauts of Apollo 12 landed near Surveyor 3, and returned parts of the spacecraft for analysis.

Surveyor 4 (14 July 1967) Flown successfully to the Moon until the final two second before retro-burn 2.5 minutes prior to landing, when it is believed to have suffered an onboard explosion, and impacted on the surface (0.4° N, 1.33° W) in the Sinus Medii region.

Surveyor 5 (8 September 1967) Landed successfully, after suffering technical failures in flight, on 10 September, in the Sea of Tranquillity at 1.5° N, 23.19° E. In addition to 18,000 images, the spacecraft returned data on lunar surface thermal reflectivity, and performed the first on-site chemical soil analysis.

Surveyor 6 (7 November 1967) Achieved a soft landing in the Sinus Medii region on 9 November (0.53° N, 1.4° W), and returned 29,500 images of the lunar surface, Earth, Jupiter, and several stars. Data were transmitted on the dynamics of the touch-down and the surface characteristics, determining the abundance of chemical elements in the lunar soil. On 17 November, 177 hours after landing, the three thrusters were fired for 2.5 seconds, lifting Surveyor about ten feet off the surface and moving it to a new landing site about 6.5 feet to the west. Surveyor 6 was the first spacecraft to be launched from the Moon, and to land twice.

Surveyor 7 (7 January 1968) Aimed for an area of the Moon with higher scientific interest than previous landings. All of the programme objectives designed to support Apollo design studies had been met by the sixth flight, and as a result, the remaining three Block II spacecraft were cancelled together with plans for a Block III Surveyor that included an automated rover. Surveyor 7 landed at 40.86° S, 11.47° W, on 10 January, near the crater Tycho, and returned 21,274 images (with several in stereo) detailing specific rocks of interest. Two lasers were aimed at the spacecraft, from observatories in California and Arizona, demonstrating the ability of using lasers for both communications and accurate measurements of the Earth–Moon distance (which also became an Apollo experiment). Surveyor 7's landing site also featured in the planning of later Apollo missions before budget cuts curtailed the manned programme.

Lunar Orbiter (1966–1967)
The final major unmanned programme flown in support of the Apollo missions was the Lunar Orbiter series, evolved from the planned Surveyor Orbiter project of 1960.

A Lunar Orbiter 2 photograph of the Marius Hills area taken across the surrounding terrain rather than vertically down, providing an impression of the approach viewed from an LM. This area was at one time a leading contender for Apollo 19 and then Apollo 17. (Courtesy NASA via the British Interplanetary Society.)

The resulting studies defined a spacecraft that could return images, with a resolution of three feet, of potential Apollo landing sites within 5° of the lunar equator and between 45° E and 45° W (see p. 191), by using a self-contained photographic system, developed by Eastman Kodak, that included wide- and narrow-angle lenses, a processing system, and a scanning and transmission system.

Lunar Orbiter 1 (10 August 1966) The first US spacecraft to be placed in lunar orbit (14 August), which was eventually reduced to a perigee (closest approach) of only 31 miles. The photographic system returned 207 images covering 25,500 square miles of candidate Apollo landing sites, and further images of the lunar far side. The tracking of the spacecraft also revealed variations in the gravitational field that affected the orbital parameters of the spacecraft, providing valuable data in planning the Apollo orbital and landing profiles. This phenomenon was attributed to mass concentrations (mascons) in the lava-filled lunar mare regions. Lunar Orbiter 1 focused on southern region landing sites, and from images of large blocks on the surface, provided further evidence that the surface would support the LM. At the time, the Ocean of Storms, near Surveyor 1, was a leading area for the first astronauts, as it contained about 20% fewer craters than other areas photographed. Lunar Orbiter 1, like those that followed it, was deliberately impacted on the lunar far side (6.7° N, 162° E), during its 577th orbit on 29 October, to prevent its becoming orbital debris of danger to later spacecraft, including Apollo.

Lunar Orbiter 2 (6 November 1966) Arrived in orbit on 10 November, and continued the photography of potential Apollo landing sites, including the surface debris that the Ranger 8 spacecraft ejected from the impact crater which it made in the Sea of Tranquillity. This time, northern candidate landing sites were targeted, covering thirteen primary (22,400 square miles) and seventeen secondary sites photographed in 211 high- and medium-resolution images. The spacecraft was crashed onto the surface at approximately 4° S, 98° E, on the far side of the Moon, on 11 October.

Lunar Orbiter 3 (5 February 1967) Arrived on station in lunar orbit on 8 February, and took 211 images, but transmitted only 182 (72%) due to a malfunction in the imaging system. Objectives included twelve of the most promising Apollo and Surveyor landing sites, as well as continued gravitational field and lunar environment studies. It impacted the surface on 9 October at 14.6° N, 91.7° W. This mission completed the programme's primary goal of photographing the major landing sites being evaluated for Apollo.

Lunar Orbiter 4 (4 May 1967) Entered orbit four days later on a mission to conduct scientific surveys of the lunar surface, and became the first spacecraft to enter lunar polar orbit. The 193 images returned included the first of the lunar South Pole region. Contact was lost on 17 July, and impact occurred during 6 October.

Lunar Orbiter 5 (1 August 1967) Entered lunar polar orbit four days later to complete the programme's photographic coverage of the surface that would exceed 99%. It was also a target for photographic tracking exercises conducted by Earth-based observatories using sunlight reflected off its solar arrays. A total of 212 images of both the Moon's near side and far side were transmitted, as well as further data on gravity, micrometeorites and radiation. The mission included the imaging of thirty-six sites of scientific interest, including five potential Apollo and several Surveyor sites. Luner Orbiter 5 crashed on the surface on 31 January 1968, during its 6,034th revolution. A planned sixth Lunar Orbiter was not flown.

Other missions
Four other Pioneer spacecraft (Pioneer 6–9) were placed in solar orbits between December 1965 and November 1968. These were designed to conduct missions aimed at gathering information on interplanetary radiation and magnetic fields that had important applications in both future unmanned and manned space programmes – such as providing up to a 15-day warning of potential solar flare activity used in planning missions outside the protection of Earth's radiation belts. Each spacecraft was placed at a different point to study the Sun during its eleven-year cycle. Discoveries included the Earth's long magnetic tail, which extends about 3–4 million miles on the side facing away from the Sun, and accurate density measurements of the solar wind. Pioneer 8, 9 and E (the only launch failure) also carried Test and Training Satellites (TETR) designed to evaluate the Apollo communications network.

Pioneer missions

Spacecraft	Launched	Solar orbit
Pioneer 6 (A)	1965 Dec 16	0.814 x 0.985 AU*
Pioneer 7 (B)	1966 Aug 17	1.01 x 1.125 AU*
Pioneer 8 (C)	1967 Dec 13	1.0 x 1.1 AU* (TETR-1, re-entered 1968 Apr 28)
Pioneer 9 (D)	1968 Nov 8	0.75 x 1.0 AU* (TETR-2, re-entered 1979 Sep 19)
Pioneer E (TETR-C)	1969 Aug 27	Lost during launch failure, 8 min 3 sec into flight

*Astronomical Unit – mean Earth–Sun distance (around 93 million miles)

Several of the Explorer series of spacecraft were used in a variety of scientific missions to study the Earth's environment, terrestrial, solar and interplanetary space in support of unmanned satellites, and lunar and planetary programmes including Apollo. Most closely related to Apollo development was the series of Interplanetary Monitoring Platforms (IMPs) designed to study magnetic fields, solar wind and cosmic rays outside the magnetic field of Earth. The information provided over the eleven year solar cycle, together with that from the solar Pioneers, was also used during the Skylab programme.

Robotic explorer summary
The programme of unmanned robotic explorers aimed at the Moon during the ten years preceding the first manned Apollo missions, and the series of interplanetary research studies conducted through the mid-1970s, provided valuable information for the planning and execution of Apollo and Apollo Applications programmes, including preliminary studies on sending astronauts to Mars.

Details of the composition of the lunar surface helped finalise the design of LM landing gear and, in part, the EVA suits worn by the astronauts and the equipment they were to use or deploy on the surface. In addition, information on the effects of solar radiation, cosmic rays, solar wind and gravitational fields around the Moon were also valuable in the planning and timing of lunar missions, the flight profile of extended-duration missions in Earth orbit, and proposed voyages to Mars. Photography from orbit and from the surface produced high-resolution images which were important for the manned landings and, in part, in the training of the astronauts. The popular impression of the lunar surface prior to the space programme was one of jagged peaks and a soft dust layer several feet in depth. The unmanned robotic explorers proved that although the surface was far from smooth, the dust was only inches thick, and though dust clouds might affect the final approach to the surface, it would support the LM mass upon landing – which also reassured the astronauts who were being prepared to go there.

Although separate programmes, the unmanned lunar spacecraft were an integral

Explorer missions

Designation	Spacecraft	Launch	Re-entry
IMP-A	Explorer 18	1965 Nov 26	1965 Dec 30
IMP-B	Explorer 21	1964 Oct 4	1966 Jan 30
Failed to achieve its intended orbit, but returned useful data before it re-entered			
IMP-C	Explorer 28	1965 May 29	1968 Jul 4
Placed in a highly eccentric orbit to predict solar flares for Apollo missions			
IMP-D	Explorer 33	1966 Jul 1	High eccentric Earth orbit
An excess velocity of its launch vehicle prevented entry into its planned lunar orbit but a high eccentric earth orbit of 9,900 x 270,500 miles.			
IMP-F	Explorer 34	1967 May 24	1969 May 3
170,000 hours of data during period of bright solar flares before re-entry			
IMP-E	Explorer 35	1967 Jul 19	Lunar orbit
Inserted into a 50 × 4,800-mile lunar orbit three days later to measure solar plasma flux, energetic particles, magnetic fields and cosmic dust; it also determined that there was no detectable lunar magnetic field			
IMP-G	Explorer 41	1969 Jun 21	1972 Dec 23
IMP-H	Explorer 43	1971 Mar 13	1974 Oct 2
IMP-I	Explorer 47	1972 Sep 22	High eccentric Earth orbit
IMP-J	Explorer 50	1973 Oct 25	High eccentric Earth orbit; last of series

part of the Apollo effort, which supplemented the data being gathered from the Saturn development, launches and spacecraft ground test programmes. By the summer of 1968, the next major step towards the Moon lay in placing both men and machines in orbit to test the hardware and procedures designed to place Americans on the Moon by the following year.

APOLLO TRAILBLAZERS

Following the pad fire and the cancellation of the Apollo 2 mission, a new mission schedule, leading to the lunar landing mission, had to be devised. With only a little over two years left in the decade, the Manned Spacecraft Center was working hard on a development plan that featured the unmanned and manned development flights required to achieve that goal.

On 17 April 1967 – about ten weeks after the pad fire – a meeting held at MSC between representatives of the Mission Operations Division and the Flight Crew Operations Directorate included discussions that identified certain requirements to be considered prior to attempting the first landing.[2] These discussions resulted in several opinions in several areas of mission development, including a consensus that a demonstration of CSM rescue of a stranded LM or EVA transfer from the LM to the CM would not prevent a manned LM Earth orbital mission that included undocking from the CSM. However, prior to any manned flight with the LM, there should be a demonstration of various unmanned ascent and descent engine burn profiles, and there was a growing question over the ability of LM-2 to be manned. The inclusion of a 'rendezvous exercise' on the first manned Apollo (101) with a 'rendezvous pod' (similar to the experiment carried out on Gemini 5 in August 1965) would be worthwhile attempting after the first day of the mission had been completed, and in addition to this single CSM flight, three manned Earth orbital flights of the CSM and LM in joint operations should be the minimum number of missions allocated in the programme prior to the first potential lunar distance mission. However, although a lunar orbit mission would not be considered a step in the primary programme, such a profile should be a contingency option if the CSM achieved lunar mission capability before the LM. According to the memo: 'The gain in operational experience [was] considered sufficient to justify the risk of such a mission'. Finally, the Saturn V should be man-rated as soon as possible.

This was followed on 9 June by a memo from Robert Aller, at NASA OMSF, to the Apollo Program Director Sam Phillips, highlighting the considerable analysis completed at MSC to the most effective sequencing of missions following the first manned flight in Earth orbit. The current official assignments included three high-apogee Earth orbital manned CSM/LM missions to evaluate joint operations, lunar simulation, and finally the lunar capability. MSC had suggested that there were considerable operational advantages in changing the third high-altitude Earth orbital flight from a lunar simulation to a mission placing Apollo in lunar orbit. Phillips considered these options, and suggested the inclusion of a lunar orbital mission as a prime or alternative profile in the Mission Assignments document being developed over the ensuing three months.

Seven steps to the Moon
On 20 September 1967, MSC proposed a sequence of seven steps, developed primarily by Owen Maynard, Chief of MSC Operations Division, that could be assigned to a mission or series of missions in order to progress step by step to the landing mission:

Mission	*Objective*
A	Saturn V and unmanned CSM development on missions to about 10,000 miles altitude on flights of approximately 8.5 hours from launch to splash-down.
B	Saturn 1B unmanned LM development flight to test propulsion and staging systems in low elliptical orbit for about six hours' duration from launch and without spacecraft recovery.
C	Saturn 1B manned CSM evaluation and crew performance in low Earth orbit for up to 10–11 days (the longest period considered for Apollo missions at that time).

D   Saturn V manned CSM and LM development with combined crew
             performance and operations on a Saturn V or dual Saturn 1B launches in
             low Earth orbit and again up to 6–11 days in duration.
E   CSM and LM operations in high Earth orbit using a Saturn V for up to
             eleven days.
F   Deep-space evaluation of CSM/LM to include lunar orbital mission
             operations using Saturn V and a flight duration of about 8–10 days.
G   Initial lunar landing mission, to include one EVA of about three hours
             duration, deployment of experiments and collection of samples and
             return to Earth during a mission of about eight days.

This original sequence was subsequently expanded to include:

H   Lunar landing missions of up to ten days in duration, to include two
             EVAs on foot, deployment of the Apollo Lunar Surface Experiment
             Package, collection of geological samples, and future landing-site
             photography from lunar orbit.
I   Lunar orbit survey missions' launches to be determined if manifested.
J   Extended-duration 'super-scientific' lunar landing missions to include
             three periods of surface EVA, additional surface modes of transport,
             expanded surface and orbital experiment packages, and geological survey
             and collection from landing sites of high scientific interest.

It was this step-by-step approach that formed the framework by which all Apollo
missions would be carried out through 1972.

Apollo designations

Grissom's mission was officially termed Apollo–Saturn 204, although the astronaut
office had some time earlier identified the first three manned missions as Apollo 1,
Apollo 2 and Apollo 3. Shortly before the accident, Grissom's crew had also received
official approval for the designation and patch bearing the number. Following
several requests from the widows of the three astronauts, NASA decided not to use
the Apollo 1 designation again, and reserved it for the unflown mission. For a while
the next scheduled manned launches were still identified as Apollo 2 and Apollo 3,
but documentation was already being distributed showing that the next mission to be
launched – the first unmanned flight of the Saturn V – was being called Apollo 4,
which added to the confusion. Immediately following the fire, all named crews were
stood down while the investigation proceded, and with them the formal designations
of the earlier missions Apollo 2 and Apollo 3 would never again become official
mission designations.

According to the NASA statements, the Apollo Applications Program flights
would be designated separately and in sequence (AAP-1, AAP-2), and in future
Apollo mission number designations would not differentiate between unmanned or
manned missions, nor Saturn 1B or Saturn V launches. However, all hardware
would retain its own designations (for example, CSM-101, LM-1), and launch

vehicles would be identified with the 200 series number for Saturn 1B (AS-200) launches, and a 500 series number for Saturn V (AS-500).

Between 25 March and 24 April 1967, the OMSF conducted correspondence with the manned spaceflight field centres in Texas (MSC), Alabama (MSFC) and Florida (KSC), on the approved designation for future missions. George Low, MSC Deputy Director, offered the suggestions that the three Saturn 1B missions flown in 1966 should be redesignated, for historical purposes, in line with the official Apollo 1 designation, thus: AS-204 (Grissom's Apollo 1), SA-201 (Apollo 1A), SA-202 (Apollo 2), and SA-203 (Apollo 3). Then, on 24 April OMSF further instructed the centres that Apollo 1 should be recorded as the 'first manned Apollo Saturn flight – failed on ground test', that the unmanned Saturn flights launched in 1966 would not be renumbered in the 'Apollo' series, and that the next flight would be known as Apollo 4. A flood of correspondence followed this announcement, but as the official history of Apollo observed: 'The sequence of, and reasoning behind, mission designations has never been really clear to anyone'.[3] On 4 May, NASA decided to formally designate AS-501 as Apollo 4, AS-204/LM-1 as Apollo 5, and AS-502 as Apollo 6, and these designations began appearing in news releases and internal documents.

UNMANNED APOLLO

By 4 November 1967, NASA had developed a programme of launches for 1968 and 1969 that would follow the first unmanned launch of the Saturn V (501 – Apollo 4), scheduled five days later. A schedule of launches and alternative plans allowed for a limited number of CSM and LM missions, configured for lunar landing, which could be used on test flights towards the Moon by 1969.

The planned 1968 missions were to be flown in the following order, and as quickly as safety and launch preparations could achieve:

AS-204   First unmanned test of LM (LM-1) in Earth orbit.
AS-502   Second unmanned flight test of Saturn V and CSM in high Earth orbit.
AS-503   Third unmanned test flight of Saturn V and CSM if required.
AS-205   First manned Apollo flight, ten-day CSM qualification mission in Earth orbit.
AS-206   Second unmanned test flight of LM (LM-2) in Earth orbit if required.
AS-504   First manned Apollo flight on Saturn V. This would also be the first time that manned operations would be carried out with the CSM and LM in Earth orbit, including crew transfer between vehicles, rendezvous and docking.

During 1969, five manned Saturn V missions were planned (AS-505 through AS-509), with the first four being considered as 'lunar mission development flights or lunar mission simulations' that would take astronauts out towards and around the Moon but without attempting a landing. The news release (67-282) revealed that NASA had considered the possibility that the first landing could be attempted on AS-509, but that it was equally possible that it might be delayed 'until one of the remaining six Saturn V flights'. This statement gave notice that current Saturn V production would end with vehicle 515 if further procurement was not forthcoming. Therefore, if AS-204 became 'Apollo 5' as expected, 515 would end the programme with Apollo 21, if all hardware was used on lunar Apollo missions and not transferred to AAP operations.

It will be recalled that at the same time as developing the lunar landing, NASA was also pursing the Apollo Applications Program with 'surplus' Apollo hardware. On 24 May 1967, a new launch schedule had been issued, realigning AAP schedules within the Apollo framework as a result of the Apollo 1 accident and programme delays. The new schedule stated that the AAP programme would include twenty-five Saturn 1Bs and fourteen Saturn V launches, with two Saturn 1B-launched workshops and two Saturn V-launched workshops, with the first launch planned for January 1969. To help support this, NASA had awarded Boeing a contract for long lead procurement for two additional Saturn Vs in support of the AAP programme, on the same day that NASA Administrator James Webb testified at the NASA authorisation bill hearing, during which the Senate Appropriations Subcommittee informed him of a substantial funding cut in the NASA budget requests for 1968 and 1969.

By 29 August, internal memos between AAP managers featured how restrictions in funding would affect the AAP.[4] The AAP management recognised the likelihood of serious funding cuts in FY 1968 and FY 1969, which would affect how much hardware AAP might inherit from Apollo, which in turn affected future planning. MSC recommendations for the next phase of AAP mission planning proposed an Earth orbital manned mission during 1969, two manned visits to the OWS (28 and 56 days) during 1970, a manned ATM mission in 1971, long-duration manned flights (2–12 months duration) flown during 1971 and 1972, and 'manned lunar landing missions (including surface operations) in the post-Apollo period.'

Evidence of the effects of the budget cuts was apparent in the revised NASA HQ AAP schedule issued on 3 October 1967. The new schedule revealed only two Saturn 1B and one Saturn V workshop launches remaining with three ATM missions/launches. AAP comprised of only seventeen Saturn 1B missions and seven Saturn V launches, of which only four would venture to the Moon after the mainstream Apollo missions, with the first workshop not flying before March 1970 – if, of course the hardware was available to support this plan. The budget restrictions were clearly already affecting what was hoped would follow the first lunar landing, two years before NASA attempted the mission. Despite this as (described the following chapters), planning for extensive operations in Earth orbit and on and around the Moon continued for some time, but without firm commitment of hardware to fly them, or funds to support them. Meanwhile, NASA pressed on with testing the existing hardware to progress towards the primary task of sending Americans to the Moon within the following two years.

Apollo 4 (A mission)

The next planned mission (the first of the new designations – an A mission) was perhaps the most important launch in the Apollo test programme. Apollo–Saturn

On 25 May 1966 the Apollo Saturn V Facilities (500-F) Test Vehicle and Launch Umbilical Tower (LUT) on top of a Mobile Launcher Platform (MLP) begins its three-mile journey to Pad 39A by means of one of two crawler transporters. The combination emerges from the Vehicle Assembly Building (VAB) and in front of the Launch Control Centre (LCC).

mission 501 – now officially called Apollo 4 – had the primary objective of demonstrating that the Saturn V worked as advertised on its maiden launch, but also that the mission would provide NASA and the American public with a renewed boost of confidence in the progress of the nation's space programme, which was still recovering from the tragic loss of the Apollo 1 astronauts. The media began calling Apollo 4 the 'Big Shot', with reports featuring the 'gee-whiz' data that became a standard way of discussing anything about the Saturn V in the press over the ensuing years. Indeed, it was a 'big' – mission not only in statistics and dimensions, but also in what NASA was trying to achieve. Prior to adopting the 'all-up' concept, there were to be four missions flown to achieve what Apollo 4 was planning to do on one launch: the testing of all three stages together. The mission featured the first launch from the new Pad A at LC 39, the first flights of both the first and second stage which throughout 1966 had a difficult year of ground testing to reach this point, the first flight of an LM test article, and the first use of aircraft (Apollo Range Instrumentation Aircraft – ARIA) instrumented with a 7-foot antenna, to communicate with the Apollo in orbit, supplementing the range of the ground station and providing maximum contact with the spacecraft.

Apollo/Saturn 500-F at LC 39A at KSC on 17 June 1966. Saturn 500-F was never intended for a launch, but was used to verify LC 39 launch facilities, train the launch crews, and develop test and check-out procedures prior to operational launches.

Apollo 4 represented the first full lunar stack launched with the three stages of Saturn, the instrument unit, launch adapter and spacecraft with the launch escape tower. The LM was a mock-up (LM Test Article, or LTA-10) that included a flight-qualified descent stage, but without any landing gear and fuel tanks filled with water/glycol (simulated fuel) and Freon (simulated oxidiser), while the ascent stage was a ballasted aluminium shell that contained no flight systems. The CSM was a Block I version (CSM 017) that featured many of the Block II systems, including the first flight hardware of the new CM hatch design incorporating changes as a result of the fire.

At 07.00 am on 11 November 1967, the resident wildlife of the normally peaceful Florida Wildlife Game Refuge received a sudden and tremendously loud awakening as the five F-1 engines of the S-1C stage exploded into life, creating a huge man-made earthquake and shock wave that vibrated for several miles from the launch pad as the first Moon rocket left Earth. Even though the force of the launch was known from numerous ground tests, the impact of seeing, hearing and feeling the Saturn leave the pad was unforgettable to those who witnessed the event from the safe distance of three miles. Even the most experienced space reporters were awed by the event, as 7.5 million pounds of thrust lifted the vehicle off the pad and into the

Floridian skies. Covering the launch for CBS was veteran journalist Walter Cronkite, who emotionally described the experience of seeing the first of what would be only thirteen launches of a Saturn V: 'The ... it ... the building is shaking ... b .. boy, it's terrific. The building is shaking. The big glass window is shaking as we're holding it with our hands ... look at that rocket go!' The author still recalls the emotion of watching the launch of Apollo 8 on black-and-white television, and the excitement that it added to the whole adventure of sending men to the Moon. Those lucky enough to experience the event at the Cape have always described it as spectacular.

As the first two stages performed flawlessly, telemetry was transmitted from the LM test article to the ground for the first twelve minutes of flight as it was held at thirty-six points in the spacecraft adapter. These instruments recorded the vibration, acoustics and the structural integrity of the test article hidden inside the adapter as it rode on top of the Saturn. These recordings provided useful data for analysis of what the subsequent LMs, intended for a manned crew, might experience when leaving the Earth and entering orbit. The first stage was jettisoned after about 2 minutes 30 seconds of powered flight, followed by almost seven minutes burn of the second stage. The escape tower, no longer being needed, was jettisoned as planned, just over three minutes into the ascent. Once the second stage had achieved its goal, the third stage, in a 2-minute-45-second burn, took the Apollo into a 115-mile orbit, where it shut down, although it was not separated from the Apollo.

During the next three hours, onboard systems were checked and verified, and then, during the second orbit of Earth, the third stage was reignited to boost the spacecraft out towards deep space, on a second five-minute burn, duplicating the manoeuvre that manned missions would undergo to begin the three-day journey to the Moon. On Apollo 4, however, it was to be a short test-flight, and at an altitude of 10,500 miles the CSM was separated from the S-IVB. The SPS was then fired to boost the spacecraft a further 1,000 miles into space to test the thermal characteristics of the spacecraft in a four-and-a-half-hour cold soak, during which the CM was orientated so that half of it would bake in the Sun, while the other half froze in deep shadow. Data on the CM internal temperatures and capacity of the environmental control system revealed that the system could provide a habitable environment inside had a crew been on board. The data revealed no deterioration in the cabin environment

Once the apogee had been achieved, Apollo 4 began the long 'fall' back to Earth, but it was not a gradual decline in its orbit. With the CM nose pointed towards the planet, the SPS was fired again to increase the speed of the spacecraft to simulate return speed from lunar distances. The CM was separated from the SM and rotated to point the all-important heat shield for the planned re-entry test. When Apollo 4 entered the 400,000 feet atmospheric re-entry zone it was travelling at 24,800 mph, and as the vehicle descended into the atmosphere the maximum temperature recorded on its outer surfaces was more than 5,000° F, doubling the entry loads experienced during Mercury or Gemini flights; but again, this was within tolerable limits.

After the vehicle skipped through the atmosphere the parachute recovery system performed flawlessly, and the spacecraft splashed down in the Pacific Ocean about 500 miles north of Hawaii, and less than ten miles from the prime recovery carrier *USS Bennington*, after a flight of 8 hours 37 minutes. NASA felt elated that everything seemed to work so well, and confidently announced that once again Apollo was on the way to the Moon. The recovery from the pad fire, though never to be forgotten, had overcome its first big test. Next was the last major element to be tested in flight: an unmanned production series Lunar Module

Apollo 5 (B Mission)

This flight, designated Apollo 5 (B class mission) would be launched by the smaller Saturn 1B (in fact, the rocket that should have launched Apollo 1 – AS-204). As it was only the relatively light LM that was to be tested, there was no need to use a Saturn V, and for this mission there would be no CSM or LES. It was therefore a shorter and stubbier Saturn 1B, topped with a nose cone, that left LC 37 on 22 January 1968 at 5:48 pm. Ten minutes later, the S-IVB placed its payload, LM-1, in a low 138-mile orbit.

Less than 45 minutes later, the RCS of the ascent stage on LM-1 separated the vehicle from the S-IVB stage to begin its test programme. This LM had no landing gear, and was to be burned up on re-entry after completing the tests, which included two burns, each of both the ascent and the descent stages. After a coasting flight of four hours, engineers commanded the first burn of the descent stage, planned as a 38-second burn, but just four seconds into the firing the LM guidance system sensed that the LM was not travelling fast enough, and promptly shut down the engine. On manned flights this 4-second cut-off due to low velocity was a planned feature, allowing the astronauts time to analyse their situation and to decide whether to proceed or to abort the burn. On normal missions the burn would have started with full tank pressurisation, and would have easily reached design velocity in the four seconds, but on Apollo 5 the tanks had been only partially pressurised, and six seconds were needed to reach full pressurisation, thus triggering the cut-off. Flight controllers decided to change the flight plan and adapt an alternative mission providing minimum mission requirements, deciding that the engine had worked as directed and that the safety systems had worked as planned. This change of flight plan was a feature of all missions, allowing controllers (and later astronauts) to adapt to new situations to obtain the very best from the mission.

Each Apollo flight was planned to proceed through several levels, during which it could be decided whether to carry on or abort (Go/No Go), or which other planned missions could be followed that were not the primary mission but the best alternative options given the situation (contingency missions). These contingency missions could range from the Apollo being unable to leave orbit and head out to the Moon but capable of completing an Earth orbit mission, to the LM being unable to descend to the surface (forcing a lunar orbital mission), to shortening the surface stay time to restrict EVA time, or indeed allowing only one astronaut on the surface instead of two. In all of these options, the safety of the crew was a priority, and if anything threatened this mission rule, then the whole mission was aborted, and the crew took the safest (but not necessarilyy the quickest) option to return home – a profile dramatically demonstrated on Apollo 13 in April 1970.

Although the first firing of the LM descent stage in space 'failed' in its primary goal of a 38-second burn, it did 'burn, and triggered safety procedures which called in a contingency mission option, allowing the controllers to call upon trained for, but not expected, alternative flight plans, and invaluable flight experience.

The alternative flight plan consisted of two brief burns of the descent engine, of 26 seconds and 33 seconds, before separation of the ascent and descent stages, with a one-minute firing of the ascent engine and a second burn of six minutes. All other aspects of the test performed satisfactorily. Despite the one malfunction, Apollo 5 was deemed a success. Despite burning for only just over one minute of the planned nineteen minutes, the three good brief burns had determined that the system worked, and that most of the primary objectives had been met. The main mission of Apollo 5 was completed 7 hours 50 minutes after launch, but the stages of LM-1 would not re-enter and burn up until until 12 February.

Earlier that month, on 9 January, George Mueller, NASA Associate Administrator for Manned Spaceflight, in a letter to MCS Director Robert Gilruth, had summarised key Apollo programme decisions that were required to instil a sense of urgency on the need to certify Apollo for manned flight. The first of these assumed a successful flight of LM-1 that would certify the lander for manned flight on AS-503. With the official declaration that Apollo 5 was indeed a success, the need for a second unmanned test flight of the lunar lander (LM-2) was deleted. The next time that an a LM would fly in space, astronauts would be on board.

Apollo 6 (A mission)
NASA had indicated as early as 1 December 1967 that if the second unmanned test flight of the Saturn V (502) was successful, the next vehicle (503) would become the first manned Saturn V launched, eliminating the need for a third unmanned test flight. This decision – depending on the success of Apollo 5 and Apollo 6 – would release both a complete Saturn 1B and a Saturn V from the unmanned test programme to support later manned operations. At the time NASA was trying to find ways of fulfilling the launch manifest with existing hardware, under the restrictions of the current budget requests for the next two years that seriously threatened further procurement of hardware well before any astronaut flew an Apollo into space.

The success of Apollo 4 provided confidence that the Saturn V would safely support a manned launch, but NASA pressed on with the second test, primarily to provide launch teams with an additional cycle of launch preparations prior to including a human crew in the system. Apollo 6 was essentially a re-run of Apollo 4 on a high-apogee Earth orbit and with a high-speed re-entry test for the CM.

The payload for Apollo 6 consisted of CM-020 and SM-014 (the one originally planned for Schirra's Apollo 2 mission), and LTA-2R simulating the LM. For a second time, the ground around KSC shook as AS-502 left the pad at 07.00 am on 4 April 1968. Witnessing the monstrous rocket leaving the pad, journalists began to wonder whether the power needed to make the Saturn V move upwards might cause Florida to sink a little!

Planned to supplement data provided by Apollo 4 and to prove that the first launch was not a random success, the second Saturn V actually provided a wealth of new information from which many new things were learnt about launching and flying such large vehicles. During the first two minutes of powered flight on the first stage, thrust fluctuations caused the whole vehicle to bounce forward and backward like a pogo-stick for about thirty seconds (thus afterwards known as the pogo effect). Low-frequency modulations were recorded as high as $\pm$ 0.6 g in the CM, exceeding design criteria, and well above the 0.25-g upper limit permitted for Gemini launches. The buffeting also dislodged a panel on the adapter fairing.

When the second stage took over, two of the five J-2 engines ceased functioning, forcing the other three to burn longer to compensate for the loss. The stage therefore failed to reach its designed altitude before the fuel was exhausted, and the third stage had to burn longer to finally place the Apollo in a 110×228 mile orbit instead of the planned circular orbit of 99 miles. For the next two orbits, controllers left the vehicle in a parking orbit, allowing for a complete systems check and attitude manoeuvres before proceeding with the mission. During this period, 70-mm automatic cameras in the CM windows recorded excellent Earth terrain photographs in evaluating the camera system for use in future cartographic, topographic and geographic studies. Similar experiments were to be carried out on manned Earth-orbiting Apollo missions – partly in preparation for the larger surveys expected to be conducted during AAP OWS missions.

When the J-2 engine on the third stage was commanded to reignite, nothing happened. With a change in the flight plan, the CSM separated from the third stage, and a test firing of the SPS engine for 442 seconds, instead of a planned 280 seconds, provided sufficient velocity for simulated translunar injection. This length of burn far exceeded the maximum requirements on lunar mission profiles of SPS engine operations, and further qualifyed the SPS for operational use. Apollo 6 ventured to an apogee of 13,800 miles, and then began its long return towards the Earth. The altitude was sufficient for a good simulation of an Apollo returning from the Moon, but the SPS did not have a sufficient level of fuel remaining to provide the vehicle with the desired speed of 24,400 mph for entry. The CM reached a velocity of just over 22,000 mph, and missed the planned splash-down point by about fifty miles. Apollo 6 landed safely in the Pacific Ocean, 10 hours 23 minutes after launch.

Post-flight analysis recorded a successful mission, but there were still several unanswered questions, indicating that a third unmanned Saturn V flight might be necessary. The pogo effect was recorded in a much smaller level on Apollo 4, and had appeared during the Titan qualification for the Gemini programme. NASA performed investigations into the effects on Apollo 4 and Apollo 6, and determined that the increased weight differences on the flown LM test articles changed the mass distribution on Apollo 6 and enhanced the pogo effect. It was also found that two of the five engines on the first stage were tuned to the same frequency (instead of different frequencies to prevent two or more engines pulling the Saturn off balance). Further tests determined that bubbles of helium inserted in the fuel lines ten minutes before launch served as a shock absorber against the pressure surges and prevented build-up of the oscillations. Vent holes in the adapter panels prevented the build-up of internal pressure that caused the problem on Apollo 6, and strengthened propellant lines and modifications to the design prevented the vibration that was thought to cause the failures in the two engines.

At the beginning of 1968, a year after the pad fire, NASA was still optimistic about flying astronauts on Apollo 'in the last quarter of 1968', By June, after analyses of the flights of Apollo 5 and Apollo 6, combined with that of Apollo 4 and several ground simulations, this still seemed a distinct possibility.

ASTRONAUT ASSIGNMENTS

At this point it is worth recalling the assignments of the first Apollo crews after the pad fire. Immediately after the accident, all mission training was suspended, followed a week later by an official NASA announcement, on 3 February, that all future manned flights were postponed indefinitely pending the outcome of the review board. At the time of the accident there were six three-man Apollo crews in training (three prime and three back-ups), but now there were five, with the Commander of the third mission, Frank Borman, also directly involved in the investigation into the accident. Slayton was faced with the task of reassigning the crews he already had in training, and taking the opportunity of introducing new astronauts to form crews for later missions. In addition, he had to maintain the momentum of training with generic Apollo training across the group, both for the lunar missions and the AAP, and reflect the changing launch manifest. It is these changes that influenced later crewing for missions leading to the lunar landing and for all assignments through 1975.

In early 1967, the crew commanded by McDivitt had been preparing to fly the first manned LM for about a year, and Slayton did not wish to waste that experience. In late March therefore, he approached Schirra to head up the first flight, with Eisele and Cunningham – a task which he knew Schirra was prepared to take up as his last astronaut role before retirement. On 9 May 1967, the names of the first crew were revealed to the US Senate Committee for Aeronautical and Space Sciences, followed by the public release of the assignment on 18 May.

Mission	Position	Prime	Back-up	Support
C	Commander	Schirra	Stafford	Evans (CSM specialist)
CSM-101	CM Pilot	Eisele	Young	Swigert (CSM specialist)
	LM Pilot (no LM)	Cunningham	Cernan	Givens (CSM specialist)

(On 6 June Givens was killed in an off-duty car accident, and was replaced on the support crew by Bill Pogue, who was also a CSM specialist. In his autobiography, Slayton wrote that had he lived, Givens would have been one of the first from his selection (1966) to be assigned to a flight crew as a CMP.)

The mission was to be the manned qualification of the CSM in Earth orbit, including a rendezvous with the S-IVB stage, several firing of the SPS engine, star-sighting and navigation tests, and photographic experiments S-005 and S-006. No EVAs were planned, and an ocean recovery, after ten days in orbit, was planned. Prior to this flight, other members of the astronaut team would perform a number of ground test exercises. The original experiment programme (announced on 28 June 1967) for the C mission included M-006 (Bone Demineralisation), M-011 (Cryogenic Blood Studies), M-023 (LBNP), and S-005 (Synoptic Terrain Photography) S-006 (Synoptic Weather Photography), with D-008 (Radiation in Spacecraft) as a further option. On 21 July, only S-005 and S-006 were assigned to the C mission, with M-006, M-011 and M-023 defined as pre- and post-flight medical evaluations, and not experiments. D-008 was deleted. Schirra and his crew would not be burdened with an extensive programme of experiments, and instead were to concentrate on the higher-priority engineering qualification objectives.

Simulation crews

While preparations for launching the first manned Apollo continued throughout 1967 and most of 1968, and while other assigned crew prepared for follow-on missions, other teams of astronauts performed important simulations and tests of elements of hardware and procedures that were important milestones in preparing Apollo. These teams of astronauts performed what most astronauts are employed for when not training for or flying in space: as engineering test subjects. They formed teams never intended to train as a flight crews, and are often the 'forgotten Apollo crews' who remained on the ground.

CM-007A Sea Trials Crew A full post-landing system test was conducted in the spring of 1968 in an indoor water-tank at MSC and in the Gulf of Mexico. The spacecraft used was CM-007A, and the crew comprised Jim Lovell (Commander), Stuart Roosa (CMP), and Charlie Duke (LMP). The first part of the trials – a 48-hour sea test, completed on 5–7 April – consisted of the CM being lowered into the Gulf of Mexico, south of Galveston, initially with the apex of the CM in the water (Stable II position), to test the three inflation bags to right the spacecraft to the upright position (Stable I). The crew then remained in the CM for the next two days, testing recovery beacons and communications equipment with recovery forces and para-rescue divers. The astronauts were to evaluate survival equipment crew comfort and systems onboard the CM. The sea remained relatively calm (depth 18 fathoms, waves 3–4 feet, and winds 12–14 knots) and none of the crew reported sea sickness or any other medical problems. The second part of this test programme was conducted in a water tank at MSC, for 27 hours, on 7–8 May. This time the crew evaluated post-landing ventilation systems up to design limits, with simulated wave motions in the tank. In the earlier test it was set at a low mode during the day and only turned on over thirty minutes to purge carbon dioxide from the atmosphere. In this test it was kept on continuously, and temperature and humidity levels were induced to test the system to its capacity. Again the crew reported no problem, and the tests were deemed successful.

2TV-1 Thermal Vacuum Test Crew An eight-day CSM 'mission' was conducted in June 1968 with the spacecraft configured to the Apollo 7 CSM located in Chamber A of the Space Environment Simulation Laboratory at MSC, Houston, where it was exposed to simulated space vacuum conditions. Crew-members were Dr Joe Kerwin (Commander), Vance Brand (CMP), and Joe Engle (LMP). Their objective was to verify a number of systems under spaceflight conditions in the vacuum chambers, including the CM heat shielding structure under cold-soak conditions, changes being incorporated inside the CM following the fire, and ensuring that the stowage of the Earth recovery system (parachutes) was compatible with the simulated space environment. In addition, the 'mission' was to demonstrate the capability of the CSM systems and sub-systems in the simulated space environment, both in a pressurised and unpressurised mode, gather thermal data for later use in analysis of the heat loads across the spacecraft, analyse the capability of the electrical power system radiators to cope with heat transfer from the fuel cells, and determine the water consumption rate of the environmental control system. At the end of the test, all these results had been met, adding to the qualification of the CSM to support a crew in space.

LC-39 Pad Escape System Crew From early in 1968, simulations were conducted of the Apollo Emergency Ingress/Egress and Escape Systems. These were designed to evacuate astronauts and the launch crew from the White Room on the top of the launch tower area to protected areas, in the event of impending explosion prior to sealing the crew inside the spacecraft and the launch team vacating the pad, after which the launch escape tower could be used. It could take just thirty seconds to descend more than 300 feet by elevator from the White Room level down to the base of the launch tower, to exit the pad area by protected recovery vehicles. Depending on the time available, the other method of White Room evacuation was by means of slide-wire baskets on the launch tower structure (still provided for Shuttle crews) to a parked armoured personnel vehicle that would take the crew and pad team to a bunker 1.5 miles away. The third option was the use of escape tubes that led from the top of the tower to a blast room mounted on coiled springs to absorb the explosion of the Saturn below the pad area. In these tests, astronaut Stuart Roosa tested the slide-wire system, while Charlie Duke evaluated the escape tube, both in unsuited and partially suited modes.

These ground support assignments were necessary milestones in qualifying the CSM and Saturn V for manned flight before a crew left the ground.

D and E mission assignments
In preparation for the mission that would follow Schirra's C mission, Slayton refined his choices for the next two flights, now designated D (LM manned Earth orbital test using two Saturn 1Bs) and E (first manned Saturn V with high-apogee orbital mission). Slayton assigned the McDivitt–Scott–Schweickart crew to the D mission, and moved the Conrad–Gordon–Williams crew up a flight to serve as their back-ups. He then created a new crew to serve as back-up to Borman's crew: Neil Armstrong (CDR), Jim Lovell (CMP) and Buzz Aldrin (LMP). Before the official announce-

ment, CC Williams, LMP on Conrad's crew, was killed in the crash of his T-38 on 5 October 1967.

Upon the suggestion of Pete Conrad, Slayton brought in Al Bean from heading up AAP to replace Williams as back-up LM on the D crew. On 20 November 1967, a few days after the flight of Apollo 4, the next two crews were announced:

Mission	Position	Prime	Back-up	Support
D	Commander	McDivitt	Conrad	Worden (CM specialist)
CSM-103	CM Pilot	Scott	Gordon	Mitchell (LM specialist)
LM-3	LM Pilot	Schweickart	Bean	Haise (LM specialist)
E	Commander	Borman	Armstrong	Mattingly (CM specialist)
CSM-104	CM Pilot	Collins	Lovell	Carr (LM specialist)
LM-4	LM Pilot	Anders	Aldrin	Bull (LM Specialist)

Slayton now had eighteen astronauts in training as prime and back-up, which would provide the core for the first six manned Apollo missions that were expected to test hardware and procedures and, possibly, attempt the first landing, depending on the pace of the programme on the fourth, fifth or sixth mission. Slayton warned the astronauts that they were not necessarily to stay as complete crews or fly in the order indicated, but the 1967 assignments stayed relatively unchanged to the end of 1969 and the second lunar landing.

The D mission

The original mission's McDivitt crew was scheduled to complete the launch of the CSM-101 on a Saturn 1B (now 205) approximately 24 hours before the launch of the LM on a second (208) Saturn 1B. During the first flight-day, McDivitt and his crew would check out their spacecraft in preparation for the rendezvous and docking with the LM atop the S-IVB stage on FD2. After extracting the LM, the crew would prepare for the transfer of McDivitt and Schweickart to the LM, leaving Scott in the CSM to begin the solo LM operations of testing both the descent and ascent stages before being rejoined by McDivitt and Schweickart to complete the Earth orbital mission in the CSM.

During the preparations for the mission in 1967 and 1968, several key events were to change this profile. Foremost were the delays in man-rating the LM, due to increased overall weight, wiring and corrosion problems, descent engine qualification difficulties, and failures of the spacecraft windows under pressure testing. LM-1 was assigned to Apollo 5 as an unmanned test flight, and there was an option to fly LM-2 as a second unmanned test with the McDivitt crew taking LM-3 into space. In July 1967, MSC Director of Flight Operations Chris Kraft raised the question as to whether the LM-2 could be adapted to first perform an unmanned burn, and then rendezvous with the CSM for a crew to board it and test the lander manned as an additional mission, which might allow flexibly later in the schedule.

By October 1967, the decision was made to leave LM-2 unmanned and on the ground unless the Apollo 5 mission failed. In addition, on 28 October 1967 an EVA transfer was added to McDivitt mission, resulting in the only pre-lunar-landing mission space walk. The plan was for an EVA transfer from the LM to the CM in a demonstration of the contingency method of crew transfer, in the event that the internal use of the docking system and the transfer tunnel prevented internal transfer.

McDivitt would take LM-3 and CSM-103 into orbit on Apollo 8, which in the planning schedule of 4 November indicated that this might also involve the third launch of the Saturn V – the first with a crew – replacing the duel Saturn 1B launch based on the results from Apollo 4 and Apollo 6. Meanwhile, ground tests for qualifying the LM for manned flight continued with a programme of thermal vacuum tests, with the crew of Jim Irwin (Commander) and John Bull (LMP) using LTA-8. All the test and planning data became official on 27 April 1968 when NASA decided that the third Saturn V would be manned and that the McDivitt crew would ride it into space. All this depended on having the hardware available to fly the mission. The combination of solving the pogo problems and processing the vehicle at the Cape was taking too long. An unmanned 503 mission might be possible by July 1968, but a manned flight would not occur before November of that year. However, the progress of preparing the LM-3 for flight was slow, and it appeared that it would not be ready before the end of 1968.

The E mission
In early 1967, Borman, Collins and Anders were training to become the first crew to take a Saturn V off the pad. Following the high-altitude unmanned tests of the first two (or three) Saturn V tests, Borman's crew would take a CSM 104/LM-4 to an apogee of 4,000 miles, and in doing so set a new manned world altitude record, (surpassing that of Gemini 11 (850 miles)), where Borman and Anders would transfer to the LM and separate to complete similar LM tests that the McDivitt crew had completed in low Earth orbit, but far deeper into space.[5]

By April 1968 this mission was designated Apollo 9, and was the second manned Apollo flight on the fourth Saturn V. However, preparations for preparing the LM for flight included difficulties with the rendezvous radar, and it was expected that the vehicle assigned to the E mission (LM-4) would not be ready before the spring of 1969, as difficulties with LM-3 indicated that it would not be ready before December 1968. The following month, Chris Kraft was expressing concerns to the ASPO Manager George Low on the escalation of mission E goals, which seemed to him extremely complex and ambitious for the third manned flight, with a 'low' chance of achieving everything in the flight plan. However, Kraft added that by giving first priority to the objectives, then 'if we are fortunate, then certainly the quickest way to the Moon will be achieved.' It would not be long before the E mission would be adjusted to a far greater objective than that about which Kraft had expressed concern.

The members of the Apollo E mission on 5 March 1968, during the crew compartment stowage review at Grumman for LM-4, which they were to fly to a high elliptical orbit later that year. The vehicle used in this test was Lunar Test Article (LTA) 1, configured to resemble the LM-4 flight vehicle.

Astronauts assigned to Apollo E discuss the progress of the stowage review at Grumman on 5 March 1968: (*left*) mission Commander Frank Borman, (*centre*) CMP Mike Collins, (*right*) support crew astronaut and LM specialist Jerry Carr. Also participating (but out of frame here) was LMP Bill Anders.

Circumlunar Apollo

The idea was to send Borman and his crew on a high elliptical orbit that looped around the back of the Moon and back towards Earth – a circumlunar profile. The hardware included two CSMs ready to fly (Schirra's 101 and McDivitt's 103) on their Saturn Vs, but no LM would be ready before the end of the year. In July, the latest news from the Cape was that no LM would fly manned in 1968, which helped evolve the decision to put an Apollo in lunar orbit.

During August 1968, management discussions were held in Washington, Houston and Huntsville on the proposal to send Apollo 8 to lunar orbit on a new mission called C-Prime – a CSM-only flight – by December 1968, and thus delay the launch of the first LM into 1969. The LM would be replaced by LTA-B to simulate the mass of the Lunar Module. Doing so could eliminate several steps in the test programme, and the first landing might be attempted by either the fourth manned Apollo (now Apollo 10), more probably the fifth (Apollo 11), or, if need be, the sixth (Apollo 12). The general consensus was to try for the Moon with Apollo 8. As von Braun observed once the decision was made to man Saturn 503: 'It did not matter how far you went.'

Plans were drafted to follow Apollo 7 in October with Apollo 8 to orbit the Moon in December 1968, and to test the LM on Apollo 9 in early 1969, and the hardware prepared at the Cape. There remained only the question of who were to fly the revised missions. The Schirra crew and their back-ups were so far advanced in their training for Apollo 7 that the change to the next two missions did not affect them. However, changes had to be made to the crews before any public announcements.

In April 1968, astronaut John Bull – who was under consideration for an LM Pilot position on an early crew, and was involved in the thermal vacuum testing of the prototype LM – was grounded with a pulmonary disease that forced his retirement from the astronaut programme in July. At the time he was also support crew-member on Borman's crew, and so Vance Brand took his place.

In July, Collins had been taken off flight status to undergo surgery for a bone spur between two vertebrae, and although he was expected to return to flight status, it would be several months before he could be assigned to a crew, and he would certainly miss Apollo 8. Slayton reassigned the experienced Lovell from the back-up crew to the prime crew, reasoning that his previous experience with Borman on the 14-day Gemini 7 mission would compensate for the lack of time in retraining. Buzz Aldrin was 'promoted' to replace Lovell and to serve as back-up CMP for the flight, while Fred Haise – one of the best members of the fifth selection, and support on the McDivitt crew – moved across to fill in as back-up LMP for Apollo 8, with his support crew place on McDivitt's flight filled by Jack Lousma. The changes were announced on 8 August, but nothing was said of the intention to orbit the Moon.

Before the official decision to send Apollo 8 to the Moon, on 10 August Slayton discussed the idea with Jim McDivitt who, as Commander of the following mission (Apollo 8), needed to be told what was planned for his mission. Slayton knew that the LM would not be ready for some months, and thought that since the McDivitt crew had trained for nearly eighteen months then he should reassign them as the prime crew for Apollo 9, although they could take the Apollo 8 flight if they wished.

McDivitt has since commented that the Apollo 8 mission was there if he wanted it –
but it was never really offered. He agreed with the decision to not waste the training,
and his crew would therefore take Apollo 9. Borman's crew had also trained for an
LM, but less so than had McDivitt's, and when Borman was asked if he wanted to
take the command of 'eight', he accepted the challenge.

The final assignments were:

	Position	Prime	Back-up	Support
Apollo 8, December 1969				
CSM-103	Commander	Borman	Armstrong	Carr (LM specialist)
First Saturn V launch	CM Pilot	Lovell	Aldrin	Mattingly (CSM specialist)
Lunar orbit mission	LM Pilot	Anders	Haise	Brand (CSM specialist)
Apollo 9, February 1969				
CSM-104 /LM-3	Commander	McDivitt	Conrad	Mitchell (LM specialist)
Second Saturn V launch	CM Pilot	Scott	Gordon	Worden (CSM specialist)
Earth orbit mission	LM Pilot	Schweickart	Bean	Lousma (LM specialist)

These changes, announced on 19 August 1968, eliminated the E designation from the
mission schedule. The assignments also proved important in the future careers not
only of the prime crews but also the back-up crews. Dave Scott, on McDivitt's crew,
was unhappy about losing CSM-103, having worked with the vehicle since 1966; and
Bill Anders, on Borman's crew, lost his chance of piloting an LM. Indeed, it has been
observed that it was perhaps the Conrad crew who lost the greatest prize. As the
back-up crew to the original Apollo 8 flight, Conrad and Bean would have rotated to
fly Apollo 11 and become the first and second men on the Moon – but now they were
in line for Apollo 12. The 'Eleven' mission seemed most likely to be the first landing
flight, though by no means certain, as Apollo 8–10 still had to fly to prove the
hardware and profile. It will be noted that the support crew also changed with their
prime and back-up crews as a three-tier unit, and actually were more suitable to the
new missions. Borman's Apollo 8 support crew had two CSM specialists for the
CSM-only flight, while McDivitt's support crew included two LM specialists to help
prepare the first manned LM flight.

The final crew assignments announced in 1968 were those on 13 November for
Apollo 10. This was the F Mission, the flight plans of which ranged from Earth orbit
to lunar orbit tests of the CSM and LM spacecraft, depending on the success of
Apollo 8 and Apollo 9, and was announced after the return of Apollo 7 and formed
by the back-up crew for that mission.

	Position	Prime	Back-up	Support
Apollo 10, second quarter of 1969				
CSM-106/LM-4	Commander	Stafford	Cooper	Engle (LM specialist)
AS-505	CM Pilot	Young	Eisele	Irwin (LM specialist)
Lunar orbit	LM Pilot	Cernan	Mitchell	Duke
mission				(LM specialist)

The F mission (Apollo 10) was being planned as a full dress rehearsal of the lunar landing mission in lunar orbit, up to the point of initialising the descent engine at nine miles above the surface. As the flight approached its launch date in early 1969, there was some discussion about allowing Apollo 10 to make the first landing, but the LM-4 on which the crew had trained was too heavy to support a landing, having been originally planned for a high orbital test flight, and the crew was not trained to perform a surface EVA (that crew would begin training in January (see below)). The choice of back-up crew was interesting. Under Slayton's rotation system, this was the crew in line to fly Apollo 13, which was a high probability of a lunar landing crew; but Slayton had already indicated that assignments from Apollo 13 were by no means carved in stone, and that they were prone to change once the first landing had achieved its goal. And change they did.

MAN-RATING APOLLO

The missions of Apollo have been well documented over the past three decades, and it is not intended to discuss those events in great detail once again. What follows, therefore, is a brief review of the lessons learned on the four missions preceding Apollo 11, and how these applied to future plans and operations beyond the first landing, together with a summary of what the Apollo 11 and Apollo 12 crews experienced that would influence the future planning of lunar surface operations.[6]

These four missions were:

Apollo 7 (C mission) A CSM-only Earth orbital mission launched by Saturn 1B and flown in October 1968 by the crew of Schirra, Cunningham and Eisele. This mission of eleven days featured an extensive evaluation of the CSM systems including eight firings of the SPS engine, rendezvous with the spent S-IVB stage, and regular television transmissions by the crew. It was essentially the Block II version of the Block I missions planned for Grissom's crew on the Apollo 1 mission and the original Apollo 2, but with fewer scientific experiments. The mission has been described as a 101% success, and it finally qualified the CSMs for manned spaceflight.

Apollo 8 (C-Prime mission) The second CSM-only Apollo mission, but this time launched by the Saturn V for the first time, and sent to orbit the Moon ten times during the Christmas holiday of December 1968. Astronauts Borman, Lovell and Anders continued the qualification of the CSM, this time at the lunar distance, and are also remembered for broadcasting their famous Christmas message to the world and for views of the Earth floating in space.

Apollo 9 (D mission) In March 1969, the delay resulting from the crew's recovery from mild colds returned astronauts to Earth orbit operation. McDivitt, Scott and Schweickart tested the complete Apollo stack of Saturn V/CSM/LM for the first time with a crew on board. The important evaluations of the LM, the rendezvous and docking with the CSM, and the first Apollo EVA, were completed early in the mission, allowing several days of Earth Resources observations and photography in the second half of the flight.

Apollo 10 (F mission) Flown in May 1969, this was in all respects a complete lunar landing mission profile, but without the landing on the Moon. Astronauts Stafford, Young and Cernan completed a dress rehearsal of the landing mission to the point of firing the descent engine to begin the final approach for landing, and from separation of the LM stages (in lunar orbit) to splash-down.

The learning curve
The primary objective of these four missions was to complete qualification of all hardware, procedures, systems and infrastructure prior to supporting the first manned lunar landing and a series of follow-on missions in Earth orbit and at the Moon. This consisted of the man-rating of the CSM and LM spacecraft and pressure garments, the Saturn 1B and Saturn V launch vehicles, operational use of the launch preparations and ground tracking network, crew training and mission control teams, spacecraft recovery, and post-flight debriefing and analysis procedures, both in Earth orbit and lunar distance missions.

In addition, crew comments and experiences formed part of the habitability studies designed to evaluate the systems and procedures onboard the spacecraft – designed not only to complete the assigned mission, but also to support the crew in reasonable comfort and safety. There had been years of ground studies in this area during the development of the spacecraft, with many astronauts taking active roles in providing personal input into the evolution of crew stations and facilities. This was not new to Apollo, as similar involvement and evaluation occurred during the Mercury and Gemini programmes, and it was also part of the AAP being developed at the same time as the lunar effort.

The results from these four missions, therefore, were not aimed at merely proving the Apollo concept for the first landing. They had a much wider application far beyond the Moon.[7]

Crew training As none of these flights included a lunar landing, the training programme concentrated on detailed reviews of CSM and LM procedures and systems, with the astronauts from both the prime and back-up crews participating in almost every phase of spacecraft testing and check-out. The reason for this was that the vehicle systems' performance in either normal or aborted flight was neither well documented nor understood. A significant number of hours was spent in simulators supporting spacecraft test programmes, backed up by numerous briefings and speciality training. The crews spent more than 11,000 hours in the simulators for these four missions, and in addition logged over 4,000 hours on special-purpose training, 8,000 hours going through procedures, 5,900 hours in numerous briefings, and 2,500 hours in tests of the spacecraft. This totalled almost 32,000 hours for the four flight crews (excluding the back-up crew training time). The Apollo 7 crew conducted eighteen integrated mission simulations with flight controls in the CM simulator, while the Apollo 8 crew logged fourteen, the Apollo 9 crew completed ten, and the Apollo 10 crew completed eleven simulations. In addition, the Apollo 9 crew completed twenty-two integrated crew simulations in the LM and eight in the combined CSM and LM simulators, while the Apollo 10 crew completed seven CSM/LM combined simulations.

The benefit of this training was evidenced in the success of the first four missions in completing almost all of their objectives with no serious setbacks or failures. In addition, the training experience of the back-up crews of those missions was applied to the first landing missions (noting that the Apollo 10 crew was the Apollo 7 back-up crew). Further members of the crews from Apollo 7 through Apollo 10 provided experienced crew-members on the flights of Apollo 13 to Apollo 17. In comparison, the Apollo 11–14 crews completed 26,800 hours of training, while the series J crews of Apollo 15–17 logged 25,320 hours, excluding geological field training and lunar surface activity simulations. Others that benefited from the early training cycles were the support crew astronauts, who would rotate to later missions and the flight control teams.

Launch through docking No significant problems were encountered on this phase for Apollo 7 through 10. The Apollo 7 crew found that one of the spacecraft lunar adapter panels had not deployed to the full 45° and remained at only 25°, but since no LM was carried, and as there were plans to jettison the panels from future flights, this was not a serious concern, reflecting the earlier decision not to keep them attached for AAP S-IVB operations, which could jeopardise docking activities. The Apollo 7 crew discovered an unexpected phenomenon that could have also have had application to AAP Earth orbital missions had they been flown as planned. Between a perigee of 90 and 120 miles, the CSM was affected by the gravity gradient (pulling) effect, while on Apollo 9, with the docked LM, the crew reported that in drifting flight the longitudinal axis of the combined spacecraft (with the LM closest to Earth) was aligned with the orbital plane, and that the autopilot was very effective in rotating the spacecraft about any axis while still holding attitude in the remaining axis. This became a major feature in the accurate pointing of instruments from the SIM-bay on the J series of missions, freeing the CMP for other duties, and would have helped qualify systems and procedures for proposed AAP Earth orbital missions such as the cancelled AAP-1A.

On Apollo 7, the crew rendezvoused with the S-IVB stage, while full LM docking extraction, undocking and re-rendezvous and docking operations were performed on

Apollo 9 in Earth orbit and on Apollo 10 in lunar orbit. The first crew to use the internal transfer tunnel – on Apollo 9 – reported that it took 5–7 minutes, and that the mass handling (84-lb tunnel hatch and 80-lb probe) was not a problem in zero g.

Translunar coast The first crew to accomplish this was the Apollo 8 crew, who evaluated the passive thermal control mode (bar-b-que) at a rate of about 3 revs per hour. For Apollo 8 the RCS usage to control this was quite high, and thruster firings caused an interrupted sleep periods. From Apollo 10, the RCS control was relaxed somewhat, thus saving on fuel consumption and allowing for uninterrupted sleep of the crew. This gave the crew on '10' more time to adjust to the environment, and they could rest in preparation for the heavier workload in lunar orbit. During this phase, star and horizon navigation sightings were evaluated and conducted, and it was found that the auto-optics control occasionally failed to correctly position the sextant in relationship to the target star, and required manual control of the spacecraft, resulting in additional use of crew time and propellant. This period of the mission was used to check out equipment, including the LM, on Apollo 10, housekeeping, and television transmissions, and allowed time for the astronauts' bodies to become adjusted to the spaceflight environment – the added volume of the combined CSM and LM giving rise to demonstrations of the weightless environment during the television shows.

CSM operations Verification of the SPS system came with eight firings during the Apollo 7 mission. On Apollo 8 there were five burns of the SPS, including the four-minute burn to place the spacecraft in lunar orbit, the three minute burn to take Apollo 8 out of lunar orbit, and three minor course corrections. This successful demonstration of the SPS was a milestone in the qualification of the Apollo system. In CSM-only flight in lunar orbit there would be no back-up, and if this engine failed, the astronauts would have been stranded with no hope of rescue. Repeating the event on Apollo 10 with the circularisation of the lunar orbit added confidence for the lunar missions. In addition, Apollo 9 SPS performance in Earth orbit added to the database for later lunar operation and for planned AAP operations in Earth orbit. On Apollo 7 and Apollo 9, significant Earth landmark and terrain tracking and photography was accomplished in preparation for expanded Earth Resources operations on AAP (Skylab). In addition, a single stand-up EVA was accomplished by the CMP (Scott) on Apollo 9 in Earth orbit to photograph the LMP on EVA and to retrieve samples from the exterior of the spacecraft. Scott found that his access was made easier by stowing the central couch, provided more volume in which to work.

LM operations Apollo 9 and Apollo 10 verified the in-flight activities and operations of the Lunar Module in both Earth orbit and lunar orbit. Apollo 10 developed operational procedures that became standard operations on all lunar flights, and these induced communication and telemetry checks and stowage transfers during the Earth coast and initial lunar orbits prior to preparation for undocking. Apollo 9 also verified the Lunar Module power-up and check-out procedures in Earth orbit, while the Apollo 10 crew repeated these operations in lunar orbit about six hours before undocking. RCS firings were quieter than expected in the LM than in the CM, and the Apollo 9 and Apollo 10 crews found that it took about ten minutes to vent the tunnel before undocking. The Apollo 10 crew was unable to vent the tunnel completely before jettisoning and also recalled a loud pyrotechnic action at undocking, suggesting that future crews should wear pressure suits with helmet and gloves to prevent against unexpected venting of the cabin atmosphere due to shock from the pyrotechnic firing in the docking system.

One of the most valuable exercises carried out on Apollo 9 was the manual throttling of the descent engine used in both the docked configurations as back-up to the automated systems and for SPS engine failure and undocked configuration, again for auto-system redundancy. This was used to great effect during the aborted Apollo 13 landing mission, during which the LM descent engine was used for course corrections in the manual mode to return the crew to Earth. On Apollo 10, an example of a lapse in crew coordination and training was demonstrated in which switch sequencing was out of synchronisation, resulting in confusion to the flight systems and wild gyrations of the LM ascent stage during simulated lunar ascent.

During Apollo 9, a single EVA was accomplished by the LMP (Schweickart), who exited from the front hatch onto the porch and secured himself in the 'golden slipper' foot restraints, while wearing the lunar surface EVA garment and back-pack. Although the full transfer from the LM to the CM by EVA was not completed due to his earlier bout of Space Adaptation Syndrome (SAS), he attempted a partial traverse on the front of the LM by using the handrail intended to support contingency transfer in the event that the tunnel and hatches were inoperable. In addition, this system was also previously evaluated for use in flying the AAP ATM on the LM and in experiment data retrieval from LM Lab missions (discussed earlier), and was demonstrated as an adequate procedure on Apollo 9.

Lunar orbit activities On Apollo 8, the crew found that defining their ground track while flying over the far side of the Moon was much more difficult than expected, due to the imprecise nature of the preliminary maps of that region obtained from the Lunar Orbiter probes. They were therefore not completely sure where they would enter more familiar features on the near side of the Moon, resulting in a few moments of reorientation as they came back around the Moon. The situation improved for Apollo 10, due to experience and photographs from Apollo 8, and these missions established an ever-growing photographic database of far-side features for later crews. This also included the creation of additional lunar geology observations for the CMP on later flights to assist in gathering photographic and visual observation 21 data of lunar features and potential landing sites. The crews of Apollo 8 and Apollo 10 were also the first to encounter the mascons discovered during the Lunar Orbiter flights, and provided useful flight parameters of the CSM and LM in docked and undocked orbits around the Moon.

Rendezvous and docking activities The rendezvous of the Apollo CSM with the S-IVB stage on the first manned mission was also the first rendezvous exercise conducted on Apollo. On this flight the crew did not have reliable back-up ranging data from the LM to check the CM data, and the manual controlled braking manoeuvre was 'very discomforting'. Apollo 9 completed the first CSM/LM rendezvous manoeuvres in Earth orbit, to be duplicated by Apollo 10 in lunar orbit. After jettisoning the empty LM, the Apollo 9 crew tracked the ascent stage visually by using the sextant to about 2,500 miles, while on Apollo 10 the crew evaluated the maximum range performance of LM rendezvous sensors with the CMP, visually tracking the approaching LM through the sextant against the lunar surface at 125 miles in daylight and above the horizon, in daylight to 275 miles, and in darkness to 230 miles. A practise initiated on Apollo 9, to become standard operating procedure on each CSM/LM rendezvous, was the protection of the LM RCS firings by 'mirror image' readiness on the CSM. In case the ascent stage RCS thrusters should fail, the CMP could initiate opposite firing to effect the same result. The LM was tested as the active rendezvous vehicle on Apollo 9, and then the crew used the LM as the active component in docking and found that it was much more difficult to control. In addition, the accuracy was not so great as when using the CSM as the active vehicle, and it took the crew more time to initiate the docking from the LM than from the CM. The major factor in this was that the LM Commander had to lean back and look 'up' through the overhead window while his flying controls and displays were set at 90° to his line of sight, whereas the Command Pilot had direct line of sight of control and displays and the rendezvous window from his crew position. As a result, CSM active docking became standard on subsequent missions.

Trans-Earth coast The most quite period of each lunar mission was usually between the CSM's return from the Moon and the point of preparing for entry and landing. On Apollo 8 and Apollo 10, the crews began what became a feature of all Apollo lunar missions: television broadsheets showing the receding lunar disc and the approaching planet Earth.

Habitation issues Most Apollo astronauts adjusted well to zero g in the first few hours of a mission. Some of the crew-members suffered brief periods of disorientation or SAS, but none of this seriously affected mission operations, apart from the delay of the Apollo 9 EVA by the LMP, and shortening of the exercise; and the Apollo 7 crew suffered from head colds that irritated their sinus cavities. The crew usually reported favourable habitation issues including sleeping and eating, although food preparation took longer than expected, and gas bubbles injected into the food packages when rehydrating caused intestinal discomfort. Another important facility, evaluated and used on all four Apollo missions, was the waste management system. The Apollo 7 crew reported the excess of IVA handgrips that were perfect for EVA but pointless for internal activities, as there was no problem in moving around the CM. Traversing through the tunnel to and from the LM did not pose serious problems on Apollo 9 or Apollo 10.

Entry and recovery None of the crews experienced any difficulty during this phase.

The Apollo 8 crew became the first to return from the Moon, at a speed of 36,221 fps; and Apollo 10 set the unofficial speed record at 36,314 fps – a record that still stands more than thirty years later, and is not likely to be exceeded for many years to come.

Overall, the crews of Apollo 7-10 fully evaluated and proved the systems and procedures planned for Apollo and AAP, except for lunar landing and extensive lunar surface, lunar orbital and Earth orbital research programmes. These flights were the final steps towards the ultimate test – landing on the Moon – and waiting to fulfil that goal were Apollo 11 and Apollo 12. The first landing crews were announced on 9 January 1969 (Apollo 11) and 10 April 1969 (Apollo 12).

Position	Prime	Back-up	Support
Apollo 11, third quarter of 1969			
CSM-107/ LM-5 Commander	Armstrong	Lovell	Mattingly (CM specialist)
AS-506 CM Pilot	Collins	Anders	Evans (CM specialist)
First lunar landing LM Pilot (G) mission	Aldrin	Haise	Swigert (CM specialist)

Haise stepped down from the Armstrong crew to make way for Collins, who had returned to flight status as CMP. Aldrin returned to LMP duties. Anders' assignment was a dead-end position, as he had already indicated that he would be leaving the agency after the flight, and Slayton therefore assigned Mattingly, effectively as the second back-up Command Pilot, who would rotate with Lovell and Haise to fly Apollo 14.

Position	Prime	Back-up	Support
Apollo 12, fourth quarter of 1969			
CSM-108/ LM-6 Commander	Conrad	Scott	Gibson
AS-507 CM Pilot	Gordon	Worden	Weitz (CM specialist)
Lunar orbit LM Pilot mission	Bean	Irwin	Carr (LM specialist)

Ed Gibson was the first of the scientist-astronauts to receive a support crew assignment, although Kerwin and Schmitt had performed Apollo support assignments during simulations in 1968. The back-up crews for this mission were in line to rotate to Apollo 15 – the fourth and last of the H-series missions.

PLANNING THE FIRST LANDINGS

In addition to deciding how to reach the Moon, and in developing the hardware and the ground infrastructure to support the effort, the question of where to aim the first landing crew had been a topic of debate well before the commitment to go there.

Landing sites

The easiest area on which to land would be on the lunar near-side (always facing the Earth), and one of the primary considerations was the capability of the spacecraft designed to achieve that landing.[8] By simply looking at the Moon from the Earth, the *mare* (seas, or lowlands) appear darker than the lighter uplands. Telescopes reveal the undulating surface of craters, channels (rilles) and mountain ranges, and even careful observation can lead to dramatic descriptions of the features (usually based on terrestrial features), which helped to provide artistic licence in the dramatic jagged peaks and deep chasms in the interpretations of the classic 1950s artwork and writings. By the early 1960s this information was being supplemented by spacecraft actually reaching the Moon and answering some of the fundamental questions about its structure and appearance.

Geologists divided the face of the Moon into areas of landfall (highlands and lowlands), geological features (craters, mare and basins), and areas of special interest (specific craters, rilles, or formations) in their study of where it would be scientifically useful to send a package of instruments or from where to obtain samples for study. When Apollo was targeted for the Moon it was realised that detailed geological maps of the surface would be required, and that the capability of the launch vehicles would determine the best place for early landings.

Launching a spacecraft to the Moon from the Cape Canaveral launch site would be defined by the window of opportunity, allowing correct alignment the Moon when the spacecraft arrived there a couple of days later. It was not a case of launching directly at the Moon, which travels in its own orbit around the Earth. The probe therefore had to be sent to where the Moon would be when the probe intercepted the lunar orbit – the principle applied to Apollo. In addition, it had been known since the late 1950s that spacecraft launched from the Cape would require less energy to fly over the equatorial regions of the Moon than over the polar regions. Some of the earliest plans for the Ranger probes called for targeting on the Lansberg region (16° N to 16° S, 10°–50° W) in the western equatorial zone. The initial detailed maps therefore covered this area, and this allowed planning the first spacecraft missions to the surface – which then determined which feature in that area would be of most interest.

Following this, a major study of lunar equatorial areas eliminated unfavourable areas from prospective landing sites for both Surveyor and Apollo. Aided by powerful telescopes, areas of great crater impacts large boulders, rilles and highlands were discounted, leaving the relatively flat mare regions. Using Earth-based observations (later confirmed by close-up Ranger photographs), studies of crater rays and steep slopes revealed bright features under solar illumination, and also smaller craters and undulating feature along these rays. The darker areas were less impacted and thus younger in nature, and by using the theory that the darker areas had far fewer unresolved crater impacts and other rougher features, a list of candidate sites for Surveyor landings and Lunar Orbiter photography established; and from these areas, the Apollo landing sites would be selected.

Other important features in selecting the landing sites were in the capabilities of the LM, the training of the astronauts, and what was expected on each landing. The scientists' desire to deploy sophisticated scientific instrument packages and to carry out detailed geological field trips – perhaps with a trained geologist on at least the first landing – invariably gave way to the Astronaut Office and NASA's desire for mission success and crew safety. The first landing soon became a brief trip, to be followed by others that would extend the duration and increase exploration coverage, although the inclusion of a geologist on the landing crews was by no means certain, at least under the mainstream Apollo.

By 1965 the targets of the first manned landings were within an area known as the *Apollo Zone* – 10° wide and 90° long, located between 5° N and 5° S latitude, and 45° E and 45° W longitude, which equated to a narrow strip measuring about 190 miles wide and 1,700 miles long. Within this basic rectangle were the main landing areas for Apollo, although the actual boundaries were not geometrically defined, as the landing area was a little wider in the west than in the east, and narrower in the middle, and changed a little for each launch window once a month throughout each calendar year.

The constraints on this type of area evolved from the amount of fuel required to reach the landing site, the requirement for the main spacecraft to remain on a free-return trajectory if the main engine failed, accessibility throughout the year from TLI burn over the Atlantic Ocean or the Pacific Ocean, direct radio communications with the crew on the surface, eliminating the need for a comsat relay system in lunar orbit, navigation and landmark training for guidance, and the lack of knowledge of features at each limb of the Moon, the poles or the far side. In addition, the requirement for bringing down the astronauts onto water and in daylight at the end

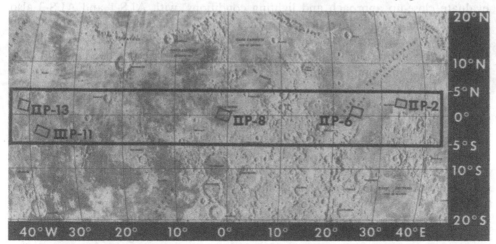

A 1967 NASA photograph of the lunar surface, showing some of the potential Apollo landing zones of interest, based on Lunar Orbiter photographs.

of the mission led back to specific dates and times when the Saturn could be launched, arrival times at a suitable landing approach with correct lighting conditions, and protection of the recovery constraints. If the launch opportunity of several days each month were to be missed, then the Earth and Moon would shift in their orbits, and Sun angles would change and not be optimum for another four weeks.

From this Apollo Zone, a list of thirty-two prime landing areas clustered in groups were identified and photographed by the first three Lunar Orbiters. In addition, another nine less intensely photographed but geologically interesting sites were photographed, and together these were identified as 'Set A'. The initial landing site would be selected from these twenty locations. Details of these sites were exhaustively studied and debated as the time came to choose the first landing area. This information narrowed to a group of five primary targets. MSC wanted an eastern site so that if the launch was scrubbed for a few days then the opportunity to land in a western site still remained in the same launch window. This led to focusing on two sites in Mare Tranquillitatis – specifically ALS-1 and ALS-2; therefore, as the debates continued, Apollo 8 had the task of detailed photography and observation of these sites for making the final selection. ALS-1 was described by Jim Lovell as resembling a geological training area that he had visited in New Mexico, with special features and characteristics that was felt 'too exotic though possibly interesting' by the geologist for the first landing. ALS-2 was forty miles from the impact point of Ranger 8, and sixteen miles from the landing site of Surveyor 5, and so was not a totally new area. ALS-3 in Sinus Medii offered a two-day recycle alternative site, and would be included in the astronauts' training; while ALS-5 in Ocean Procellarum was the back-up to ALS-2, even if ALS-3 could not be reached in the launch window. ALS-4 was out of consideration until December 1969 or January 1970.

In May 1969, ALS-2 was used by Apollo 10 as the dress rehearsal for Apollo 11 to evaluate close-up approach and lighting conditions, with ALS-1 and ALS-3 also receiving their attention visually and photographically. The reports from the Apollo 10 crew produced greater confidence in selecting ALS-2 in the Sea of Tranquillity as the primary landing site for the first Apollo, although once this decision had been made, discussions continued on where to send the next three or four missions.

Experiments

When NASA issued proposals for a separate lunar landing module in 1962, it included the capacity for delivering a 'scientific information system' that would study the lunar atmosphere, surface and interior after the crew returned to Earth. Over the next four years the NASA Office of Space Sciences and Applications evaluated the experiment package defined by the constraints of the LM to deliver it to the surface. Preliminary studies with Grumman indicated a total mass of 230 lbs for experiments in a 15-cubic-feet location in the descent stage. In addition there would be an additional 80-lbs and 3 cubic feet in the ascent stage and CM for sample return containers and exposed film. The method of power for the Lunar Surface Experiment Package required it to be self-contained, support experiment operation for one year, and be capable of sending telemetry to Earth. This focused on using heat generated by radioisotope fuel capsules converted into electricity by a power module.

Discussions between MSC and scientists focused on the initial experiments flown. These would feature a passive seismometer to record so called 'moonquakes', an active seismometer to record the effects of explosive changes detonated on the surface, a lunar gravimeter used in conjunction with tidal effects to determine the internal structure of the Moon, heat flow measurements, radiation and micro-meteoroid detectors, an lunar atmosphere analyser and studies of the magnetic field properties and solar wind .

Following the identification of the shortlist of proposed experiments, Phase I definition studies conducted during 1965 evaluated potential contractors to fabricate the hardware. On 16 March 1966, NASA selected Bendix Corporation's Systems Division as the prime contractor for the design, manufacture, test and operational support of initially four Apollo Lunar Surface Experiment Package (ALSEP) units. Bendix would provide the central station that transmitted data to Earth, integrate the whole package and liase with subcontractors on the experiment hardware.

By the late summer of 1968, a demonstration of pressure-suited astronauts deploying the Apollo Lunar Surface Experiment Package was scheduled to evaluate the proposed deployment sequence. The astronauts selected to perform this test were physicist Don Lind and geologist Jack Schmitt, who completed two simulated EVAs on 26 and 27 August 1968. The first of these was to deploy the ALSEP, and the second was to be an evaluation of other surface activities such as collection of samples. The tests revealed a heavy workload during the first demonstration, while the second was void of any clear procedures. The test was deemed successful, but highlighted the fact that the proposed workload for the first landing could be too excessive.

In light of this simulation, on 11 September a meeting of the MSC Management Council established that the first ALSEP should be deleted from the initial landing due to safety and time constraints, and be replaced with a smaller abbreviated set of instruments: an Early Apollo Surface Experiment Package (EASEP) consisting of a passive seismometer, a laser ranging retro-reflector, and a passive solar wind collector. To assist in evaluating the EVA planned for Apollo 11, astronaut Don Lind was assigned to develop procedures and timelines, and to provide support and guidance for Armstrong and Aldrin.

Equipment

One of the other priorities of the landing was the collection of samples of lunar material. To achieve this, a range of sampling tools, collection bags and containers was designed for the astronauts to use while wearing a pressure suit. In addition, attachments were made to carry a camera on the front of the suit, a cuff-mounted check-list time-lining the surface activities, and a lunar surface close-up camera to be placed on the lunar material to take photographs of the soil *in situ* without requiring the astronaut to bend down to reach the surface – which was a difficult task in the initial (Apollo 11–14) lunar surface suits (designated A7L, and developed by the International Latex Corporation) due to lack of limb mobility in the suit design.

The determination of lunar characteristics

	Topography	Geodesy	Gravity Field	Terrain Characteristics	Geological Formations	Sub-surface Characteristics	Petrology	Compositional Characteristics	Geophysics	Biology	Particles and Fields	Lunar Atmosphere	Lunar History	Landing Mission Support
Gamma-Ray					x			x	x		x		x	
X-Ray								x			x		x	
Ultraviolet				x				x						
Infrared Spectrum				x			x							
Infrared Scanner				x			x							
Microwave Spectrum				x			x		x			x		
Microwave Scanner				x		x		x	x			x		
VHF Reflectivity						x			x					
Radar Imaging	x			x	x	x			x					
Radar Altimeter				x	x									
Metric Cameras	x	x		x	x									
High-Resolution Cameras	x			x	x									
Ultra-High-Resolution Cameras	x				x									
Multi-Band Synoptic Camera	x				x									
Geochemical Sensing							x	x					x	
Micrometeoroids											x		x	
Gravity			x						x					
Probes														
Hard Landing	x					x								x
Soft Landing	x		x	x		x								x
Survivable Capsule	x			x		x					x			x

Achieving the goal

On 20 July 1969, Apollo 11 astronauts Armstrong and Aldrin achieved the primary goal of the programme in landing on the Moon. They conducted a 2-hour 31-minute EVA at Tranquillity Base, collected samples, took photographs, and deployed the EASEP during a stay time of 21 hours 36 minutes. This was followed on 19–20 November by Apollo 12 astronauts Conrad and Bean, who completed two periods of EVA (3 hrs 56 minutes and 3 hour 49 minutes) at the Ocean of Storms landing site, gathered a second collection of samples, deployed the first ALSEP, and examined the unmanned Surveyor 3 lander during a surface stay time of 31 hours 31 minutes. President Kennedy's commitment to send Americans to the surface of the Moon and bring them safely home by the end of the decade had been achieved twice.

The two landing missions had demonstrated that the chosen method of reaching the Moon by Apollo worked – and worked well. The hardware, systems, procedures and astronauts all performed relatively flawlessly, and post-flight debriefing of the two crews indicated that the other eight missions then undergoing planning as part of the mainstream Apollo would be achievable. However, at the time of Apollo's greatest success, dark clouds were looming to threaten not only Apollo but also the future of manned lunar exploration, and the whole direction of American manned space exploration over the rest of the century. The steps to the Moon had been achieved, the tragic fires had been overcome, and the improved systems had been demonstrated as workable. The future operations of Apollo focused on the landings that would immediately follow Apollo 11 and Apollo 12, and on the application of Apollo hardware and systems in Earth orbit, at the Moon, and perhaps even further ... towards Mars.

1 *NASA Historical Data Book Volume II, Programs and Projects 1958–1968*, Linda Neuman Ezell, NASA SP-4012 (1988), pp. 300–331); also *Solar System Log*, Andrew Wilson, Jane's Publishing Company Ltd, London, 1987.

2 Meeting with FOD on Flight Program, Memo for file, C. Perrine, MOD, 17 April 1967, from *Apollo Chronology*, Volume IV, p. 123.

3 *Chariots for Apollo*, Brooks Grimwood, Swanson, NASA SP-4205, 1979, pp. 231–232 (footnotes).

4 *Skylab Chronology*, pp. 104–125 (various 1967 entries); *Apollo Spacecraft: A Chronology* pp. 68–189 (various 1967 entries).

5 *Carrying the Fire*, Michael Collins, Farrar, Straus and Giroux, New York, 1974, pp. 267–269.

6 *Apollo by the Numbers, a statistical reference*, Richard Orloff, NASA SP-2000–4029, October 2000.

7 *Apollo Program Summary Report*, JSC-09423, LBJ Space Center, Houston, NASA, April 1975; also crew post-flight technical debriefing transcripts, Apollo History Collection, formerly at NASA JSC History Office and Rice University Fondren Library (researched by the author, 1988–1994).

8 *To a Rocky Moon*, Don E. Wilhelms, Arizona University Press, 1993. A detailed analysis of the evolution of lunar exploration in the 1960s and 1970s (highly recommended for further reading).

Lunar logistics

Project Apollo could be regarded as four programmes in one. First there was the effort to develop the hardware and procedures necessary to achieve the lunar landing goal, and then there was the Earth orbital programme that applied Apollo hardware to the creation of a space station. Both of these initial efforts would in turn support the long-term exploration of the Moon and initial journeys to Mars. Developments in the first two phases have been discussed, and while these were underway, NASA and aerospace contractors were evaluating how Apollo hardware could be used to create a reliable lunar logistics system than could be used to expand the exploration of the Moon and support the first lunar bases.

Once the lunar orbital rendezvous system had been selected in 1962, and after the decision to begin the development of the separate lunar landing module (LM), dozens of plans, profiles and hardware concepts were generated, exploring the idea of lunar exploration after the initial concept of using Apollo to land men and machines on the Moon had been qualified. In addition, at several conferences, scientists and engineers discussed the most appropriate way to approach exploration of the Moon after 1969 – the most notable being the summer conference on lunar exploration and science held at Falmouth, Massachusetts, during 19–31 July 1965, and, two years later, the 13 July–31 August 1967 studies of lunar science and exploration held in Santa Cruz, California.[1]

LUNAR LOGISTICS STUDIES, 1962

On 2 August 1962, NASA issued a Request for Proposals (RFP No. 10.35) which announced (news release 9 August 62-181) the start of in-house and industry contract studies to determine whether if an unmanned lunar logistics system would be required to support manned lunar landings that would improve the capabilities of the crew on the surface. The three-month studies consisted of two elements: a spacecraft bus that could carry support payloads to the Moon, and a variety of payloads that could be soft-landed near the manned LM. The plan was to use the Saturn 1B and later the Saturn V with minimum modification as launch vehicles. For the purpose of the study, seven payloads were identified to evaluate how a crew surface stay time could be extended, in both pre-planned exploration and emergency situations, by additional hardware that would also expand the scientific exploration programme.

The logistic support payloads were identified as crew stay-time extension, crew shelter, roving vehicle, lunar surface modification equipment, power station equipment, communications station equipment, and multiple-purpose payloads. Twenty-four applications were received for the spacecraft bus proposal request, and twenty-eight applications were submitted for the payload request proposals, with several companies applying for both contracts, which would be issued on 1 September and completed on 1 December 1962.

By 5 September, NASA reported (Release 62-190) that three companies would conduct the study. Space Technologies Laboratories would study the spacecraft bus concept, while Grumman (which was also about to be awarded the LM contract, on 7 November) and Northrop Space Laboratories would study the payloads that might be carried. The guidelines were based on the concept of a 9,000-lb vehicle carrying about 1,500 lbs of equipment and launched by Saturn 1B, with future application to the development of a 90,000-lb vehicle with a 20,000 payload envelope launched by a Saturn V. Meanwhile, NASA field centres would be studying optimum trajectories and adapting the Saturn launch vehicles, alternative spacecraft propulsion systems, lunar landing dynamics and the use of roving vehicles.

In a memo from Joseph Shea, Deputy Director for Systems, OMSF, to Robert Seamans, NASA Deputy Administrator,[2] it was stated that such a system would, following the initial Apollo landing, not only expand the capabilities on the surface, leading to semi-permanent and permanent lunar bases, but also 'provide long lead propulsion stages for a direct flight back-up capability' to LOR. Shea reminded Seamans that in the C-5 Direct Ascent mode recently rejected in favour of LOR, the concept included a modified Service Module with the capacity to provide direct return from the lunar surface to Earth. Shea suggested a deferment of Direct Ascent until mid-1964 to retain its availability to support LOR operations, and to await experience in rendezvous concepts with Gemini. However, Direct Ascent was not the primary mode of sending Apollo to the Moon, although, with EOR, it could still play an important role in supporting exploration after the early landings.

When the reports were submitted they were, as expected, positive in their belief that such a system could be produced.[3] They also noted some areas that required further study – such as the operation of equipment and systems during the two-week lunar night, designs of effective thermal radiators, the extreme variations of temperatures during the lunar day and night cycle, the use of solar energy for power during the day, and storage batteries (or nuclear fuel cells?) for use during night operations. The capability of a shelter was evaluated as providing a pressurised shirt-sleeve environment which included power supplies and life support systems. The studies also identified the benefits of developing a non-pressurised lunar roving vehicle of a simple and reliable design, using small rechargeable batteries or solar power, and offering a twenty-mile round trip expedition from the landing base. Anything further would require much larger, complex and pressurised vehicles.

GRUMMAN AND BOEING STUDIES

In the Grumman study paper,[4] the approach was similar to other industry proposals; but it also hinted at the possibly of using the LM itself as a unmanned logistics transport system, and perhaps incorporating an airlock and shelter into the design of the LM, which itself was still being evolved as the contracts were signed with NASA. In effect Grumman was hedging its bets with the basic LM design, and indicating that it would (initially) not require extensive modification in adapting the basic lunar lander to other roles (the principle of what became the Apollo Applications Program). The studies also indicated the capability of a second LM for crew rescue from a (LM?) shelter, delivery of system spares for a disabled LM, or resupply of consumables and exhausted equipment and systems, prolonging the surface stay time well beyond that proposed for the initial landers.

Grumman had recently completed twenty months of study of in-house concepts for unmanned and manned (unpressurised and pressurised) rover concepts, and suggested that a 1,000–1,500-lb LRV would be useful in a dual unmanned and manned role, possibly being relocated to a new site once the previous crew had left the surface. In these rover studies, begun in January 1961, Grumman estimated that a payload of 500 lbs with an 800-mile range (unmanned) and a 1,250-kWh emergency capacity would be well within Apollo/Saturn payload design limits. In late 1961, Grumman expanded these studies with an advanced rover (launched on a Saturn C-5) with a 500-mile range, pressurised, and capable of supporting 'a crew of *three* in a shirtsleeve environment for two weeks'. Weighing about 14,000 lbs, it included a bulldozer blade for surface operation, and two manipulators (similar to the Shuttle RMS, but with grapples) with a 5,000-lb lifting capacity. This might seem to be a drastic method of sample collection, but was aimed more at future lunar base construction than scientific geological surveys.

A report by the Boeing Company[5] illustrated the suggested sequence of exploration over the next twenty years (see the illustration on p. 200) and how the exploration of the Moon was but one element of a major exploration of manned exploration of the Earth–Moon system and Mars, using Apollo–Saturn-developed hardware.

Proposed LLS/C-1B mission concepts included mapping the lunar surface from orbit, measuring lunar surface properties and environment, providing a landing beacon for site identification, emergency shelters and escape vehicle or repair kits and early extended-stay facilities for up to twenty-eight days.

Those proposed for LLS/C-5 missions included shelter and life support facilities while surveying the area to establish a permanent base (three-men growth to twenty-one men, with partial crew rotation on three-month and six-month cycles), (nuclear) power stations, communications stations, lunar surface vehicles, scientific equipment, a manned LM transported unmanned for emergency return capability, and the 'special materials equipment and supplies for permanent base construction and use'.

This survey (extending to approximately 175 pages) also evaluated performance trade-offs, system consideration such as flight profiles, pinpoint-landing targets, unmanned support vehicles (such as those proposed for the Lunar Orbiter

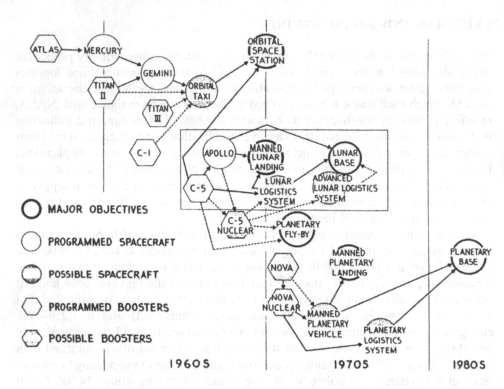

This 1962 Boeing diagram reveals the long-term planning for manned space exploration in successive objectives for the twenty years after Apollo was assigned to the manned lunar goal (1960s–1980s). (Courtesy Boeing Aircraft Corporation.)

programme), and estimated costs over a variety of options. The concept also evaluated the landing of a three-man Apollo CSM by direct ascent from the surface. Boeing estimated that a pure cargo payload could also replace the manned CSM and supplement the normal two-man lunar lander. The weight breakdown for using a three-stage Saturn C-5 was quoted as:

Apollo CM	10,000 lbs
Apollo SM	9,300
SM propellants	7,700
LLV	19,000
LLV cryogenic propellants	44,000
Total Earth-escape weight	90,000 lbs

The propellant requirements for SM Earth return was quoted as 28,600 lbs, with refuelling on the lunar surface (20,900 lbs) met by an additional LLV/C-5 tanker flight.

An important element reviewed in these studies was the launch vehicle capabilities and, more importantly, availability over a 10–15-year period. Although these studies were intended to be only conceptual, they focused attention on what could be achieved within the framework of Apollo, which in turn might lead to a larger and more permanent establishment of a manned lunar base similar to the polar research stations in Antarctica.

GRUMMAN LM UTILISATION STUDIES

Grumman was awarded the LM contract on 7 November 1962, and by the end of June 1963 most of the LM design had been frozen, allowing sub-contractors to be selected and the design of sub-systems to begin. By July, Grumman was proposing an unmanned logistics version of the LM (as indicated during their 1962 studies) that would supplement the manned versions, drawing on its systems but not directly interfering with the main development programme. The concept envisaged a logistics vehicle (LM Truck) to increase surface payload delivery, and to also support the installation of a habitable shelter–laboratory (Shelab, or LM Shelter) for two men to extend their time on the surface.[6]

The study followed the guidelines of a maximum use of LM hardware, under development, and if changes were required there could be no modification to the SLA. This resulted in a commonality between all designs to ensure smooth production flows of mainline and utilisation vehicles. These gave rise to a series of utilisation studies under NASA contract (NAS 9-3681) of LM vehicles for what was then termed the Apollo Extension System (AES) mission, but which became the Lunar Apollo Applications Program (LAAP).[7]

LM Lab for lunar orbital missions

The use of an LM as a laboratory with the CSM were completed for both Earth orbital missions and lunar orbital missions. The lunar orbit concepts were identified as Configuration C, and depended on an Apollo X CSM in which the LM Lab supplied payload equipment requirements but depended on the CSM for life support for up to fourteen days; or Configuration D, in which a Block II CSM was used with the LM Lab which with modifications extended the CSM capabilities of the sub-systems to 30-day and 45-day missions. In both cases the CSM was relied upon for G&N, communications, atmosphere control, and crew habitability, and neither was totally compatible for lunar orbital missions.[8]

The LM Lab would operate in polar orbit at 80 nautical miles for 27 days of a 33-day mission duration. A three-man crew would be required with no more than two astronauts in the Lab at one time to operate experiments. In addition objectives were listed as studies related to the oblateness of the Moon at the poles, location of the centre of mass, surface feature mapping and observations, gravitational anomalies, magnetic field studies, and the physical environment in the immediately vicinity of the Moon. Biomedical experiments and observations were to be carried on all flights, but since the lunar polar orbit mission was not expected to be the first extended flight (exceeding fourteen days), there were no medical concerns with this flight profile.

The report stated that each astronaut would follow a 10-hour work/14-hour rest cycle, with a least one astronaut awake at all times (although this would have probably changed to a simultaneous sleep cycle for all three, due to in-flight experiences on Apollo 7; see p. 137). All three astronauts would be on duty for each 3.5-hour EVA with PLSS (tethered), or each one-hour EVA (with 15-minute breaks) wearing a manoeuvring unit (tethered or untethered).

Studies of providing a manned orbital reconnaissance platform to select Apollo landing sites were some of the earliest studies of using Apollo hardware for non-landing roles after the lunar programme was assigned to Apollo in 1961. One of these studies[9] featured the replacement of the LM with landing capability with a simple Lunar Reconnaissance Module (LRM) that was essentially based on the Experiment Module that featured in the AAP-1A mission to carry the early LM&SS hardware, and was further developed by Grumman into the LM Lab concept

Studies determined that the use of the cartography camera system every sixth orbit was both efficient and operationally desirable, with supplementary photographs filling the gaps. There were also plans to fly an infrared mapper, multispectral and ultraviolet cameras, and a high-resolution panoramic camera, as well as the cartographic cameras. It was estimated that 3,500 high-resolution stereo photographs of about 450 areas could result in net coverage of 531,990 square miles at a scale of 7,750 feet per inch, with cartographic photography of the entire lunar surface at 162,000 feet per inch. UV photographs of the surface at 324,000 feet per inch, multispectral (three-colour) photographs providing full-colour reproductions at 324,000 feet per inch, infrared scans at 250,000 feet per inch, and radar maps of one half of the Moon at 80,000 feet per inch, as well as the magnetometer, thermometer, X-ray spectrometer, radar altimeter and gravity gradiometer recordings being continuously taken.

Configuration studies featured the placing of instruments up the CSM/LM long axis, allowing orientation in any of three principle directions relative to the Moon/orbital plane system. This 'forward–side–end' flyer was a descriptive term for the placement of the experiments on the LM, although there were attempts to reduce the different locations and requirements for excessive orientation changes for data-gathering. Several designs were considered for this configuration: a fuelled descent stage with propulsion but without landing gear, and associated equipment offering the least structural changes and greater abort capability; a descent stage Lab without propulsion that deleted the descent engine but retained the tankage; a Lab with a low-profile descent stage which replaced the normal descent stage, providing an experiment support rack and little else, making the greatest payload weight savings.

For surface missions after the initial landings, Grumman proposed a fleet of LM derivatives that included:

Uprated LM (ULM) Minimal changes to the LM systems allowed the basic 36-hour stay time to be extended to about 72 hours, and allowed for a slight increase in payload mass to the surface, providing for up to four EVAs over the basic LM capacity for two EVAs. (These evolved into the J-series of LM landers.)

An artist's impression of the Grumman LM Truck, capable of delivering a 9,000-lb payload to the Moon. Atop the descent stage is a 760-cubic-foot fixed crew living quarters, with provisions for a 14-day surface stay, and a 4,800-lb, 550-cubic-foot scientific payload. The astronaut is unloading a Lunar Roving Vehicle. (Courtesy Grumman Aircraft Engineering Corporation.)

Extended LM (ELM) This depended on the relaxation of flight rules, would add an additional 1,500 lbs of consumables and equipment, and would extend surface stay time to about seven days, dependent on the CSM reaching the Moon.

Augmented LM (ALM) Improvements to the descent stage by increasing the size of propellant tanks and strengthening its structure ('beefed up'), allowed delivery of 2,800 lbs to the lunar surface, and an extended surface stay time of fourteen days for the two astronauts.

Lunar Payload Module (LPM) This was the original 'stripped LM' version with all ascent propulsion systems, crew controls and displays and life support sub-systems removed, to be replaced by automated landing systems and a capacity to support a payload on the lunar surface for up to three months. It could deliver up to 8,200 lbs based on the basic LM, and up to 10,200 lbs based on the ALM design.

Grumman also studied improved versions of the basic design.

LM Shelter

This concept was designed to provided an unmanned landing on the Moon and to serve as a two-man shelter on extended lunar exploration for up to 14 days (lunar daytime), launched in conjunction with a second vehicle (LM Taxi) designed to transport the two astronauts too and from the surface.[10] Essentially, this design was a 'stripped LM' with the ascent capability removed, and was able to land unmanned in an automatic mode to deliver 8,438 lbs to be utilised for storage of stay-time expendables and scientific equipment, offering the provision of initial expansion of a lunar base site.

As well as the 14-day active support capability, the study evaluated the possibility of storage of the LM Shelter, either in lunar orbit or on the surface, for a period of up to three months. The resulting report and configuration diagrams were not defined designs, but concepts showing how the various sub-systems and experiment payload could have been adapted into a flight version LM Shelter. Three concepts were put forward – the first requiring minimum modification and the addition of further expendables to increase the lifetime. In addition, an unpressurised LRV was included to extend the foot range of the astronauts beyond the 1,000-feet (1965) planning limitation of early landings. This vehicle was to be used during the lunar daytime offering 295 hours of use, but without storage capability during periods of lunar night.

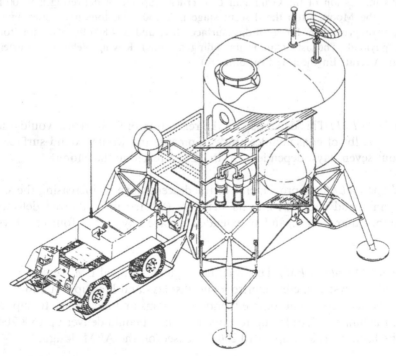

An NASA (MSFC) concept for an advanced LM Shelter/Laboratory concept called Shelab is shown with deployed Lunar Roving Vehicle.

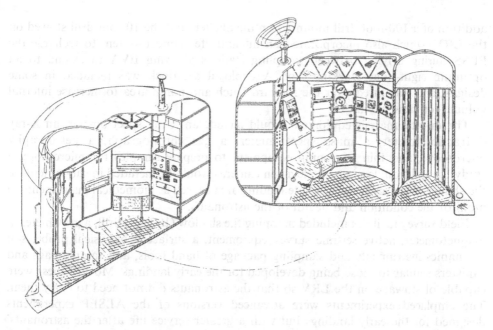

An exploded cut-away illustration of the proposed internal layout of the Shelab, showing airlock access to and from the module.

The other two configurations represented partial or fully closed cycle life support systems, reducing the need for stored consumables and increasing the payload capability for a larger LRV or flying vehicles to venture 'out and beyond the visible horizon'. Configuration 2 contemplated CSM-type fuel cells with cryogenic reactants, a fourteen-day manned capacity, and 1,000 hours (42 days) of storage time, while Configuration 3 used Gemini-type fuel cells with ambient storage of reactants and an airlock, offered a fourteen-day stay time with up to three months' storage capability. There were also two modes of landing the shelter. First, the LM Taxi was landed by the crew, who 'remotely guides and lands the Shelter near by'; or second, the manned Taxi completes a lunar surface rendezvous with the previously landed unmanned Shelter.

Typically, there would be two 'missions' for the Shelters. Mission 1 would extend exploration beyond the 1,000-feet radius of early landers, and exceed the thirty-four-hour limitation of mainline Apollo LMs (essentially beyond the H series of landings). It would include equipment inside the ascent stage for preliminary on-site analysis of lunar samples prior to selection of those to be returned to Earth, field survey equipment, instrument packages for long-term automatic monitoring of lunar environment (advanced ALSEP), and psychological–behavioural and biomedical experiments and equipment. There would be a drill capable of boring to ten feet, and a surface vehicle 'of minimum size' allowing a walk back to the Shelter in case of failure. Maximum EVA supported by rechargeable PLSS would be limited to three hours during each excursion.

Mission 2 would extend the exploration radius to about 4–6 miles with the addition of a 100-foot drill mounted on the Shelter, and the 10-foot drill stowed on the LRV, which also incorporated an integral life support system to recharge the PLSS during driving between sampling stations allowing EVA to extend to an optimum eight-hour working day. A deployable airlock was included in some designs, and was fitted over the forward hatch and porch area to increase internal volume for crew habitation

The on-site analysis equipment would include an X-ray spectrometer, an X-ray diffraction meter, a mass spectrometer, a gas chronograph, a petrography microscope, and thin sectioning equipment to prepare samples for microscopical study. In addition, lasers and television cameras would be included in sub-systems in the shelter to survey the immediate landing area as well as biomedical instruments to monitor the condition and health of the astronauts.

Field survey facilities included mapping the shallow (10-foot) drill, a gravimeter, a magnetometer, active seismic survey equipment, a surface geophysical probe, soil mechanics instruments and sampling package of hand tools, and sample bags and canisters similar to those being developed for the early landings. Most of these were capable of stowage on the LRV so that the astronauts did not need to carry them. The emplaced experiments were advanced versions of the ALSEP experiments designed for the early landings but with a greater service life after the astronauts' return home.

Most of the LM hardware would be retained in its original location whenever possible, with additional stowage capability for LiOH canisters for the cabin atmosphere, and back-packs. Secondary structural components would require minimal modification to accommodate equipment or experiments. The 'stripped' LM, without the ascent facilities and crew displays, controls, restraint harnesses and associated equipment, would be reconfigured for the particular mission that the Shelter would perform. The flight configuration depended upon the amount of modification required in replacing the skin panels with redesigned thermally protected skin (for protracted surface exposures during 14 days' operation or storage), or addition of new external structures to support the LRV and/or 100-foot drill and support structure

LM Taxi

This version offered the most fundamental design for adapting the LM: transportation of two astronauts from lunar orbit to the surface, the performance of a pinpoint landing near a Shelter (the crew's remote landing capability of a shelter from the surface was not mentioned), the capability of fourteen-day power-down storage (moth-balled), the provision of transport from the surface for two astronauts to rendezvous with an orbiting CSM, and the delivery of up to 250 lbs of ascent payload from the surface.[11]

Most of the changes in this design reflected adjustments to the systems, with the capacity of powering down the vehicle for storage during the surface stay time, but still retaining an 'any-time' abort capability. Alternative power supplies (Gemini-type fuel cells, or ALSEP-type RTG) additional micrometeoroid and insulation protection materials, and transponder landing capability, resulted in eight configurations to encompass these requirements. The environmental parameters resulted in a design guideline of 40–130° F ascent stage limitations, and 50–90° F for the optimum mission profile range. The study determined that an LM Taxi could remain on the surface for 336 hours (fourteen days) in sunlight (lunar daytime) and between 163–336 hours (6.8–14 days) adequately protected during the lunar night.

The sequence of events, outlined by Grumman, after LM Taxi touch-down (with windows facing the previously landed Shelter) were also itemised, and required the Commander to leave the taxi and proceed alone to the Shelter – estimated to be a 20-minute foot-traverse distance away – leaving the Systems Engineer (LMP) in the Taxi. There was an estimated one-hour inspection of the Shelter, with a 35-minute contingency timeline to confirm that the Shelter was 'Go/No Go' for occupation. Once this was achieved, the Systems Engineer would spend 15–20 minutes deactivating the Taxi into the storage mode, and then join the CDR in the Shelter, where they would base their operations for the next 13.8 days, after which they would deactivate the Shelter and return to the Taxi for the ascent from the surface.

Changes to the LM structure included additional micrometeoroid protection, more batteries for descent (four to seven units), deletion of scientific payload installation of RTG power units (two for a day/night mission, and one for a day-only mission).

LM Truck

This was a descent-stage-only configuration with modifications to make an unmanned automatic landing from lunar orbit with a large payload. The Truck included the required ascent stage sub-systems to support the landing, integrated into the configuration without the pressurised ascent cabin structure, crew LSS and provisions, and ascent engine sub-systems.[12] In this design, certain guidelines were observed during development, incorporating a configuration that included a landing area that had to be relatively clear of large obstacles and severe slopes. The guidance and control system would be used to control descent and touch-down, with the landing gear operating within the design envelope to support the landing. Relative status monitoring of onboard systems would be required at MCC from CSM separation to touch-down, to provide failure analysis. There would be no electrical support from the CSM, and the Truck would be deactivated at landing. Its final design and performance depended on the required mission assigned to it.

Study results indicated that the truck was (in theory) capable of delivering payload of 2,100 cubic feet and up to 10,300 lbs, the major modification being the relocation of selected ascent stage systems to the descent stage. Ridge attachment of RCS quads from the ascent to the descent stage provided for RCS propellant storage. In this study the 'payload' would require a CSM/LM docking system facility to allow extraction of the Truck from the S-IVB and to transport it to lunar orbit. In addition, either an CSM/Truck or an MCC/Truck data command link would be required to provide real-time updates to the onboard G&N system on the Truck. The manual optical alignment telescope on the manned LM was replaced by an automated version, a remote link for IMU alignments being performed either through the CSM or MCC. Apart from the added sub-systems, the basic descent stage, landing gear and SLA attachment structures remained, with the quadruple tubular truss at each end of the cruciform main structure of the descent stage carrying a system of pulleys ramps, or a crane which could remove payload from the top of the Truck to the surface; and since the stage would not be used again, these did not need to be of an elaborate design, or to be capable of lifting mass back onto the top of the Truck.

Another version of this design was the Early Lunar Shelter (ELS), which featured an LM Truck, fitted with a horizontal cylindrical habitation module in place of the ascent stage, and capable of supporting two men for up to fifty days on the surface. The Truck could also automatically deploy a Stay-Time Extension Module (STEM), which was an expandable (inflatable?) flexible structure that provided additional living space for two men for between 8-14 days and included a small airlock offering an internal shirt-sleeve environment.

HARDWARE DEVELOPMENTS

In order to support the above proposals certain hardware elements needed to be developed or improved in addition to the Grumman Lunar Module (mentioned previously).

CSM requirements

Earlier discussion focused on the upgrade of the Block II CSM to the Block III design, intended for AAP in Earth orbit but never authorised or fabricated. However, in the AAP lunar mission (described above) there has been little focus on the CSM operations other than the planned solo lunar orbital mission. Most studies indicate that during many of the extended surface stays, the CM Pilot would remain alone for up to two weeks, conducting orbital surveys from orbit. Without an Earth return capability the ascending LM could not return the astronauts to Earth, and thus the manned CSM was required on station to return the landing crew to Earth.

During the J missions the CM Pilot conducted solo SIM-bay operations and observations while the crew explored the Moon. These amounted to more than 66 hours (Apollo 15), 71 hours (Apollo 16) and 74 hours (Apollo 17), totalling 211 hours over three mission by three different CMPs. What the lunar AAP missions proposed involved the CMP flying alone in lunar orbit for missions of up to 336 hours. The longert time that any person had remained alone in space was in June 1963 when Soviet cosmonaut Valeri Bykovsky spent 119 hours on Vostok 5. On each of these missions, the crew was occupied with the flight programme of varying degrees of complexity and participation and none of the participants appear to have reported any adverse experiences due to their unique spaceflights. There was no question that fourteen days in space was not medically possible, as Gemini 7 had proved in December 1965; however, physiological separation was another matter.

With only telecommunication contact with MCC and the exploration crew, even a television link and a chat with the family could prove inadequate, and was sometimes more difficult to deal with (as discovered on Skylab and Salyut during long-duration missions). Constant rotation of work–sleep–housekeeping–work was not the most productive way to maintain crew morale and peak performance levels (as was also discovered on Skylab). A day or half day 'off' – if the lunar AAP CM Pilots were given one – would not be an escape from the routine of daily life in space, especially in the cramped confines of the CM. Even though the view out the window would always be amazing, it could not be so visually (or emotionally?) stimulating as looking down at the constantly changing weather patterns and varied land-masses of planet Earth. The availability of a lunar orbital space station or augmentation module could have helped solo operations in lunar orbit over fourteen days. What is clear is that an extended solo flight around the Moon would have been a human endurance challenge that none of these studies appears to have approached.

In 1967, BellCom – the Apollo planning contractor – completed an additional study of CSM consumable requirements to support the extended lunar missions, to supplement the reports that focused on positioning orbital sensors and experiments for orbital operations, using NASA-published 30 October 1966 CSM design mission refinement as a baseline[13] (see the Table on p. 210). The parameters for the Apollo 17 flown J-series mission are included to compare how these estimates evolved into the final flown Apollo lunar mission.

Six-man CSM
During the 1966 Congressional studies on the future of the national space objective,[14] NAA also revealed the concept of a six-man CM protracted as an advanced Earth orbital logistics spacecraft. Studies revealed that relocating some equipment and changing the couch could accommodate a support structure for four men; and in addition, the development of external impact cushioning (possible by means of a deployable heat shield and crushable struts) would allow much of the same structure and dimensions of the Block II CM to be retained. This allowed for between four to six men for logistic, lunar or interplanetary applications. This second generation CSM would greatly enhance the capabilities to meet the demands of proposed AAP programme in Earth orbit, the Moon and, initially, for manned planetary missions based on Apollo/Saturn hardware. Such applications (already described feature in the early lunar base designs, with six astronauts landing on a direct ascent trajectory in the late 1970s (see p. 211).

Lunar Flying Vehicle (LFV)
During the 1960s a wide variety of one-man and two-man small Lunar Flying Vehicles was proposed. At this time the US Army was evaluating power packs/belts for military survey and reconnaissance,[15] while others were unique to Apollo. The LM could land only in a relatively flat area with a clear approach path, and the more interesting craters, rilles and mountains were therefore out of range and not easily accessible by LRV. Early designs featured a platform over the power supply, controlled by a set of upright handlebars. Studies revealed that this was prone to tipping over, and so the designs evolved into a seated flying vehicle with a splayed undercarriage, allowing the astronauts to strap themselves in, as with the Apollo LRV.

Lunar AAP CSM extended mission consumable requirements

Mission profile	Baseline ref ('66)	28 days orbit	3 days surface	8 days surface	14 days orbit	12 days surface	14 days surface	J-series (1972) (Apollo 17[16])
Trans-lunar coast (hrs)	79	125	125	125	125	125	125	083
CSM in lunar orbit (hrs)	49	672	86	206	350	302	350	147
CSM lunar orbit crew	1	2 (no LM)	1	1	2 (LPM)	1	1	1
LM on surface (hrs)	49	n/a	72	192	n/a	288	336	74
Trans-Earth coast (hrs)	88	110	110	110	110	110	110	067
Mission duration (hrs)	216	907	321	441	585	537	585	301

Estimated total CSM consumables required, including contingency allocations

Oxygen (lbs)*	640	1,976	834	1,060	1,341	1,240	1,331	990
Hydrogen (lbs)	56	193	75	99	128	118	128	87
Food (lbs)	46	152	66	76	98	84	88	70†
LiOH (CO₂ absorption) (lbs)	80	264	115	132	170	147	154	125†
CSM dry mass (lbs)	23,562	27,900	24,000	24,600	25,900	25,100	25,200	26,324

* Including breathing supply, cabin repress after EVA, leakage, and power generation in fuel cells for producing electricity and water
† Estimated weight in absence of official data at time of publication

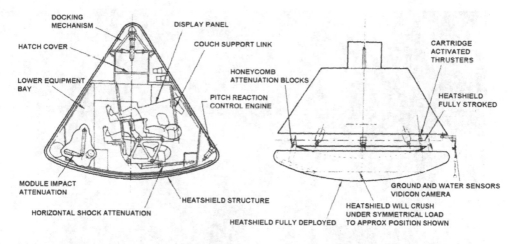

An NAA drawing of a conceptual six-man Apollo CSM to support advanced Apollo Applications Program activities in Earth orbit and towards the Moon. (Courtesy North American Aviation.)

The fabrication of the LFV would, it was suggested, use existing hardware and systems wherever possible, be light, and capable of folding in the payload envelope of the LM, and be deployable by a pulley system similar to that evolved for the LRV. Studies placed the dry weight of about 300 lbs with an additional 300 lbs of propellant, 370 lbs of payload, and a 380-lb suited astronaut. The total mass was approximately 1,400 lbs. Designs featured replaceable RCS tanks for resupply sorties, four throttleable engines using hypergolic propellants transferred from the redundant LM descent stages (an estimated 800 lbs surplus), and a gimbal steering system. Deployed from Quadrant IV of the LM, the astronaut would lower the unit when preparing for flight. The studies indicated that the unit could be carried or dragged to a launching area about fifty feet from the LM – the launch 'pad' being a deployed fabric matting which would prevent surface debris from damaging the flyer or astronaut when the engines were ignited.

Fuel transfer was enabled by connecting two hoses (oxidiser and fuel). A test flight to 200-feet altitude was planned for each new vehicle. The payload capacity included twin payload racks for scientific instruments or return samples. To return to the lander, the astronaut had to deploy a second launch 'pad' and drag the vehicle onto it to begin his return trip. Studies indicated that a cruising height of 2000 feet and a round trip of ten miles was feasible. Studies of using a 'hopping' trajectory by firing the engines to fly a modified ballistic trajectory, revealed that very little propellant was saved over direct flights, and if the pilot's approach was wrong he risked 'a high-velocity impact.' One of the last studies for an LFV for Apollo[17] estimated that a FY 1970 start (1 October 1970) would provide an LFV for an April 1972 lunar flight (Apollo 18 – originally Apollo 16) and would cost $37.5 million (1969–1974).

One concept for a Lunar Flying Vehicle was this 1967 'pogo' rocket-powered platform, seen here under evaluation at NASA's Langley Research Center in Virginia. (Courtesy NASA via the British Interplanetary Society.)

An 180-lb one-man Lunar Flying Vehicle fuelled by transferring 300 lbs of residual LM propellant from the descent stage after landing. It would have been capable of carrying one man and 370 lbs of scientific equipment or, on a rescue mission, two astronauts without the equipment. (Courtesy Grumman Aircraft Engineering Corporation.)

Lunar Cargo Vehicle (LCV)

In 1968, studies were conducted on the concept of a low-cost unmanned cargo delivery system that improved the payload delivery mass to the surface over the family of proposed LM vehicles from 1975.[18] Early designs featured an automated landing vehicle launched by either an Intermediate Saturn (two stages) or upgraded Saturn V, the diameter being the same as the S-IVB (21.7 feet), with capacity for 35,140 lbs of cargo. Using several Apollo and Saturn sub-systems to complete the trajectory to the Moon, enter lunar orbit, and then land automatically, this would be the first of a new generation of lunar landers representing the next generation after the LM. From 1980 there was a proposal to add a six-man CM with a 23,900-lb Earth return stage and service support package to a 45,900-lb landing stage and a 33,500-lb lunar orbit stage. This would provide a landing vehicle for six astronauts, and storage on the moon for 180 days. An unmanned version of this, also for use from 1980, featured a one-way unmanned lander, based on the manned version, but with 52,750 lbs of cargo instead of the CM and return capability.

Multiple EVA operations

Apollo surface exploration provided for just one EVA for the first landing, two EVAs for the H series, and three EVAs for the J series. For AAP missions beyond those planned through Apollo 20 (the K series?), multiple EVAs of four to ten excursions were suggested. (The planning, execution benefits and disadvantages of such a large EVA programme are beyond the scope of this work, but are covered in more depth in the author's forthcoming book, *Giant Strides: the Techniques of EVA*, Springer–Praxis, planned for 2003.) With the data contained in the present volume, it must be considered how these extended EVA programmes might have been approached. Ten two-man EVAs (6–8 hours) would have been a strenuous work load during a two-week stay on the Moon. During Apollo, three EVAs back-to-back on successive days were tiring – not as a result of the reduced gravity, but because of the minute detailing of the timeline, defined objectives and the ever-present lunar dust. In just three days, the clearing of dust from equipment added an extra task to the astronauts' already tight schedule. The dust was everywhere – including inside the LM cabin.

Therefore, on these AAP multiple EVAs, the provision for a flexible EVA time-line allowing for changes in real time would be advantageous, and would be similar to changing the flight planning on Skylab's last mission, during which there were items that had to be done for mission safety and experiment performance parameters, although there were others that were a 'shopping list' of secondary tasks that would be achieved as time allowed. The Skylab crews also found that personal time and housekeeping required uninterrupted periods of activity and this would also probably be reflected during extended stays on the lunar surface. During a stay of 10–14 days there would be no rush to achieve everything in the early phase of the mission. There would be a defined end to the mission, priorities, and contingency – but as the Soviets learnt on Salyut and Mir, there could still be time in the near future to complete the assigned task.

If the crews were to analyse samples on site then this would need additional time,

Various modes of transporting astronauts across the lunar surface were evaluated for the Apollo programme. Here a suited engineer evaluates a lunar motorbike in Building 29 at MSC. He is attached to a hardness device that removes ⅚ of his Earth weight and that of the bike, allowing an accurate ⅙-g evaluation of the device. Other tests were performed in flying the KC-135 aircraft in parabolic curves to reproduce a ⅙ gravity environment in short 20–30-second bursts.

and it is unlikely that EVAs would be carried out back-to-back by the same astronauts. The provision of a rest day, or rotating EVA teams are procedures followed on Shuttle EVAs, and this would be no different on the Moon to maximise work and rest cycles to ensure that all objectives are achieved first time. The provision of an airlock, and spares for failed equipment, would also help alleviate the dust problem – to some extent.

Mobile Laboratory (MOLAB)
In a move to expand the range of exploration on the lunar surface, NASA initiated the Apollo Logistics Support System that featured the Apollo CSM/LM, the LM versions (Truck/Shelter/Taxi), the LFV, and a large pressurised roving vehicle called MOLAB. Several aerospace companies participated in two contract studies worth $8 million each during the period 1964–1966. A team headed by Bendix

A January 1967 evaluation of the Mobility Test Article – a full-scale model of the mobility system for the proposed Mobile Laboratory (MOLAB). Tests were conducted at the Aberdeen Proving Grounds and the US Army Proving Grounds at Yuma, Arizona, for NASA's Marshall Spaceflight Center. The vehicle was built by the Bendix Corporation for evaluating the basic mobility concept. (Courtesy NASA via the British Interplanetary Society.)

Corporation, Lockheed Missile and Space, (which designed the crew cabin) and Hamilton Standard Division of United Aircraft (life support systems) produced one design, while Boeing undertook a similar study. The study contract also evaluated possible missions that such a vehicle might conduct across the surface by two men in a fourteen-day 250-mile traverse, and have the contingency capacity of 21 days' maximum life support without resupply. The upper weight limit was set at 6,500 lbs, allowing it to be delivered by the LM truck, but also allowing for enlargement so that it could to be delivered by future logistics vehicles in the late 1970s or early 1980s. These vehicle were, in fact, designed for post-Apollo extended surface operations in conjunction with early lunar bases.[19]

The Bendix design featured a horizontal cylindrical cabin, 12.5 ft long x 7.3 feet in diameter with a forward crew compartment of about 410 cubic feet and a rear airlock facility of 125 cubic feet. The internal atmosphere was 100% oxygen at 5 psi, with 450 lbs of hydrogen–oxygen for LSS and 650 lbs for the fuel cells, with a roof mounted radiator that dissipated heat from the ECS and fuel cells and also served as a thermal and micrometeoroid shield. Primary power would be derived from the three 2.5-kW fuel cells, providing 700–800 kWh for the duration of the trip. Secondary power designs included RTGs for periods of unmanned storage and batteries system redundancy. Crew access to and from the vehicle would have been via two rear hatches through an airlock system. The rear section would also contain stowage for surface equipment, two LFVs, and modular ALSEP experiments. An overhead docking system would be used for the transporting and docking manoeuvre on the way from Earth, but was not planned for use during normal operations on the surface, although it did provide for an escape hatch.

An early model of the 1964 Bendix MOLAB design on top of an LM descent stage and with its wheel system stowed for landing. It would have used a deployable ramp, to be offloaded by remote control several months before the astronauts arrived in a second LM. (Courtesy Bendix Corporation Systems Division.)

Grumman's design for a Mobile Laboratory, depicted in this Lunar Base Simulator, featured flexible spring wheels for negotiating rocky terrain, and a rear stowage compartment that contained scientific experiments and survey equipment, and space for collected lunar samples, thereby allowing more volume inside the pressurised living compartment. (Courtesy Grumman Aircraft Engineering Corporation.)

This allowed for a shirtsleeve environment in the crew compartment, which would feature side-by-side driving controls and seats. Depending on available power and volume, crews provisions would also include sleep stations, possibly a small galley over the freeze-dried option flown on Apollo, a waste management system, communications and navigation sub-systems, a remote television control station, a science station for analysis of samples taken during the trip, and pressure suit maintenance and stowage facililties. There would also have to be room for exposed films, retrieved samples, and crew waste products. The scientific payload capacity was quoted as 750 lbs.

About a thousand combinations of design were evaluated before the final one was selected, based on a rectangular chassis 16.5 feet long 8 feet wide, with four metal wheels of about 80 inches diameter and with inner metal rings supported by metal spokes. Studies revealed that the six-wheel and eight-wheel designs offered no better surface adhesion or mobility than did a four-wheel design (the Apollo LRV would have four wheels, and the Soviet Lunokhod would have eight wheels), and the track systems were too complex to adapt to this design. The fully deployed wheels had an overall length of 30 feet, and the height of the vehicle, from the surface to top of the radiator, was about 12 feet.

Observation was by means of four large windows (7 feet off the ground), with

MOLAB studies were terminated due to budget cuts in the late 1960s, but not before this strange LM design was proposed. An unmanned LM ascent stage would have had wheels attached, and would have landed in advance of the exploration crew, who would use the Mobile LM (MO-LM) for several weeks' surface exploration. (Courtesy Grumman Aircraft Engineering Corporation.)

As the MOLABs were being phased out, a simpler and unpressurised design – the Local Scientific Survey Module (LSSM) – was under evaluation. This 1967 Bendix concept of the one-man vehicle could carry 600 lbs of scientific equipment to up to five miles from the LM Shelter, and had a capability of up to 125 miles travel around the site. A second astronaut could ride in the rear cargo space if required. (Courtesy NASA via the British Interplanetary Society.)

additional television cameras at the front – one over the rear door, with internal monitor screens, and three inside for crew monitoring. There was a steering capability to turn the wheel to change direction, with each wheel being driven independently by means of electric motors in the wheel hub. If this failed, then differential steering would be used, as with a tracked vehicle. If a power-pack failed, the system could route power from the other wheels, and up to two wheels could fail before the MOLAB became undriveable.

Due to budget cuts in 1965, MOLABs never progressed beyond the design concept. What emerged in 1966 was the non-pressurised Local Scientific Survey Module (LSSM)[20] which offered a deployable vehicle from the LM or LM Truck measuring 15 feet in length and 8 feet in width, with four wheels each 45 inches in diameter. Weighing 1,100 lbs, this design by Bendix also incorporated much of the MOLAB design technology in traction and steering, but removed all of the crew compartment pressurised shell and habitation facilities. The LSSM could carry 2,000 lbs of equipment locally to the LM, and not undertake long traverses that were planned for the MOLAB, although unmanned applications were designed to allow for relocating the vehicle between manned landings. LSSM would be battery-powered, and resembled more of a dune buggy than a mobile laboratory. This type of design was the genesis of the Boeing LRV used on Apollo 15–17.

Moon base

The concept of creating a permanent or semi-permanent base on the lunar surface is as old as the desire to explore the Moon, and has produced countless renditions, studies, plans and concepts for well over fifty years. Concepts proposed during the Apollo era are far too numerous to detail in this current volume, but the early bases were planned to be simple structures – possibly inflatable, and certainly connected with the range of options open from adapting the LM – and were only temporary habitats designed to last for about fourteen days to a couple of months. Related studies of adapting Saturn technology also included the provision of a soft-landing system for an S-IVB after it performed TLI, instead of crashing on the Moon. The stage could remain upright for habitation, but artwork usually depicted the stage being lowered to the surface horizontally, and an automatic vehicle covered it with lunar material for thermal and particle impact protection. The arriving LM crew would then outfit and use the S-IVB in much the same way that the original wet-stage OWS was planned to be used in Earth orbit.

Other options included the use of the Saturn LM Adapter as a temporary structural shell for a lunar outpost – called a Mini Base – instead of jettisoning the panels early in the mission. Here they formed the uppermost part of an adapted LM descent stage carrier vehicle.[21] For manoeuvring prior to descent orbit insertion, the top of the configuration also featured an inverted SM, but without RCS quads or high-gain antenna, and offering 20–30 tonnes of propellant for payload delivery to lunar orbit. Separating from the SM module, the SLA Mini Base descended to the surface to await the arrival of the two- or three-man crew, depending on the mission. The onboard systems in the Mini Base could, it was calculated, support two men for fourteen days or three men for ninety days. Longer missions (perhaps up to 120 days) could be completed using the structure, which stood 26 feet tall fully deployed on the modified descent stage. A pressurised environment was provided at 5 psi with consumables available for the supply of 8 lbs of water, 2.3 lbs of food, and 2.04 lbs of oxygen per man per day. Apollo-type fuel cells provided additional water and heating, and studies were also completed on adequate solar flare radiation shielding. A duel Saturn V launch profile was requireed for this system–one for the Mini Base and lunar orbital station module, the other delivering the CSM, LM and 3-6 astronauts. In this concept six dual missions could be launched over a ten-year period from 1975 at a cost of $5.25 billion. When this concept was finally published in September 1970, the demise of Apollo was all but finalised, and the comment in the text that 'over half of the total program cost is associated with the procurement of the twelve Saturn V launch vehicles required' is notable in that it was at the time when NASA halted Saturn V production to reduce budgets.

Lunar orbit space station

A version of the SLA Mini Base was adapted for use in lunar orbit. On missions up to 14 days it was expected that the lone CM Pilot could remain in the CM awaiting the return of his colleagues, but on longer surface excursions it was advisable to dock the CSM to an orbital outpost to provided additional facilities for the crew, hibernate the systems, or conduct expanded orbital observations to supplement the surface programme. It was planned to send an S-IVB OWS-type wet-workshop to lunar orbit early in the AAP, but this gradually disappeared, to be replaced by smaller LM-scale orbital labatories and modules supported by the CSM for temporary occupation rather than for an extended mission, with crew rotation in an S-IVB cluster – essentially a lunar-orbit Skylab!

Uprating the Saturns
Describing the numerous and various versions of the Saturn family of the 1960s would require the writing of another book, and here we shall describe only the range of improvements to the Saturn 1B and Saturn Vs (illustrated on p. 221). Other vehicles studied to support Apollo Earth orbital lunar and interplanetary missions were the Titan III-C and III-D, and the Agena and Centaur upper stages and, eventually, nuclear upper stages.[22] In the mid-1960s the Titan/Saturn 1B payload delivery capability was about 20,000–35,000 lbs to low Earth orbit (LEO), while the standard configuration Saturn V could deliver 250,000 lbs to LEO. It was estimated that a post-Saturn booster (undefined) could deliver 500,000 lbs or more to LEO,[23] and it was therefore proposed to uprate the Saturn and the Titan to 'fill the gap' between 40,000 lbs and 250,000 lbs, and perhaps go as high a 350,000 lbs.

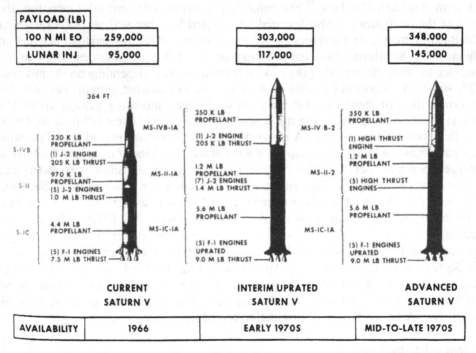

Saturn V growth concepts.

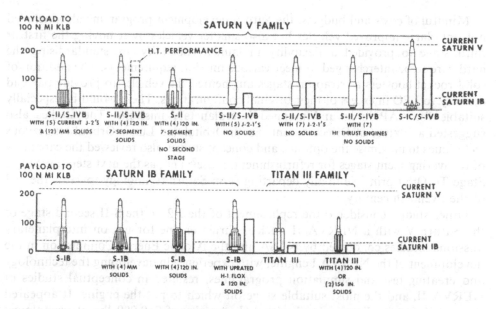

Candidate medium payload launch vehicles, mid-1960s.

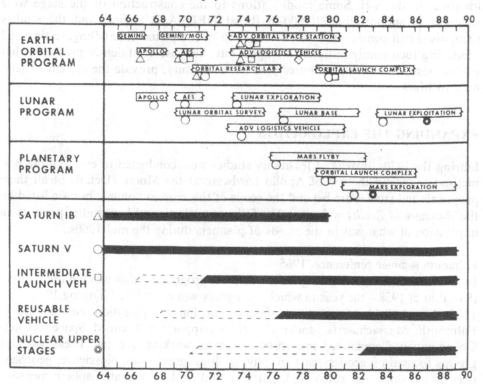

A 25-year integrated launch vehicle programme planning chart.

Mindful of costs and budgets, the Saturn development programme also included not only large improved vehicles but also smaller vehicles by removing the first or third stage to provided a flexibility in launch capability, by standardising the hardware to be interchanged to meet various mission requirements. The addition of solid rocket boosters as strap-on stages augmented the vehicles to provide payload delivery to LEO in excess of their original configurations. These would be especially suitable for AAP lunar and manned interplanetary missions. The studies also suggested a mixture of Direct Ascent, Earth Orbital and Lunar Orbital rendezvous techniques to maximise the options, and concept studies also reviewed the capability of recovering spent stages for refurbishment, reuse[24] and, as the next step, the Single Stage To Orbit principle finally departing form Saturn technology towards the end of the twentieth century.

Other studies considered the replacement of the J-2 on the S-II second stage of the Saturn V with a NERVA II nuclear thrust engine for use on interplanetary missions. Studies conducted by NASA and the Atomic Energy Commission in the development of the NERVA I engine, with experience in developing the technology and creating test and simulation programmes, resulted in conceptual studies of NERVA II, and the most suitable stage on which to put the engine. It appeared that at 33 feet in diameter and with LH capacity of 250,000 lbs, it was almost identical to the S-II. Some modifications to the construction of the stage were required to support the 230,000-lb thrust NERVA II engine, and the studies recognised that considerable work needed to be completed in handling, testing and preparing such configurations for flight; but they also confidently predicted that such a design would, in the latter part of the century, provide the system to send men to Mars.

EXPANDING THE EXPLORATION

During the 1970s, dozens of feasibility studies were conducted to evaluate the best method to extend the use of Apollo hardware at the Moon. (Details of all these proposals and studies are beyond the scope of this current volume, but are listed on the *Romance to Reality* web site.) The following selection of key studies presents an impression of what was in the minds of planners during the mid-1960s..

Falmouth summer conference, 1965
By 1965, NASA was benefiting from a funding programme that was ten times larger than that of 1958 – the year in which the agency was established. During 19–31 July 1965, NASA held a summer conference for lunar exploration and science in Falmouth, Massachusetts, under the NASA-appointed Manned Space Sciences Coordinating Committee, comprised of seven working groups. These working groups focused on geodesy/cartography, geology, geophysics, bioscience, geochemistry (mineralogy and petrology), particles and fields, lunar atmosphere measurements, and astronomy, and each group presented a report that was summarised by a coordination committee and published as NASA SP-88 later that year.

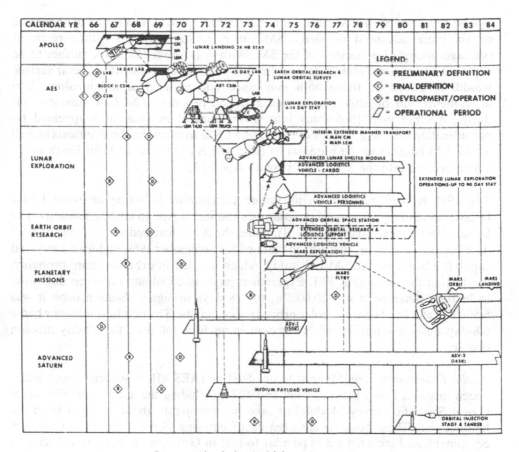

Integrated mission/vehicle programme.

This report identified major requirements for the lunar exploration programme for the decade following the initial landing by Apollo astronauts. These studies – some of which overlapped – resulted in five 'mission' types:

Apollo missions These were identified as the initial lunar landings that achieved the programme goal set in 1961. They featured a surface stay time of one or two days, and focused on an area immediately local to the landing area, with the first priority being the collection of 'the greatest number and variety of samples', and the second to set up a small experiment package (then termed the (Apollo) Lunar Surface Experiment Package – LSEP (later ALSEP). This evolved into the H series of Apollo missions flown as Apollo 12–14.

Lunar Orbiter These were both the unmanned Lunar Orbiter missions and the manned Apollo lunar orbital missions, although carrying only 'simple diagnostic experiments', and supplementing data collected from lunar surface experiments. The manned orbital programme would be expanded by the Apollo Extension System.

Apollo Extension System Manned Lunar Orbiter (AES-MLO) consisted of about five or six more advanced manned CSM lunar orbital missions, with experiment packages mounted in Sector I of the SM. With only about 1% of the surface to be explored 'in the near future', major collection of data across the whole lunar surface would be gained by flying both unmanned and manned orbital missions. The instruments and objectives were similar to the proposed LM Lab missions, but during a subsequent 1966 study into these types of mission there appeared no consideration for using Sector I of the Apollo CSM to carry these instruments as well as the LM (similar to the proposal to carry the ATM in the CSM in Earth orbit, which evolved into the SIM-bay fly-by on Apollo 15–17).

The 1966 report[25] revealed an integrated programme following on from Lunar Orbiter and Survey unmanned programmes and initial Apollo landing missions. The Lunar Orbit Survey Mission (LOSM), under the AAP, featured seven flights of 5–28 days' duration, with the first launch planned for September 1969, and with a payload capability based on the LM Lab study, which was left in orbit (or more probably, crashed on the surface) at TEI. Experiment payload capability for seven days low-inclination lunar orbit was 20,000lbs, for 28 days in high-inclination orbit it was 5,500 lbs, and for 28 days in polar orbit it was 4,600 lbs. The first four missions had a 6.8-day mission duration, while the remaining four followed the 28-day mission profile.

Apollo Extension System Manned Lunar Surface (AES-MLS) featured about half a dozen missions flown between 1969 and 1974, including the use of a smaller rover (Local Scientific Survey Module) to take the astronauts about 10 miles from the landing site. These rovers were capable of carrying about 1,300 lbs of support equipment, and included a drill (similar to that in Grumman's study) to collect core samples from 10 feet below the surface. Unlike the early Apollo missions, these flights could deliver more than one pack of experiments to the surface, and could return about 650 lbs of lunar samples. These missions would feature two Saturn V launches to deliver the manned LM and unmanned LM carrying support hardware that could also feature a lunar flying vehicle. Local mapping would be a dominant feature, together with sample collection and analytical selection of material returned to Earth. For these longer excursions away from the landing site accurate automated positioning systems would be required to accurately pinpoint the landing vehicle and the exploring astronauts, who would be tracked by automated camera at each station stop (as was the case during the J missions).

Post-AES missions These covered a five-year period from 1975 through 1980, and required a three-man landing crew to support 500-mile traverses across the equatorial region of the Moon, in expeditions lasting up to two months. To achieve this, a more advanced and pressurised vehicle would be required – termed Mobile Laboratory (MOLAB) in several contractor studies. This featured the capability of supporting a minimum of 500 miles traverse, with provisions and facilities to support a three-man crew, and did not require a return to the starting point (indicating an expedition between major lunar base sites). Major construction featured the fabrication of a lunar base, allowing permanent expedition crews to remain for a year on the surface in rotations of perhaps three to six months, large drilling equipment to the depths of about 1,000 feet, and the operation of a large radio telescope.

Exploration plans, November 1966
In February 1966, NASA established a Lunar Exploration Working Group, which over the ensuing nine months established a set of candidate programme options that identified objectives and system requirements for a protracted programme of lunar exploration over 15–20 years. Although this report was published before the major budget cuts which would loom over the following four years, and also before the tragic pad fire of January 1967, it forecast a difficult road ahead; but the summary stated that the findings did not represent official policy or firm intent to proceed, although an effective exploration plan could be gained by a careful matching of 'unmanned systems and Apollo derivatives within a reasonable budget'. These 'exploration plans' featured a four-phase uninterrupted programme in the post-Apollo period.

Phase I Apollo, Block II/III Lunar Orbiter and *Block II Surveyor* (1968–1975) At the time, the upgraded Lunar Orbiter was to map the entire surface of the Moon between 1969 and 1971, supplementing the date with manned Apollo orbital surveys. A series of ten Block II Surveyors would land in areas too remote or too hazardous for the LM.

Phase II Lunar AAP (1970–1973) This featured one lunar AAP each year, employing two Saturn Vs The first launch would consist of a two-man CSM to deliver the unmanned LM Shelter and return to Earth, and a second Saturn V would then launch a three-man crew towards the Moon. Two would land in the LM Taxi near the Shelter for fourteen days of surface operations, while the lone third astronaut would conduct a fourteen-day orbital mission before all three crew-members returned to Earth.

Phase III Mobile Exploration (1974–1976) A combination of two Saturn V launches – one standard and one uprated Saturn V (also known as the 188% Saturn V) – featuring stretched stages and four solid-propellant strap-on rockets for the first stage. The uprated Saturn would deliver a large mobile rover (in this study called MOBEX – MOBile EXploration) that could weigh as much as 10 tons. The standard Saturn would launch three astronauts to lunar orbit, where they would mothball their CSM and transfer to an uprated LM Taxi capable of carrying all three astronauts to the surface to land near the MOBEX on a 90-day excursion across the surface, with station stops of several days instead of the hours performed on Apollo, and end the traverse back at the Taxi landing site to rendezvous with the orbiting CSM to reactivate it for the return to Earth.

Phase IV Temporary Station (1977–1980) This revived the Direct Ascent approach, with a new lander replacing the retired original CSM and LM system, and using three uprated Saturn V launches that would deliver the MOBEX and landing stage. The second would deliver an unmanned two-deck Lunar Laboratory Module (LLM), while the third would launch the crew on a Direct Ascent trajectory in a modified six-man CM capable of seven-day independent support, and a modified 'service module' featuring two stages to be used for the return to Earth, launching the CM from the surface after the exploration phase had been completed. Upon landing, the crew would transfer to the LLM habitation module and use the MOBEX for exploration by three of the astronauts on two or more traverses. Total stay time was planned at 180 days, and featured Saturn/Apollo-derived hardware based on the planning for initial Mars spacecraft. This further application of Apollo hardware was used in an attempt to emphasise a continuing programme using a common stage design and keeping costs down, while also establishing a temporary lunar station that could evolve into semi-permanent bases.

Summer study, 1967
During 31 July–13 August 1967, NASA held a further conference on lunar exploration and science at the University of California at Santa Cruz. The objective was to help define sites for the Apollo landings, divided into three 'Early Apollo' (EA) sites and nine AAP missions, of which two were manned CSM-only lunar orbital missions. The plan was to fly the AAP lunar missions over a five-year period after the first three early landings. The latter missions also included a two-man LSSM rover (launched by a separate Saturn V) allowing excursions of 12–30 miles, and one-man Lunar Flying Units (LFU) (0.5–1.5 mile range). The LSSMs would be used automatically for sampling excursions on long traverses in between manned missions. There would also be deployment of modular ALSEPS and sub-satellites ejected from the orbiting CSMs.

AAP Mission 1 (Year 1) and *Mission 7* (Year 3 or 4) were the orbital survey missions; *Mission 2* (Year 1 – Copernicus) used an ELM with two LFUs but no LSSM. The three-day surface duration included four LFU flights, two of which were to explore the central peaks of the crater Copernicus; *Mission 3* (Year 2 – Davy Rille) was similar in profile to Mission 2 but would explore a rille feature, while *Mission 4* (Year 2) visited the rim of Copernicus. On *Mission 5* (Year 3 – Marius Hills) an LMT would first be landed, followed by an ELM which carried the first LSSM and fuel for two LFUs. After crew departure the LSSM would conduct an automated sampling traverse towards Aristarchus, and on *Mission 6* (Year 3 or 4 – Aristarchus) a new pair of astronauts would land near an LM truck, which would deliver a fresh LSSM. They would retrieve samples collected by the LSSM from Mission 5, and then use the new vehicle for three traverses and the LFUs for five sorties. At the end of the manned phase, the LSSM would set off for Hadley Rille. *Mission 8* (Year 5 – Alphonsus) included advanced landers with payload modules, about a week on the surface, and eight LFU and five LSSM excursions, after which the LSSM would complete an automated traverse towards Sabine and Ritter *Mission 9* (Year 5 –

Sabine and Ritter) would not have a new LSSM, and would rendezvous with the Mission 8 vehicle to collect and retrieve samples and to follow a profile similar to Missions 5, 6 and 8.

The conference suggested fourteen scientifically interesting sites – some of them offering further study before they were attempted. These included sites at both poles, the crater Tycho, and Hadley Rille, both later assigned to Apollo, with only Hadley actually visited by a crew, in 1971. The report into the conference (NASA SP-167) noted the advantage of assigning a crew-member who had professional geological field training. Interestingly, one participant at this conference was scientist-astronaut and professional geologist Jack Schmitt, who five years later achieved that goal as LMP on Apollo 17.

Studies during 1968

Two studies[26] during 1968 are cited for their post-Apollo planning in the early 1970s, at a time when there was still the expectation of a boost in the NASA budget to fund an official follow-on to the early Apollo series. The Earth orbital AAP programme had suffered from restrictions in the budget from 1965 – due in part to the escalating Vietnam War – and it was hoped that perhaps an early end to hostilities in south-east Asia would allow the new President (Nixon, elected in November of that year) to restore at least some of NASA's funding request from the next FY to plan what would follow Apollo.

BellCom proposed a twelve-mission four-phase plan to 'fit the gap' that it had identified between the funded Early Apollo mission and the delayed lunar AAP programme, and continued with the 'assumption that lunar exploration will be a continuing aspect of human endeavour'. In this study, *Phase 1* (1969–1971) included five missions that follow the profile that was adopted for the G and H missions, while *Phase 2* (1972-1973) included four missions with an Extended LM, the capacity for six surface EVAs, and the first LFU on Missions 8 and 9 to more scientifically interesting sites: *Phase 3* was a 1974 CSM 28-day lunar polar orbit mission with an LM&SS developed from the AAP Earth Resources package; and *Phase 4* (1975–1976) would feature the dual-launch Saturn V concept LPM with an 8,000-lb payload capacity and augmented ELM that would deliver LFUs. Two dual-launch missions (the first Saturn V with the manned CSM delivering the automated LPM, and the second Saturn delivering the ELM crew) were proposed – the first to explore Hyginus Rille or the Davy crater chain, and the second to be targeted the Marius Hills.

The second propsal – by NASA – featured three options ranging from $725 million to $1.09 billion, each of which would involve three missions flown after completion of the initial landing using extended-duration LM and hardware.

Option A ($725 million) would see three missions launched between the fourth quarter of 1971 and the fourth quarter of 1972, staying three days on the surface using an ELM and two one-man LFUs. The astronauts would deploy advanced ALSEPs, and would complete four to six EVAs.

Option B ($745 million) featured an ELM-B design lander that used solar cells for electricity, cooling radiators, and cryogenically stored oxygen for converting to breathable oxygen. These three missions, flown between mid-1972 and mid-1973, included a six-day surface stay, six to ten EVAs, three to four LFU sorties, and, on the second mission, the use of a solar telescope on the surface.

Option C ($1.09 billion) – the favoured option – was to be initiated early in 1973 by the landing of an automated LPM/Shelter with a 5,000-lb payload that included the habitation shelter, two LFUs and LRV with dual manned and unmanned capability, a 12-inch aperture telescope, advanced ALSEP, and associated surface sampling tools. The manned landing again featured an ELM–B module arriving near to the LPM shelter. The first crew also would deliver an additional 750 lbs of payload, including a third LFV in case the LPM should be out of walking range. They would stay for up to 14 days, allowing for twelve to twenty surface EVAs, sixteen LFU excursions, and eight LRV traverses, as well as conducting astronomical observations and sample analysis inside the shelter. The LRV would have been programmed for a return-to-site automated traverse between the manned periods. About six months later, the second crew would land for about a one-week stay, allowing for six to ten EVAs, four LFU flights, four LRV expeditions, and an experiment to extract water from lunar samples. Six months later, in early 1974, the third and last crew would arrive with a solar astronomy experiment. They would also stay for about a week at the site, use the shelter for a third time, conduct six to ten EVAs, four LFU and four LRV excursions, and then set the LRV on a second remote expedition before departing.

From 1969 the number of similar studies declined as the Apollo programme did not receive the expected funding to expand the programme past Apollo 20; and, as we have discussed, it was reduced by three missions to allow funding and hardware to be used in the only AAP space station missions, which became Skylab. Studies were continued into the early 1970s but although the Saturn V was featured in launch-system studies, the Space Shuttle increasingly became the focus of America's access to space to create a large space station. It was hoped that the Shuttle would become the major US launch system for routine and economical access to near Earth orbit – and the launching system for unmanned planetary missions.

1   Details of these events can be found in *Where No Man Has Gone Before: A History of Apollo Lunar Exploration Missions*, William David Compton, NASA SP-4214, 1989, Appendix 3, Lunar Exploration Planning, 1961–1967, pp. 331–348.
2   Lunar logistics planning; memo from Joseph Shea to Robert Seamans, 24 November 1962. NASA JSC Apollo collection.
3   Some of the industry proposals research were *Lunar Logistics System*, Volume 1, Technical Proposal, Systems Division, Westinghouse Defence Center, 20 August 1962 (SMD-15023); *Proposal for Payload Performance Study for Lunar Logistics System*, Aerospace and Systems Department, American Machine and Foundry Company, General Engineering Division, 17 August 1962 (AMF I.Q. 4006). NASA JSC Apollo collection.
4   *Technical Proposals for Payload Performance Study for Lunar Logistics System*, 20 August 1962 (PDR-344-1). NASA JSC Apollo collection.
5   *Interim Study Report – Lunar Logistics Vehicle*, 10 December 1962, (D2-100042). JSC Apollo collection.

6 A detailed study into the LM design considerations required to support a crew, the training programme, and flight experiences of LM astronauts, is currently under preparation by the author.

7 *LEM Utilization Study for Apollo Extension System Missions*, Final Report, 15 October 1965 (ASR-378-2), Copy No. 24 (six volumes plus cost analysis). NASA JSC Apollo collection.

8 *LEM Utilization Study*, Volume II, LEM Lab for Lunar Orbit Missions.

9 *Systems Study for Manned Lunar Reconnaissance Mission*, NASA Statement of Work Memo (internal), 26 December 1962, original signed by Charles Mathews, Space Technology Division, MSC Ref 082-54). NASA JSC Apollo collection.

10 *LEM Utilization Study*, Volume III, LEM Shelter.

11 *LEM Utilization Study*, Volume IV, LEM Taxi.

12 *LEM Utilization Study*, Volume V, LEM Truck.

13 *CSM Requirements for Extended Lunar Missions* (TM-67-1012-7, Filing Case 232), D.R. Valley, BellCom Inc, 22 June 1967. Cited on Portree's *Romance to Reality* web site.

14 *Ibid*. Chapter 2, p. 285.

15 *Apollo by the Number: A Statistical Reference*, Richard Orloff, NASA SP-2000-4029, pp. 266, 276–277; also *Apollo 17 Flight Plan Change A*, 20 November 1972, pp. 4–14, Apollo 17 Cryogenic Summary. AIS archives.

16 *The Man-Rocket*, Robert Roach, *Spaceflight*, **5** (1963) No. 6 (November), pp. 202–206.

17 *One-Man Lunar Flying Vehicle Study Contract*: Summary Briefing, Space Division, North American Rockwell, presentation materials, July 1969. Detailed on Portree's *Romance to Reality* web site.

18 *Improved Lunar Cargo and Personnel Delivery Systems*, N. Morgan and R. Johnson, NASA SP-177, 1968, pp. 173–191. Detailed on Portree's *Romance to Reality* web site.

19 *Bendix MOLAB Study*, Donald Fink, AWST, 7 December 1964, special reprint.

20 *Lunar Survey Module Design Under Test*, William Normyle, AWST, 18 June 1966, pp. 97–101.

21 *SLA Mini Base Concept for Extended Lunar missions*, SD5 70-516, Vol 1, Summary, North American Rockwell, September 1970. Detailed in Portree's *Romance to Reality* web site.

22 *Orbital Launching of S-2 Stages Studied*, Donald Fink, AWST, 24 January 1966, pp. 52–57; *Changing Concepts Key to Design of Future Saturn Family*, William Normyle, AWST, 7 March 1966, pp. 106–113.

23 *1967 Congress Report: Future National Space Objectives*, pp. 289–296.

24 *Increment Steps Towards Booster Recovery and Re-use*, Phillip Bono and T. Gordon, *Spaceflight*, **8** (1966) No. 9 (September) pp. 326–333, British Interplanetary Society, London.

25 Memo: Transmittal of Lunar Orbital Survey Missions Study briefing charts, AAP Definitions Office Missions Group, original signed by Thomas R. Kloves, 20 January 1966. From the NASA JSC Apollo collection.

26 *A Lunar Exploration Program – Case 710*, N. Hinners, D. James and F. Schmidt, BellCom Inc (TM-68-1012-1), 5 January 1968; *Reports of the Single Site Working Subgroup to the Lunar Exploration Working Group*, 22 May 1968, NASA (4 June 1968 revision). Detailed on Portree's *Romance to Reality* web site.

Exploration planning

The success of the first four Apollo manned missions (Apollo 7–10) between October 1968 and May 1969 gave NASA confidence that the fifth mission – Apollo 11 – could, at the first attempt, accomplish the goal of landing men on the surface of the Moon.

The primary goal of the Apollo programme set by President Kennedy in May 1961 had been to land men on the Moon and return them safely to Earth by the end of 1969, and it appeared that this would be achieved with about five months to spare. However, until the first landing had been accomplished, NASA had scheduled the Apollo programme for maximum safe technical progress, and at a reasonable, manageable speed. Should Apollo 11 fail to achieve the landing, then Apollo 12 and Apollo 13 were being prepared to fly later in 1969, providing a second and a third lunar landing attempt before the decade ended.

Therefore, although they were expecting success with Apollo 11, NASA had contingency plans against any failure, and had provided sufficient hardware to back up the attempt. At the same time, they had defined plans for missions that would follow the initial landing, with the hardware already fabricated or budgeted for in the programme. This effort was termed Future Apollo Lunar Exploration Planning, and involved a team of engineers headed by Joe Loftus from the Program Engineering Office, which was a branch of the Manned Space Craft Center Advanced Mission Program Office. This group was designated the Group for Lunar Exploration Program (GLEP), and was assigned the task of evaluating what could be achieved with the available hardware assigned to the Apollo programme after the first lunar landing.

The team established a series of objectives for a further nine missions after the initial landing – a programme of ten lunar landings grouped under Phase I of the proposed manned lunar exploration programme. These early objectives were established in conjunction with a variety of technical and scientific elements of NASA, its advisory committee, contractors, and the greater scientific community across the USA.

An MSC announcement dated 6 May 1969[1] stated that the initial planning for the advanced missions after the first landing had been completed, and included several 'additional flights utilising Apollo spacecraft that are essentially unchanged'. These missions (the H series) would perform a variety of experiments in both lunar orbit and on the surface, in what was generally known as the Apollo (landing) zone of the Moon. There were three or four of these missions scheduled, flying at intervals of between four and six months. A second category (the J series) included more advanced lunar missions, in which additional flexibility in the CSM and LM would allow for landings in different areas on the surface, provide for larger payloads and longer lunar duration, and provide the capability to conduct more sophisticated scientific explorations from the surface and from lunar orbit. At the time of the announcement, exact definition studies and detailed engineering work was being conducted by the primary contractors of Apollo hardware – notably North American Rockwell Corporation for the CSM, Grumman Aircraft Engineering Corporation for the LM, Bendix Corporation for the ALSEP surface experiments, ILC for the EVA suit, and Hamilton Standard for the portable life support back-pack.

The responsibility for the initial H series of missions was assigned to the Apollo Spacecraft Program Office, while the responsibility for the J series of missions under development rested, for the time being, with the Advanced Mission Program Office. Once the mission definition had been completed and implementation decisions had been reached, responsibility for the J series would shift to the ASPO. This was finally announced in a second release from MSC dated 30 September 1969.[2]

Apollo 12–20 would be focused on delivering greater scientific payloads to the Moon, with a broader scientific programme, extending the duration spent in orbit and on the surface, developing surface mobility aids – in particular, a self-propelled vehicle, more commonly termed the Lunar Roving Vehicle (LRV) – and the 'demonstration of techniques and hardware for expanded manned space mission capabilities ... to further man's progress in space.'[3]

GOALS AND OBJECTIVES IN LUNAR EXPLORATION

The overall goal of Apollo lunar exploration was stated to be 'the efficient use of Apollo hardware to achieve technical and scientific objectives under the NASA Lunar Program [detailed below], as well as achieving the overall NASA goal of maintaining US prominence in space.'

Seven technological objectives were listed for the missions:

- Increasing the scientific payload to lunar orbit and to the lunar surface.
- Developing a high flexibility in landing site selection.
- Increasing the mission duration in both lunar orbit and on the surface.
- Increasing surface mobility by the development of a self-propelled vehicle.
- Development and demonstration of advanced techniques and hardware for the expansion of US manned space mission capabilities.
- Development of techniques for achieving pinpoint landings on the Moon.
- Demonstration of a closed-loop onboard navigation capability with application for future and more advanced missions under development.

In addition, there were nine scientific objectives:

- Investigation of the major classes of lunar surface features (mare and highlands).
- Investigation of surface processes (impact, volcanism, and mountain building).
- Investigation of regional problems (mare–highland relation, major basins and valleys, volcanic provinces, major faults, and sinuous rilles).
- Collection of samples from each landing site for detailed analysis on Earth.
- Establishment of a network of surface instrumentation to measure seismic activity, heat flow and disturbance in the Moon's axis of rotation. This would determine the gross structure of the Moon, its processes, and the energy budget for the lunar interior.
- Use of high-resolution photography and remote sensing from orbit to survey and measure the lunar surface.
- Investigation of the near-Moon environment and the interaction of the Moon with the solar wind.
- Mapping the lunar gravitational field and internally produced magnetic field.
- Detection of atmospheric components resulting from neutralised solar wind and micrometeorite impacts.

SELECTING THE SITES

On 10 July 1969 – less than a week before Apollo 11 was launched to the Moon – Farouk El-Baz delivered a presentation to the Apollo Site Selection Board. The presentation was prepared jointly by the NASA Apollo Lunar Exploration Office and BellCom Inc, on the recommendations for the lunar exploration landing sites for Apollo 11 through Apollo 20.[4]

The presentation asserted that timing was critical for planning the missions that

Mission assignments based on accessibility

Landing	Apollo	Mission	Site
1	11	G-1	Apollo landing site 2
2	12	H-1	Western Mare (Flamsteed), Apollo 7
3	13	H-2	Fra Mauro
4	14	H-3	Rima Bode II
5	15	H-4	Censorinus (north-west)
6	16	J-1	Copernicus (peaks)
7	17	J-2	Marius Hills
8	18	J-3	Tycho (north rim)
9	19	J-4	Rima Prinz I
10	20	J-5	Descartes

would follow the first landing, for selecting primary and alternative landing sites, assigning the mission sequence to reach those sites, and beginning the detailed site analysis and mapping for each mission. It was stated that providing there would indeed be ten manned landing attempts, the set of initial Apollo sites should include opportunities to sample mare material from eastern and western locations, regional stratographic material such as blanket deposits (ejecta) around mare basins, a variety of types and sizes of impact crater from both mare and highland sites, morphological manifestations of volcanism in both the mare and highlands, and from areas which might provide clues into the nature and extent of processes (other than impact and volcanism) that may have acted upon the lunar surface.

From these considerations, it was suggested that both geological and geochemical objectives could be met in the early phase of Apollo lunar exploration. The desire to set up a seismic network as part of the geophysical objective would require a specific mission requirement plan to be built up, but other scientific objectives would not be compromised in achieving this. With this goal in mind, the landing sites chosen in the first step of mission planning for Apollo 11 through 20 met the scientific requirements as well as progressing lunar exploration techniques. Using a step-by-step building process, the sites were chosen largely on geophysical requirements, while ensuring mission safety, and capitalising on previous mission experience as the programme progressed. These sites were only suggestions, and from these the final candidate sites (prime, alternative and recycled mission) were later selected, as defined in the Table on p. 233.

Indeed, in the Manned Space Flight Weekly Report (MSFWR) of 28 July 1969,[5] details of the success of Apollo 11 and plans for Apollo 12 and Apollo 13 were mentioned. Apollo 12 would head for Apollo Site 7 in the western mare area (Oceanus Procellarum) near to Surveyor 3, and Apollo 13 would be targeted towards the Fra Mauro highlands. This memo also conceded that the choice of landing sites was still under discussion by GLEP but was not yet finalised – a task that Joe Loftus and his team were currently undertaking for mission planning purposes. The Loftus team also had to consider all the options between what was scientifically interesting, operationally achievable, and financially practical. According to the 28 July MSFWR, the current Apollo planning summary up to Apollo 20 was a 'tentative planning schedule' that suggested the assignments as listed in the Table on p. 235.

Target landing sites
On 23 August 1969, the GLEP met to review the list of proposals, eventually modifying the proposed candidate list presented to MSC at the 10 July Site Selection Board meeting. Apparently, there had been some concern among members of MSC as to the status of the revised list provided by GLEP, because the Apollo Site Selection Board had not processed it. The resulting memo, dated 27 August, was issued to inform MSC personnel who were working on site selection that they were to consider the revised candidate list provided by the GLEP as 'the official desire of the Apollo Site Selection Board.' The 'official' (and revised) candidate landing site list from GLEP[6] appears in the Table on p. 235.

Provisional Apollo planning

Apollo	Launch plan	Tentative landing area
12	1969 November	Ocean Procellarum lowlands (Apollo 7)
13	1970 March	Fra Mauro highlands
14	1970 July	Censorinus highlands
15	1970 November	Littrow (volcanic area)
16	1971 April	Tycho (Surveyor 7 landing area)
17	1971 September	Marius Hills (volcanic domes)
18	1972 February	Schröter's Valley (river-like channels)
19	1972 July	Hyginus Rille (linear rille–crater area)
20	1972 December	Copernicus (large impact crater area)

Official candidate lunar landing sites

Apollo	Number	Landing area	Nominal target point
12	H-1	Apollo 7	Surveyor 3 area
13	H-2	Fra Mauro	Within Lunar Orbiter high-resolution coverage
14	H-3	Rima Bode; alternative, Littrow	Within walking distance of crater and rille
15	H-4	Fresh impact crater in the highlands – candidates in order of priority: Censorinus, Lalande, Abulfeda, Davy	Within walking distance of crater ejecta
16	J-1	Tycho	Within the field of view of Surveyor 7 television coverage
17	J-2	Copernicus	1–3 miles north of central peaks
18	J-3	Descartes	Approximately 40 miles north of crater
19	J-4	Marius Hills	Valley floor between four hills
20	J-5	Hadley Appennine	Near the source of the rille

Their final recommendations, published in December 1969, are summarised in this chapter, showing what the mainline Apollo programme hoped to achieve after Apollo 12.

Apollo 11 had been targeted for a relatively flat area of a mare region, primarily due to safety criteria. The surface duration was only short, with a single two-hour EVA near to the lander (no more than 100 feet away), involving the crew deploying simple experiments and completing a basic surface exploration timeline. Analysis of the scope of the later missions reveals a broad range of locations across the lunar surface, consisting of a variety of surface features all aimed at piecing together the geological history of the Moon.

The December 1969 publication identified the primary locations that mission planners and scientists were interested in reaching on the nine subsequent Apollo lunar landing missions. In planning these missions, consideration had to be made about the resources available and the lead time for hardware and training. Results and experiences from flown missions would also have influence in future mission planning and targeting of forthcoming missions. In the Autumn of 1969, this was the long-range planning for Apollo 12–20, although there was already an awareness that the programme was under strain from reduced budgets and hardware restrictions, and that circumstances were likely to change. But with the success of Apollo 11 at the Apollo Site 2 on the Sea of Tranquillity in July 1969, the consequences of this pressure were dismissed from the planning of future flights – at least for a while.

In the event of a mission recycle, the alternative sites for Apollo 11 were listed as Apollo Site 2 (Tranquillity), 3 or 5. For Apollo 12, it was Apollo Site 5; and for Apollo 13 through 16, plus 18 and 20, the recycle site was determined as Apollo Site 6R. There was no recycle site for Apollo 17, and for Apollo 19 no recycle site had been determined at the time of the report. Brief descriptions of the proposed landing sites are listed below (and suggestions for further reading into the geological selection of Apollo landing areas can be found in the Bibliography).

H mission landing sites

Mission H-1 (Apollo 12) Landing site: Apollo 7 – Location: 23°23′31″ W 2°58′56″ S

This site was located within the relatively old Mare Imbrium material, and contained a distribution of larger subdued craters of 60–182-foot-diameter, as well as lower-density craters of the 15–60-foot-diameter range. It was also the site of the Surveyor 3 unmanned spacecraft that landed on 19 April 1967. As a result, the primary objectives for this mission were to sample a second mare location for comparison with Apollo 11 samples and to obtain photographic data and samples from the Surveyor landing site to determine the variability in soil composition and the age of Mare Imbrium. It was also an excellent opportunity to study the effects of prolonged exposure of a spacecraft to the lunar environment, providing valuable engineering data for future planning of long-duration lunar hardware.

Although the landing site of the Surveyor was accurately known, targeting Apollo 12 to within walking distance of the unmanned craft was a demanding task for mission planners, and indeed the crew, especially since Apollo 11 had landed more than four miles off target. However, on 19 November 1969, Conrad and Bean carried out a flawless landing, and touched down LM *Intrepid* just 535 feet northwest of Surveyor 3. The accuracy of the landing provided confidence for the planners, who were targeting later Apollo missions to much tougher landing sites.

Mission H-2 (Apollo 13) Landing Site: Fra Mauro Formation – Location: 17°33′ W 3°37′ S

Fra Mauro is a vast, geological area that covers central portions of the lunar surface around the Mare Imbrium. The scientific interest, and desire, to visit this site was very strong, as it would offer the opportunity to understand the nature, composition and origin of a widespread formation, which was expected to include ejecta from the Mare Imbrium area visited by Apollo 12.

Mission H-3 (Apollo 14) Landing site: Littrow area – Location 28°57'00" E 21°44'10"

This location is characterised by the number of fresh-looking 'wrinkle' ridges in the mare. The Mare Serenitatis located in this region also features a number of minute cracks, forming systems that are sub-parallel to the ridges and giving the appearance of a stage in the formation of the wrinkles. The study of this area for an Apollo landing mission observed that, in one location, part of a fresh ridge was covered by dark blanketing material, suggesting that this darker material was relatively young, and could be of volcanic origin.

Mission H-4 (Apollo 15) Landing site: Censorinus (NW) – Location 32°24'30" E 0°08'48" S

Located within, but near the edge of, a highland block in the south-eastern Mare Tranquillities is the 2.5-mile probable-impact crater Censorinus. The proposed landing site for the LM was to the north-west of the crater within the ejecta blanket. Attempting a landing in this area would be a unique opportunity to sample both highland material and features associated with a fresh impact crater early in the exploration programme. Selecting Censorinus for H-4 provided an area large enough to reveal clear signs of impact, but small enough to be thoroughly investigated by the astronauts during foot traverses close to the LM.

J Mission landing sites

The study offered sample EVA traverses for the J missions over high-resolution photography, except for the Descartes site, for which photography was not available at the time. This was to be obtained on H-1 (Apollo 12) and H-4 (Apollo 15), and so the determination of these traverses was only preliminary. The details of the EVAs proposed for these missions (from J-2, each carrying a lunar rover) are further described below, but the availability of motorised transport expanded the astronauts' range and thus the scientific target of opportunity at each site. This directly affected the selection of later landing sites. Three EVAs were planned and worked, on a baseline of both 3 mph and 6 mph for comparisons, and deployment of 'scientific equipment similar to ALSEP' during the first EVA. This affected the number of geological sampling stations. On H-series walking missions, the deployment of ALSEP early in the first EVA had a greater impact on the traverse distance due to the higher walking metabolic rate. Beginning with Apollo 17 (J-2), the use of an LRV on the J series would see the emplacement of the scientific station (an upgraded ALSEP), having the primary effect of reducing the number of geological stations examined, rather than the traverse distance, due to the lower metabolic rate encountered while riding on the LRV. Thus, station stops of more scientific interest, at further distances and in varying terrain (within the constraints of the LRV), could be planned for the later missions.

Mission J-1 (Apollo 16) Landing Site: Descartes – Location: 15°34′E 8°51′ S

The region around the crater Descartes in the southern highlands of the Moon resembles an area to the west and north-west of Mare Humorum and, in 1969, was thought to include distinctive patterns of morphological manifestations of volcanism in the lunar surface. The landing site for this mission was an area to the north of the crater, characterised by hilly, grooved and furrowed deposits. To the west was a hilly and pitted stratisgraphic unit, and to the east, a range of rugged hills which bound Mare Nectaris. Planning a mission to this region of intensive and prolonged volcanism was very important from both geological and geochemical viewpoints. The alternative target area to this site for similar sampling was identified as Abulfeda (14°00′ E 14°50′ S).

As important as this area was to mission planning, prior to Apollo 12 there were relatively few good high-resolution photographs of the region. Working mainly from Lunar Orbiter 4 photographs, the scientists and mission planners suggested that the landing crew could perform a single loop traverse north of the proposed LM landing site, which would enable examination of the hilly and furrowed material. The EVA from a landing site south of the furrowed ridge would involve a 0.6-mile walk to the ridge, a 1,500-feet walk westwards along the base of the ridge and into the furrow, and a further 1,500-feet walk towards a fresh 0.6-mile crater (expected to be blocky in nature, and with ejecta which was expected to include material from the hilly and furrowed unit), completed by a 0.6-mile walk back to the LM.

There would be no LRV on this mission, but it would include use of the Modularised Equipment Transporter (MET – similar to that used by Apollo 14 in February 1971). The second and third traverses would be devoted to the exploration of the plain material or to other 'objects of interest' that were expected to be revealed by high-resolution photographs obtained during the previous missions.

Mission J-2 (Apollo 17) Landing Site: Marius Hills – Location: 56°34′ W 14°36′ N

This landing area is a region of domes and cones located near the centre of Oceanus Procellarum, west-north-west of the crater Marius. Here, isolated hills and clusters of hills rise above the surface of the mare, forming part of a significant north–south median ridge system that extends irregularly for about 1,200 miles across Oceanus Procellarum. Many of the hills in the region exhibit the convex upwards shape suggesting laccolithic intrusions – some of them resembling terrestrial shield volcanoes on Earth. The variety of features and similarity to Earth volcanic structures strongly suggested to scientists that this area had been subjected to intense and prolonged volcanic activity. This was the first use of an LRV.

Mission J-3 (Apollo 18) Landing Site: Copernicus (peak) – Location: 19°55′ W 9°52′N

Copernicus is a crater of about 55 miles diameter, with bright rays which extend over the surrounding surface for several hundred miles. The wall of the crater is a 2.5-mile vertical section of the lunar crust, while the floor is almost circular and extends 38 miles in diameter. Almost centrally located is a small multi-peak, with large masses of material to both the east and west. The highest of these peaks rises to 2,625 feet, and is thought to have brought to the surface material that could have once originated at a considerable depth.

A mission targeted to this remarkable location would mainly be a sampling mission, with some emphasis on structural relationships. Samples of large blocks on the peaks, of the floor material, and from mounds on the floor of the crater, were expected to be of some significance to the geochemistry of the Moon. By sampling the peaks directly, samples from deep below the surface were expected to reveal new information on the internal structure of the Moon, from far deeper than either cores or drilling could reach on the early missions.

Mission J-4 (Apollo 19) Landing Site: Hadley Rille – Location: 2°27' E 24°47' N

This was one of the most visually impressive Apollo landing areas, where the Apennine Mountains form the south-eastern boundary of Mare Imbrium. Here, they form a triangular-shaped elevated highland region located between Mare Imbrium, Mare Serenitatis and Mare Vaporum. The region also features Rima Hadley – a V-shaped sinuous rille that terminates at an elongated depression in the south, and winds in a north-eastern direction parallel to the Apennine front for a distance of more than thirty miles until it merges with Rima Fresnel II in the north.

In the proposed landing area, the mountain frontage rises 4,200 feet above the adjacent mare to the west, the south-eastern region of Palus Putredinis. Also in the location of the LM landing site was a small 3.5-mile diameter crater known as Hadley C. This was identified as a conspicuously sharp and round crater that appeared to have partly covered the rille, characterised by a raised rim and ejecta blanket that covered the mare craters in the vicinity. At the time of targeting Apollo 19 to this area, the origin of Hadley C was a matter of controversy, as morphological characteristics suggested that it probably originated from mare material.

The mission was also to sample the rim area of the rille, where fresh exposures of probably stratified mare beds occur along the top of the rille walls. Photographic evidence revealed that large boulders had rolled down the walls and settled on the floor of the rille.

Mission J-5 (Apollo 20) Landing Site: Tycho (rim) – Location: 11°15' W 40°56' S

For the final scheduled Apollo mission, the target was the fresh impact crater Tycho, located in the southern highlands. At 53 miles in diameter, it is much larger than Censorinus, and provided the opportunity to sample the features common to large, fresh impact events and the associated volcanism. In the proposed landing site, the astronauts were expected to encounter several generations of lava flows, a lava pond or pool, ejected blocks (probably from the Tycho) and other ejecta features and structures. In addition, the landing was planned to be in the vicinity of the unmanned Surveyor 7 spacecraft and, as with Apollo 12, offered the opportunity to examine its immediate landing site and to retrieve samples of the spacecraft for return to Earth.

FLIGHT PROFILES

The method adopted to allow Apollo to travel from the Earth to the Moon involved two main flight paths, or trajectories, during the trans-lunar coast, which could take 40–100 hours, depending on the method chosen to cover the 250-thousand-mile distance.

Free-return trajectory

This was used on the earlier missions to the Moon (Apollo 8, 10 and 11). The spacecraft was placed on a trajectory in which, should the main engines fail and no manoeuvre or braking burns be possible, the gravitational influence of the Moon would be used to take the spacecraft behind the Moon. It could then continue in unassisted or 'free return' to Earth, during which the planet's gravitational forces would draw it back towards re-entry and ocean recovery.

A non-free-return trajectory did not allow for the spacecraft to be returned to Earth without the use of onboard manoeuvring and braking engines. This was the safest and quickest trajectory, but it limited the range of available landing sites, and also needed good lighting conditions at those sites.

Hybrid trajectory

Essentially halfway between the free-return and non-free-return, this method was adopted after the success of the first lunar distance missions, in order to allow for more flexible launch windows, a broader range of landing sites, and heavier payloads. However, it came at the expense of a quick return to Earth, and required the use of the SPS to return to the free-return trajectory, which was possible at any point during the trans-lunar coast. The manoeuvre to the hybrid trajectory was one of the mid-course corrections used to refine the trajectory to the Moon, and was followed by a further burn to a non-free return in the vicinity of the Moon prior to lunar orbit insertion.

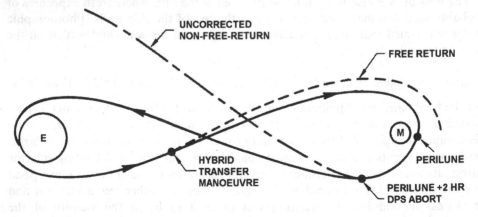

The flight profile towards the Moon, showing the free-return and hybrid concepts adopted for later missions in order to reach more scientifically interesting landing sites.

Initially, all Apollo missions were planned for free-return trajectories, with the transfer to the hybrid trajectory constrained by the abort ability of the LM descent engine to serve as a back-up should insertion to lunar orbit fail. The confidence gained from the Apollo 8, 10 and 11 missions saw the hybrid transfer trajectory used on Apollo 12 to help achieve the pinpoint landing near Surveyor 3.

The risk of following such a trajectory on the safe and early return of the flight crew was graphically demonstrated on Apollo 13, when the loss of the SPS system forced the crew to use the LM descent engine to manoeuvre to a free-return trajectory and abandon the landing attempt.

Multiple impulse LOI/TEI manoeuvres
During 1969, the Flight Analysis Branch of MPAD completed studies into the manoeuvring sequences and delta velocity levels of a multiple pulse (several short burns) of the SPS, for insertion to a high-inclination orbit. This was aimed toward the later missions and Phase II lunar missions, and was dependent upon the spacecraft configuration flown, launch windows, and mission profiles. A multi-trans-Earth burn was also studied to provide a continuous (any time) return-to-Earth capability, as opposed to the single impulse TEI burn in which a unique manoeuvre was required to bring the spacecraft home. The cancellation of the later Apollo flights terminated these studies.

CSM descent orbit insertion
The planning documents indicated that, beginning with H-2 (Apollo 13), an SPS burn while docked to the LM in lunar orbit would be termed Lunar Orbit Insertion Burn 2, designed to insert the spacecraft into a 60 x 8-nautical-mile orbit. This would greatly reduce the LM delta V required to descend to the surface (about 75 feet per second), allowing for heavier landing payloads, and enhancing precision landings, because more fuel would be available for hovering and final site selection. As a trade-off, the CSM delta V budget was increased by about 142 feet per second. After undocking, and prior to LM powered descent initiation, the CSM would return to a circular 60-nautical-mile orbit.

IMPROVING THE HARDWARE

As the landing sites with the most potential were short-listed and finally selected, and the method of reaching them defined, work continued in upgrading the flight hardware to support the expanded missions.

For the first landing – the G mission – the CSM configuration consisted of the standard Block II design and a standard-duration LM. This was categorised as the First Lunar Landing Mission. The next four H missions (Apollo 12–15) would still fly the standard Block II CSM and standard LM, but with minor modifications to support their mission category, designated Lunar Surface Science Missions. The J missions (Apollo 16–20) would fly a modified Block II CSM and modified LM spacecraft under the category of Lunar Surface and Orbital Science Missions.

In 1969, Apollo programme planning called for one G mission, the four H missions (with H-1 and H-2 hardware also available as 'back-up' for the G mission should the need have arisen), and the five J missions (which were also capable of flying I non-landing orbital missions if required). There were never any 'official' designations for any mission after Apollo 20, as the programme was terminated well before these plans developed towards an actual flight scheduling.

To support the J series of missions, significant upgrading of the hardware was required.

J series CSM modifications

Modifications to the CSM for the J missions included the capacity to fly missions up to sixteen days total duration (12–14-day mission plus two days contingency), delivery of the heavyweight LM (up to 37,000 lbs, manned) into an orbit of 60 x 8 nautical miles, and provision for the insertion and operation of a scientific instrument payload in the Service Module, together with integrated systems and the capability of in-flight EVA recovery of stored SIM-bay data.

To achieve this both quickly and safely, Rockwell ensured that the basic lunar mission capability of the CSM was retained, and that there was no degrading of safety and/or reliability as a result of the modifications. Procedural and CSM hardware changes were kept to a minimum, ensuring that the basic provisions were the same on each vehicle converted. The modifications made maximum use of existing Apollo qualified hardware, so that only new or major hardware elements would need to be qualified.

The SM modifications were restricted to Bay I (except in selected cases, depending on mission requirements) and in the CM, only in localised areas. The experiments would be installed in the CSM prior to shipment to KSC, but the option of later installation in the launch flow was retained.

The modifications (pre-Apollo 13) included the installation of an additional 50% of cryogenic storage capability (by adding one hydrogen tank and one oxygen tank to the SM), and a wire harness to associated controls and displays in the CM. As well as additional housekeeping equipment and sufficient consumables to support three men for up to sixteen days, the CM was also provided with additional stowage capability for up to three sample return containers, CM-located science and medical experiments, and returned data such as SIM-bay cassettes.

SIM-bay provisions

In order to incorporate the Scientific Instrument Module (SIM) bay in the Service Module, Rockwell had to devise a door that could be pyrotechnically jettisoned to cover Bay I for the duration of the mission, including launch, Earth orbit and translunar coast (up to 100 hours). The jettison velocity was planned to be great enough to enable deorbiting of the door from a 60 x 60-nautical-mile orbit (about 85 feet per second) to reduce orbital debris. Initially, the door would have been jettisoned after rendezvous with the LM following lunar ascent, and the SIM-bay experiments performed after the LM was jettisoned and prior to TEI. In this

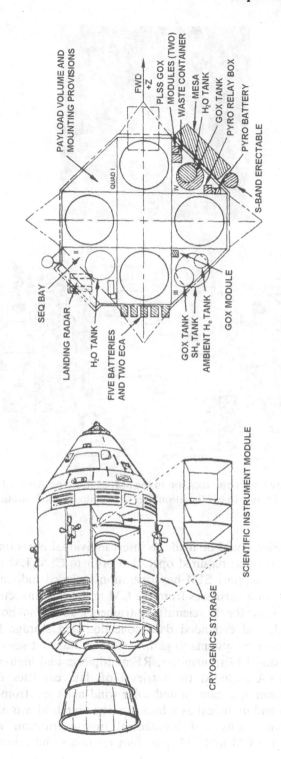

Changes in the Apollo CSM and LM descent stage for the J series of missions.

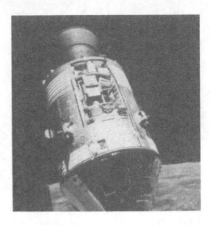

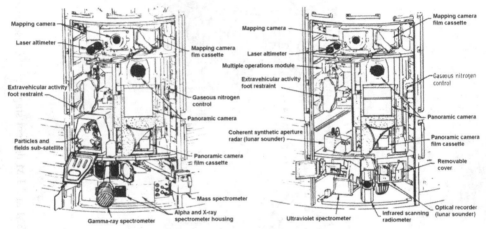

The Apollo SIM-bay layout and location in the SM, as flown on Apollo 15–17 in 1971–1972: (*left*) Apollo 15 and 16 configuation; (*right*) Apollo 17 configuration.

configuration, provision could be made to insert individual ports on the SIM door for specific experiments that required operation prior to CSM/LM rendezvous.

Other additions to support SIM-bay operation included individual instrument deployment systems, wire harnesses from the CM umbilical connections to the SM, CM displays and controls for the scientific instruments, additional housekeeping and supplies as required, and expanded data collection and storage facilities (using existing, modified and new systems to gather, store or transmit scientific data using the existing CM S-band FM transmitter, RF multiplexer and high-gain antenna).

Provisions for EVA included the retrieval of data cassettes from the lunar mapping camera, panoramic camera and solar wind mass spectrometer; secondary life support systems and umbilical as a back-up (the umbilical was also the primary carrier for communications and biomedical instrumentation with the EVA astronaut); rails on the CM and SM, plus foot restraints and other body restraint provisions; systems for transferring equipment to and from the SIM-bay area and CM side hatch; locations for additional lighting; and mounting positions for photodocumentation equipment.

Extended-stay Lunar Module
Modifications to the LM for the J-series missions required Grumman to configure the systems for a nominal surface duration of up to 54 hours while maintaining the existing structural integrity for safety. In fact, all planned consumable capacity on the LM (except electrical power) had the ability to support a 78-hour surface duration, offering further upgrade potential subsequent to the J series.

The modifications required included increasing stowage capabilities in the descent stages, with one corner quadrant and Science Equipment (SEQ) bay for payload stowage and additional stowage in the ascent stage. There were also improvements to the cabin environmental control and waste management system, to handle the extended duration and to provide the crew with the facilities to improve the habitability of the module. Finally, improvements to the guidance software would increase the landing accuracy and the capability to descend over 'rough' terrain.

To achieve this expanded capability, Grumman had to complete a number of modifications, as illustrated in the diagram on p. 243.

In the main propulsion system, the cylindrical section of the descent propellant tanks was increased by 3.36 inches while maintaining the hemispherical dome configuration, increasing useable propellant by 1,140 lbs. By adding a controlled vent bleed valve assembly, the standby time of the supercritical helium pressure increased to 190 hours. The ascent stage engine nozzle was also recontoured, and the descent stage nozzle extended 10 inches. To counter these additions and in order to save weight, the sixteen RCS thrust chamber isolation valves were removed, saving 25 lbs. Removal of the descent propulsion tank balance lines saved a further 40 lbs.

In the descent stage, the outer skin of Quadrant 1 (under the CDR crew station) was opened to accept payload stowage capability, and was provided with mountings for the Lunar Roving Vehicle 'or other payloads' (such as additional scientific payloads or the proposed Lunar Flying Vehicle). The Modular Equipment Stowage Assembly (MESA) on the adjacent Quadrant IV (under the PLT crew station) was enlarged to include palletised stowage for one-day consumable packages (such as PLSS batteries and LiOH canisters), with sufficient capability for the 78-hour surface duration (three packages – 72 hours).

An additional battery (there was a nominal four) was installed in the descent stage, and thermal protection was provided for all five batteries located on the Z bulkhead cold rails, providing up to 54 hours' surface duration. Grumman also installed another gaseous oxygen (GOX) tank and a water tank, thereby extending the capacity to 78 hours.

A new PLSS oxygen pressurisation module was installed, producing an increase in pressurisation levels from 980 to 1,280 psia, in order to support the planned extended EVA operations and number of depressurisations. The former Control Module was

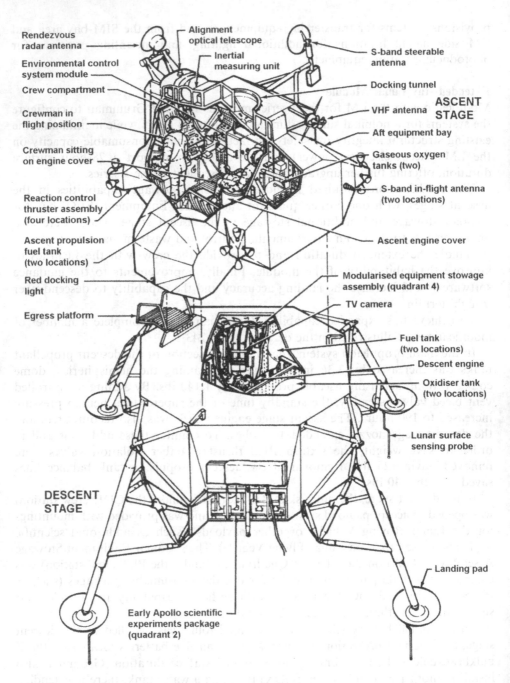

Rendezvous
radar antenna

Environmental control
system module

Crew compartment

Crewman in
flight position

Crewman sitting
on engine cover

Reaction control
thruster assembly
(four locations)

Ascent propulsion
fuel tank
(two locations)

Red docking
light

Egress platform

Alignment
optical telescope

Inertial
measuring unit

S-band steerable
antenna

Docking tunnel

**ASCENT
STAGE**

VHF antenna

Aft equipment bay

Gaseous oxygen
tanks (two)

S-band in-flight antenna
(two locations)

Ascent engine cover

Modularised equipment stowage
assembly (quadrant 4)

TV camera

Fuel tank
(two locations)

Descent engine

Oxidiser tank
(two locations)

Lunar surface
sensing probe

**DESCENT
STAGE**

Landing pad

Early Apollo scientific
experiments package
(quadrant 2)

Details of the baseline LM and its main components.

retained to provide a redundancy capability. This increased total PLSS oxygen recharge capability from 0.92 lbs to 1.5 lbs.

Crew provision stowage was modularised and rearranged to provide easier access and repackaging between EVAs. In the October 1969 planning report, it was also proposed to install an access door in the Spacecraft LM Adapter (SLA) to allow access to the science payload in the SEQ bay of the LM while on the pad, to reconfigure or attend to the instruments during pre-launch preparations.

Other crew habitability provisions planned for the J-series LM included a urine and PLSS condensate waste management system for extended duration, insulation added to the docking tunnel area to control cabin temperature levels, a reworking of the umbilical to provide independent connections for communications, liquid cooling and oxygen, and additional crew expendables and consumables to support the planned surface duration (plus contingency).

Structurally, the extended LM ascent stage was strengthened in the mid-section deck for a third sample return container, with an added support fitting for the extended MESA and for the faecal receptacle on the ascent engine cover, and pressure shell penetration points for water management and PLSS oxygen recharge lines. On the descent stage, a redesign of the descent stage structure was required to accommodate the large propellant tanks. A cold rail in Quadrant IV was provided for pyro battery mounting, and battery supports were required on the Z bulkhead. Support structures were also installed for the expanded MESA and a new deployment mechanism in Quadrant IV. On Quadrant I, the thermal insulation was relocated to provide for additional scientific payload support, and attach points and release mechanisms were installed for the rover and/or additional scientific payloads. The support and deployment systems for ALSEP-type payloads in Quadrant II were retained.

Grumman also revised and added fluid and electrical lines, and relocated the explosive device relay box, the pyro battery and the GOX control module. There were added support structures for the new water and GOX tanks and for the waste management container. Finally, the RCS plume deflectors were reworked, and exterior insulation was reduced to assist in lowering upper weight level limits.

The display and control instrumentation was modified to support all these changes, and the spacecraft guidance system was upgraded, increasing the capacity to land over rougher terrain, with the addition of the ability to update the location of the landing point. A landing radar signal pre-filter, which would smooth the variations of the lurain (lunar terrain), was also installed, along with a delta guidance capability to improve the efficiency of powered descent.

A7L suits and portable life support systems
The first five landing missions were assigned the use of the No.6 standard portable life support system and oxygen purge system. For the J series, the modified No.7 version of the PLSS and a secondary life support system (SLSS) was planned. The life support capabilities of these two configurations were:

LSS	Metabolic Rating	Operating Time Limit
No.7 PLSS	4,800 BTU	Battery: 6.25 hours
No.7B PLSS	6,000 BTU	Battery: 6.25 hours
SLSS	2,400 BTU	Battery: 2.0 hours
OPS	Not applicable	30 hours operating at a fixed oxygen flow rate of 8 lbs per hour

Modifications to the No.6 variant included the installation of a carbon dioxide sensor, located at the LiOH outlet, which provided a warning of a canister failure and increasing levels of carbon dioxide. This modification would be incorporated from Apollo 13 onwards. There was also an increased charge pressure and capacity of oxygen. The minimum charge pressure would be increased to 1,382 psia for an oxygen quantity of 1.78 lbs, compared to the No.6 design, which was at a value of 1,020 psia and 1.16 lbs.

The No.7 back-pack also had an increased water capacity, raised from 8.5 lbs to 11.8 lbs by changing the tank end domes from a structural efficient to one of volume. However, this meant that the PLSS would launch void of water, and required charging with water prior to use to minimise launch and abort weight penalties. A water quantity sensor was also installed to provide more accurate information about the state of the back-pack and the condition of the crew-member.

The SLSS was, in effect, a smaller version of the complete PLSS, using existing components where possible and repacking them into a smaller volume. This system weighed 64 lbs, and was designed to use similar mounting interfaces and occupy the same volume as the OPS. Its purpose was to furnish liquid cooling for the astronaut, and it could be used to recharge water to supply an additional 800 BTU of metabolic capability. This unit was also adaptable for use as a contingency transfer supply from the LM to the CM or for SIM-bay EVA. It did not provide variable rate cooling, communications links, telemetry data, or displays for the crew-member.

For the H series of missions, the astronauts would wear the A7L model pressure garment assembly. The A7L-B version used on the J series included improvements in mobility and visibility.

These modifications included a joint in the neck area that improved neck mobility and downward visibility for the astronaut, while a bellow joint in the waste area was installed to provide for forward bending mobility. The gas retention layer was treated with a total bladder and convolute abrasion reinforcement to increase durability. As a result of adding the waist joint, the torso zipper had to be relocated from the crotch to the chest area, which also resulted in improved leg mobility. Modifications to the shoulder area reduced the torque required to move the shoulder, and additional thumb joints in the gloves improved the ability to grasp objects, while an enlargement of the wrist ring provided the astronauts with a faster and easier donning and doffing of gloves.

As with all the equipment modified for the J series, in-flight experience would see further improvements, upgrades and changes.

Mobility aids

A limiting factor of the early landing missions of the H series was the amount of equipment that each astronaut could carry, and the restriction on gathered samples that he could return to the LM (based on the performance of his PLSS and his own efforts, and in the duration and distance he could go to still remain within operational safety and EVA time-lines). The provision of a mobility device to assist and expand the astronaut's capability greatly enhanced the scientific and operational return from later missions. This was addressed in two ways.

For missions H-3 (Apollo 14), H-4 (Apollo 15) and J-1 (Apollo 16), it was planned to utilise a Modular Equipment Transporter (MET), folded and stowed in the descent stage of the LM. This was a small tubular-wheel handcart (similar to a golf trolley, and referred to as the 'lunar rickshaw') weighing under 20 lbs, but able to carry 350 lbs of equipment and gather lunar samples

On 11 July 1969, NASA issued a request for proposals from industry for a manned lunar roving vehicle that would be flown on the J-2, J-3, J-4 and J-5 missions. NASA listed several specifications for this vehicle. Propulsion had to be via a battery-powered drive motor at each of the four wheels, and its weight had to be no more than 400 lbs (including the battery and deployment mechanisms). The vehicle had to be capable of supporting two space-suited astronauts plus 170 lbs of equipment or samples. Its top speed was 10 mph maximum and 5 mph average

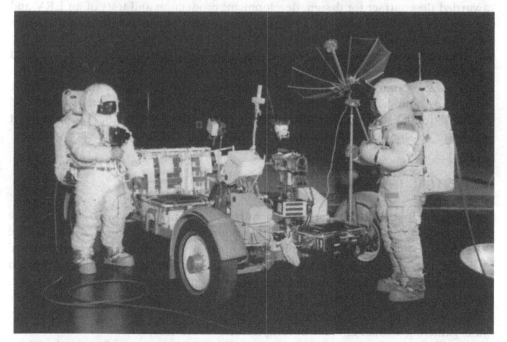

The prime crew of Apollo 15 undergoes a training session with the '1 g' Lunar Roving Vehicle trainer in Building 5, Mission Simulation and Training Facility, at MSC in Houston during January 1971. This training version reveals the final design of a Lunar Roving Vehicle used on the last three Apollo missions.

for the specifications, but operationally 6.2 mph maximum and 3.1 mph average. The range would be up to 75 miles, offering three sorties of 12.5 miles each (but still within a walk-back distance to the LM should the vehicle break down).

The LRV had to be capable of negotiating a 25-degree slope, plus obstacles 12 inches high and crevices 28 inches wide. It had to be storable in LM Quad I and locked into final configuration upon deployment, requiring minimal astronaut time to set up for operation. An EVA communications system and relay unit would provide communications. The LRV would only operate manned, and only during lunar daytime periods (although some consideration had been given to unmanned control from Earth – similar to the Soviet Lunokhod system – enabling use of the LRV between missions, and to relocate the vehicle to a new landing area). Navigation would be via gyro azimuth indication, and for the return to the LM, an odometer and computer display of the distance and direction to the lander would be available to the crew.

On 7 April 1969 an LRV Task Team had been established at MCS to conceive, design and develop a variety of proposed modes of transport for future Apollo landings. These ranged from studies developed from the LSSM/MOLAB designs, to smaller jeep-like vehicles, and even a plan to add wheels to a Grumman LM ascent stage! By 18 August this was renamed the Lunar Mobility Task Team, and included lunar flying vehicle evaluation as well as LRV studies. By 28 October, Boeing was awarded the contract for design, development, production and tests of an LRV and delivery of four flight units (although only three were delivered and flew), as well as several mock-up and training units.

The final design was of a four-wheel manually controlled and electrically powered vehicle capable of 10 mph maximum speed, and at a reduced velocity it could climb a slope of 25°. Controls allowed either crewman to drive the vehicle from the left or right crew station. The LRV was 10 feet long, 7 feet wide, and 45 inches high, and weighed from between 1530–1,600 lbs, of which 450 lbs was the LRV, the rest consisting of crewmen, tools, communication equipment, and the science payload.

SCIENTIFIC INVESTIGATIONS

During the first four manned Apollo missions, the priority was to man-rate the spacecraft, procedures and systems to support the lunar landing attempts. The scientific payloads on these missions were minimal, with most of the 'experiments' being photographic and visual in nature. Indeed, the first lunar landing G mission (Apollo 11), although featuring geological and scientific investigations, was primarily aimed at achieving the programme goal of placing astronauts on the Moon, performing a limited excursion on the surface, deploying simple experiments, and gathering a small amount of samples before returning home.

Once the Apollo system had been man-rated and the major objective had been achieved safely and successfully, then the opportunity to expand the scientific and operational returns presented itself. This was demonstrated in the objectives of the H and J series of missions, during which scientific payloads were increased, and the

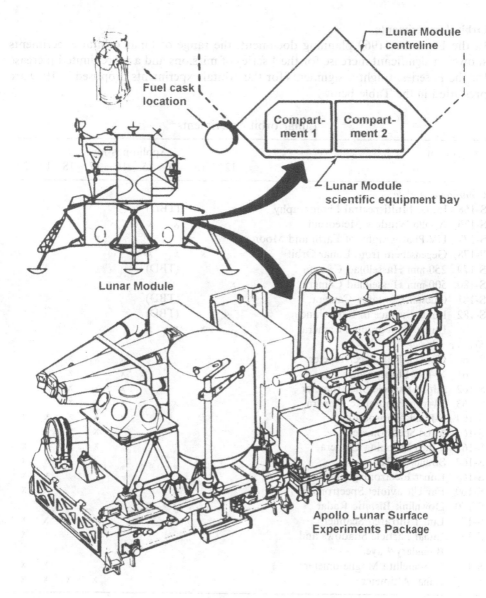

Lunar Module centreline

Fuel cask location

| Compart-
ment 1 | Compart-
ment 2 |

Lunar Module scientific equipment bay

Lunar Module

Apollo Lunar Surface Experiments Package

The location for the Apollo Lunar Surface Experiments Package carried in the descent stage of the LM, shown in its stowed form.

opportunity to study the Moon from orbit supplemented that from the surface, both at the time of the mission and after the crew had returned to Earth.

An outline of those proposals follows, based on guidelines issued in the planning documents of December 1969. (How these ideas changed as the programme itself evolved (along with what actually flew on Apollo) is discussed from p. 267.)

Orbital experiments

In the December 1969 planning document, the range of lunar orbital experiments showed a significant increase for the J series of missions and a more limited increase for the H series. Flight assignments for the orbital experiments proposed in 1969 are presented in the Table below.

Lunar orbit experiments

Experiment	Apollo mission								
	12	13	14	15	16	17	18	19	20
Command Module									
S-158 Lunar Multispectral Photography	x		(TBD)						
S-176 Apollo Window Meteoroid			x	x					
S-177 UV Photography of Earth and Moon					x	x			
S-178 Gegenschein from Lunar Orbit		x	x	x					
S-179 250-mm Hasselblad Camera			(TBD)						
S-180 500-mm Hasselblad Camera	x								
S-181 Questar Contarex Camera			(TBD)						
S-182 Lunar Surface in Earthshine			(TBD)						
Lunar Topographic Camera		x	x	x					
Service Module									
S-160 Gamma-Ray Spectrometer					x	x	x		
S-161 X-Ray Fluorescence					x	x	x		
S-162 Alpha Particle Spectrometer					x	x	x		
S-163 24-inch Panoramic Camera					x	x	x		
S-164 S-Band Transponder					x	x	x	x	x
S-165 Mass Spectrometer						x	x		
S-166 3-inch Mapping Camera						x	x	x	x
S-167 Sounding Radar								x	x
S-168 Lunar Electromagnetic Sounder 'A'								x	x
S-169 Far Ultraviolet Spectrometer								x	x
S-170 Downlink Bistatic Radar			x	x					
S-171 Lunar Infrared Scanner								x	x
S-173 Lunar Particle Shadows and Boundary Layer								x	x
S-174 Sub-satellite Magnetometer								x	x
S-175 Lunar Altimeter						x	x	x	x

Emplaced lunar surface experiments

Experiment	Apollo mission									
	11	12	13	14	15	16*	17*	18*	19*	20*
M-515 Lunar Dust Detector	x	x	x	x	x			x	x	x
S-031 Passive Seismic	x	x	x	x	x	x		x	x	x
S-033 Active Seismic				x		x				x
S-034 Lunar Surface Magnetometer		x			x	x				
S-035 Solar Wind		x			x					
S-036 Suprathermal Ion Detector		x		x	x					
S-037 Heat Flow				x		x				x
S-038 Charged Particles Lunar Environment				x	x					
S-058 Cold Cathode Gauge				x						
S-078 Laser Ranging Retroreflector	x				x	x				
– Astronomy Radiometer							x			
– Electric Field Gradiometer									x	x
– Gravimeter									x	x
– Mass Spectrometer									x	x
– Radiometer									x	x
S-080 Solar Wind Composition	x	x	x	x		x				
S-151 Cosmic Ray Detection	x					x			x	
– Low Energy Nuclear Particle Detection							x		x	
– Water Detector									x	x
– Cone Penetrometer						x	x			
– Hasselblad Camera	x	x								
– Stereo Camera			x	x	x	x	x	x	x	x
– Close-up Camera	x	x	x							
– Television Camera	x	x	x	x	x	x	x	x	x	x
– Lunar Survey Staff								x	x	x
– Apollo Lunar Hand Tools	x	x	x	x	x					
– Upgraded Hand Tools						x	x	x	x	x
– Drill (10-foot)			x			x				x
– Sample Return Container	x	x	x	x	x	x	x	x	x	x

*   Programme commitments not fully approved

Lunar surface experiments

The surface experiments package for Apollo 11 – termed the Early Apollo Scientific Experiment Package, or EASEP – was developed to accommodate the brevity of the mission and still obtain useful scientific data.

Originally, four arrays were developed for deployment on the H series of landings. In order to fly the EASEP, the central station of a fifth ALSEP was used to pre-empt the assembly of the Apollo 11 package. This was rebuilt and manifested to flight as ALSEP A-2. At the time of the 1969 planning document, the section and assignments subsequent to Apollo 15 (the J series) still had to be defined, although

ALSEP D was assigned to J-1. The Table on p. 253 shows what was proposed for these missions before final assignment.

Lunar hand-tools

To assist in sample-gathering, the range of Apollo lunar hand-tools was further developed for the H and J series. These included *special environment containers*, in which carefully selected samples of soil and the environment were sealed in small high-vacuum containers. These were not opened until they were inside the special glove boxes in the Lunar Receiving Laboratory, giving an opportunity to study the material in its original environment. Two types of collection bags were provided: the sample *collection bag*, with pockets for core tubes; the special environmental sample and magnetic shield sample, which could hold larger surface samples; and the 7½ x 8-inch Teflon *documented samples bags* in a twenty-bag dispenser.

An e*xtension handle* of aluminium alloy tubing, 30 inches long, attached to other tools so they could be used without bending or kneeling. Nine c*ore tubes* were also supplied, which could be driven into loose surface material. Each was 15 inches long, and was made of aluminium tubing, and they were designed with a non-serrated cutting edge and screw-on cap to seal the tube after extraction. The a*djustable sampling scoop* resembled a domestic garden scoop, and was used to gather samples too small for tongs or the rake. It had an adjustable pan, and was also compatible with the extension handle. The 14-inch *sampling hammer* had three functions: as a sampling hammer; as a pick; or as a mallet to drive core tubes into the soil. The *Tongs* measured 32 inches long, and were used to retrieve samples from up to 2½ inches in diameter, from a standing position. The *lunar rake* was used to collect samples of ½ inch to 1 inch in size. It was adjustable, with stainless steel tines and a 10-inch handle that fitted into the extension handle.

There was also a *spring scale* to weigh the filled rock boxes and other items prior to placing in the LM, in order to maintain the weight budget for lift-off and return to Earth. A tripod *gnomon* was used as a photographic reference. This was a weighted staff suspended on a two-ring gimbal painted grey with a colour scale to indicate local vertical, Sun angle and scale. The aluminium rack to hold these tools was the *lunar hand-tool carrier*, which could be carried by hand or mounted on the mobility aids.

Mission operations

The study also evaluated the proposed activities of each mission within the operational constraints of the hardware. The wide-ranging study included several considerations with regard to planning surface activities. The plan was to provide four launch opportunities on two consecutive months for each primary site, giving an eight-week period for launch to the target area. However, the actual launch window occurred only for a short time once each month to achieve the correct lighting conditions at the landing site and for end-of-mission recovery. From H-2, the third stage of the Saturn (in addition to the empty ascent stage after redocking with the CM) would be deliberately impacted on the surface, to stimulate seismic activity that could be recorded on placed sensors.

From mission H-3 it was also necessary to land in rough terrain, but still within an astronaut's EVA range of selenological features, dependent upon astronaut mobility, exploration requirements and safety considerations. It was determined that a landing error of ± 0.62 miles would guarantee mission success, and the techniques for this degree of accuracy would be developed on H-1 and H-2.

The execution of one EVA on landing day was the maximum allowable, and only one EVA would be performed during each successive work/rest cycle (one EVA, two meal periods, and one rest (sleep) period in as close to 24 hours as possible). Two hours' preparation and 1.5 hours' post-EVA activities were required for each EVA, and so back-to-back EVAs on the same day were undesirable, and 'would be employed only when science gain justified such a strenuous timeline'. Upon these assumptions, the H series would include two EVAs in 35 hours, and the J-series, three EVAs in 54 hours. However, the duration of each EVA could be extended, depending upon human factors rather than EMU equipment limitations.

It was determined that the first EVA would require more preparation activity than would subsequent EVAs, and therefore offered less time for scientific exploration. Estimates for ALSEP deployment activities (normally on EVA 1) were around 1 hour 45 minutes. Simulations of activities on the surface were also evaluated for the metabolic cost of walking on the surface, across different inclines, with or without equipment and samples, and for riding the LRV and then having to walk back to the LM.

In conclusion, it was determined that, for safety considerations, the end of each EVA period should allow a residual 20 minutes of additional life support, and the astronauts should remain in the pressure suit for no more than five hours. Both crew-men should also be on the surface at the same time, in constant communication with Mission Control and within sight of each other. In the event of a PLSS failure, the astronaut could (theoretically) use the back-up life support system to return to the LM, taking fifteen minutes to re-enter the LM and complete repressurisation. The LM cabin could remain unpressurised for up to six hours, but this had to be preceded by an equal period of pressurisation. In the event of an emergency return to the LM, all payloads would be discarded and the most direct route taken.

Each EVA period was divided between pre- and post-EVA time, and EVA time on the surface. Surface time was also divided into science time and traverse time, where the former involved activities at a particular point of interest or specific activity, and the latter the time spent travelling between them. Consideration was therefore also given to PLSS capacity and walk-back requirements versus time spent at each point of interest or task in determining the time line for each EVA.

FLIGHT SCHEDULE[7]

Following the success of Apollo 11 in July 1969, subsequent missions were spread out to allow more time for mission planning and post-mission analysis. Apollo 12 was moved to November 1969 to attempt a landing at the Ocean of Storms site. During the 9 October press briefing, plans for Apollo 13–20 were presented. Three options were put forward: a maximum rate, and revised Program I or Program II and III options, for long-term planning.

In all programme schedules, the rate of Apollo launches remained the same, with Apollo 13, 14 and 15 flying in Calendar Year 1970 (CY 70), three further missions (Apollo 16, 17 and 18) during CY 71, and Apollo 19 and 20 completing the initial Apollo landing exploration missions in CY 72.

The Apollo follow-on lunar exploration programme[8] (Phase II) would, under the maximum rate, fly five missions (two in CY 73, two in CY 74, and a single mission in CY 75). If a slower rate were followed under the Program I option, then, after a gap in Apollo lunar operations in CY 73, a single Apollo follow-on mission would fly in CY 74 and CY 75, followed by two missions in CY 76 and a final single mission in CY 77. If the minimum rate was chosen, then only four missions would be flown – two each in CY 74 and CY 75. Of course, all of this depended upon the new programme funding for post-Fiscal Year 1970, which for some time had not looked very promising. Nevertheless, the success of the Apollo 11 mission was seen as a possible boost to renewed efforts to extend and expand the manned exploration of the Moon, as well as to create a space station in Earth orbit from Apollo hardware. If so, then new programmes could be initiated to provide a more permanent presence not only in space generally, but also on and around the Moon, eventually leading to manned planetary (and in the distant future, interplanetary) spaceflight – the cherished dream and long-term goal of generations of 'space' theorists, scientists and engineers.

Under the 1969 planning, the earliest opportunity to put a dedicated US space station based on Apollo hardware in lunar orbit would be in 1976. In the Program I option this slipped to 1978, and in Program II/III, a lunar orbit station would not be attempted before 1981. To provide a regular supply system to support the expanded programme, a nuclear stage option was deemed feasible by 1978, in either the maximum planning rate or Program I schedules. This would slip to 1981 in the long-term planning schedule.

Finally, a Lunar Surface Base (Phase III) was estimated to be achievable by CY 78 at the maximum scheduling rate, but would slip into CY 80 under Program I and into CY 83 if the longer-term option were to be selected.

Mission assignments

Meanwhile, work continued to detail what each of the future Apollo missions could realistically achieve. For each mission, primary and detailed objectives were presented. The primary objectives were established by the NASA Office of Manned Space Flight, and were initially documented in the Apollo Flight Mission Assignments Directive (AFMAD) dated 11 July 1969 (MA 500-11, SE 010-000-1). At this time, objectives for the later missions had not been fully established, so the information presented was representative. As the later missions progressed towards flight, their primary objectives evolved accordingly.

Detailed mission objectives were developed by the Manned Spacecraft Center and documented in the Mission Requirements,[9] and again, for the later J-series missions, these were presented for information purposes only. The official NASA definition of primary and detailed mission objectives for the Apollo landing missions were as follows:

- Primary Objective: A statement of a primary purpose of the mission. When used in NASA Center control documentation it may be amplified but not modified.
- Detailed Objective: An objective generated by a NASA Center to amplify and fulfil a primary objective.

G missions

The Apollo 11 mission flew the only G mission in the programme, and had only the primary objective of a manned lunar landing and return to Earth. Subordinate objectives assigned to the primary directive included selenological inspection and sampling, and obtaining data on the capability and limitations of an astronaut and his equipment in the lunar environment.

The mission, flown between 16 and 24 July, followed a free-return trajectory and achieved a lunar landing on 20 July. The single two-man EVA lasted 2 hours 40 minutes (depressurisation to repressurisation), during which Armstrong (lunar surface time approximately 2 hours 31 minutes) and Aldrin (surface time approximately 1 hour 50 minutes) performed one excursion outside, walking no further than 0.1 km from their LM at Apollo Landing Site 2 (Tranquillity Base) and deploying the Early Apollo Lunar Surface Experiment Package (EASEP). Apollo 11 remained on the surface for 22 hours and in lunar orbit for 59.5 hours in a total mission duration of 8 days 4 hours.

The detailed objectives for this mission included collection of a contingency sample, egress from the LM to the surface, performing lunar surface operations and ingress into the LM from the lunar surface, performing surface operations with the support of the EMU, gathering data on the effects of landing on the LM, the characteristics and mechanical behaviour of the lunar surface, collecting samples of lunar material, determining the position of the LM on the surface, providing data on the effects of illumination and contrast conditions on the crews visual perception, demonstrating procedures and hardware used to prevent contamination of the Earth's biosphere (which extended from entering the LM to release from the Lunar Receiving Laboratory 21 days later), obtaining television coverage during the lunar surface stay period, taking photographs during the landing and lunar surface stay, deploying the EASEP, conducting Experiment S-059 (Apollo/Field Lunar Geology assigned to Mission G), conducting two passive experiments, and a number of television casts on the journey to and from the Moon.

And this was just the first, short-stay lunar mission with one EVA of less than 3 hours. The flight planning for the following nine missions (if funded and flown without major incidents) in the autumn of 1969 was much more involved.

H missions (Apollo 12–15)

These four missions (H-1 through H-4) followed the G mission in using standard Apollo hardware, but with a lunar trajectory of either the free return or hybrid type.

It was during these missions that the technique of descent orbit insertion using the SPS propulsion system would be developed and demonstrated. Each of the LMs would remain on the surface for an open-ended interval of no more than 35 hours, with the capability to support two two-man periods of EVA, each lasting approximately 3.5 hours (from LM depressurisation to repressurisation). A further option in the planning of these missions was the capability of using the LM top hatch for panoramic observation of the area around the landing site (which was achieved on only one Apollo landing mission – Apollo 15 – in 1971).

The maximum radius from the LM was expected to be no more than approximately 0.6 of a mile, limited by the purge capability of the life support system within the PLSS. Following LM ascent stage docking and crew transfer, the stage was to be jettisoned, and impacted on the surface to stimulate the deployed passive seismic experiments and to eliminate orbital debris. (From Apollo 13 (H-2), the S-IVB stage was also targeted to provide a larger mass and greater impact velocity for the instruments to record the impact data.).

Apollo 12 (H-1)

This was targeted for launch in November or December 1969 (and was actually flown during November 14–24), carrying ALSEP Array A for deployment at Apollo Site 7, near the Surveyor 3 unmanned spacecraft. Two EVAs were planned, each of around four hours (actually 4:01 and 3:54), and totalling eight hours (actually 7:55) on EVA and 32 hours (actually 31 hours 31 minutes) on the surface. The two astronauts would venture no more than half a mile from the LM. The scientific objective was stated to be a selenological survey and sampling of a mare region and deployment of the first full-scale ALSEP facility (achieved). Technological objectives for Apollo 12 were listed as pinpoint landing development (achieved) and surface EVA assessment (achieved). Lunar terrain photography on this flight was targeted (and achieved) for Lalande, Fra Mauro and Descartes. The first priority of all landing missions was for the Commander to obtain a preliminary geological sample shortly after stepping off the LM footpad for the first time. In the event that the EVA or surface stay had to be terminated early, at least one sample of that landing area would be secured.

This was the final Apollo mission to follow the October 1969 scheduling. The remaining eight missions encountered a changing programme of mission objectives, profiles and fates over next three years of the Apollo programme, but for the purposes of this account, the optimistic flight planning assignments detailed in October 1969 for Apollo 13–20 were as follows:

Apollo 13 (H-2)

Planning listed this mission for March or April 1970, targeted towards Fra Mauro. The crew was assigned scientific and technical objectives similar to those on Apollo 12, but with sampling away from a mare formation. Bootstrap photography from orbit by the CMP would be targeted at Censorinus and the Davy Rille. Surface activities during about 35 hours on the surface would include two EVAs, each of 3 hours 30 minutes (14 man-hours total), and deployment of ALSEP Array B (ALSEP III).

Apollo 14 (H-3)

Had Apollo 13 achieved its landing objective, then Apollo 14 would have been prepared for a launch in July or August 1970. The crew aimed to land at the Littrow formation, with the two EVAs, surface duration and objectives similar to Apollo 13, except that at Littrow, sampling would be from a highland region for the first time. ALSEP Array C (ALSEP IV) would be deployed, and future landing site photography would be attempted on the Descartes landing area.

Apollo 15 (H-4)

The original mission of Apollo 15, targeted for October or November 1970, was the final flight of the H series using the same standard Block II CSM and LM and scientific objectives as Apollo 12–14. A surface duration similar to the previous missions would have included two EVAs of about the same length, and deployment of ALSEP Array A-2 (ALSEP II) at Censorious. This allowed for sampling at a large crater-type formation. Future landing site photography would have been of Descartes.

I (alternative non-landing) missions
This was a category that could have been flown by a mission subsequent to J-1 if the potential scientific return from a non-landing mission was sufficient to balance or outweigh that from a landing mission. If a contingency had arisen with the LM prior to descent orbit insertion, this profile could have instead been followed.

The I-type flight would have been a lunar-orbit-only flight for the purposes of mapping a significant area of the lunar surface or future landing sites, and for exploring the Moon's surface with advanced remote sensors from orbit. The expected coverage of the over-flown area would have been as much a 200° of longitude, between 45° S and 45° N latitude.

Following a hybrid translunar trajectory, the CSM would have been inserted into a high inclination lunar orbit, possibly using multiple impulse technology. The mapping and scientific activities would have been conducted for no more than eight days, after which the crew might have used the multi-impulse technology manoeuvre to successfully insert the spacecraft into the trans-Earth trajectory for the flight home.

The primary objectives for this type of mission included:

- Performance of a detailed lunar orbital science survey from an orbit of high inclination.
- Obtaining metric and panoramic photographs for lunar mapping of candidate lunar exploration sites.

These objectives were aimed for missions beyond the J series (a planned but unrealised K (?) series), and although no I-type missions were ever officially scheduled, there were serious plans discussed for at least one Apollo lunar orbital mission in the early 1970s (see p. 310).

Flying an I-type mission involved a series of operational changes from the nominal landing mission. The launch window would have been defined by the assigned scientific experiments, and additional mission specific training by the crew would have been required to execute these investigations, replacing lunar surface training. Because the DPS abort capability of the LM would not exist in this type of mission, the free-return option would have been used (with the multi-impulse LOI), or the hybrid trajectory for a higher-inclination lunar orbit.

The multi-impulse LOI manoeuvre sequence consisted of a high elliptical lunar orbit, with the highest point between 2,000 and 10,000 nautical miles altitude and an orbital period of up to 24 hours. At the apocynthion (lunar orbit apogee), a plane change manoeuvre would have been executed, and a third manoeuvre would have inserted the spacecraft into a circular 60-nautical-mile orbit.

Flying such a high-inclination orbit provided the opportunity to survey the lunar surface from Littrow in the east to the Marius Hills on the west and between the maximum latitudes attainable. Further analysis by the Mission Planning and Analysis Division (MPAD) at MSC indicated that inclinations of up to 88° would have been attainable for an I-type mission profile.

Lunar orbital science was possible on six or seven days, and would have followed a similar but expanded mode of operations to that proposed for the J series of SIM-bay experiments on lunar landing flights. In the J series design of SIM-bay thermal shielding then under development, the design did not allow for orbital inclinations in excess of ±45° without changes to the manoeuvring capabilities of the spacecraft and the cycling of the instruments. This would have had to have been addressed should an I-type mission have been manifested. Flying high-inclination orbits did not allow surveillance of the lunar terminator.

Because of the excess performance requirements and the desire to return the Apollo CM to an equatorial re-entry zone (40°), a multiple-impulse TEI would probably have been a further requirement. This would have been completed by first firing the engines to insert the spacecraft into a high elliptical orbit where, at maximum apogee, a plane change/perigee increase manoeuvre would have preceded a third manoeuvre to insert the spacecraft into the desired TEI trajectory. During the flight home, EVA by a member of the crew (presumably the CMP) would have been required to retrieve data and film cassettes from the SIM-bay area, but not during solar flare activity. The opportunity to conduct experiments during the trans-Earth coast period (which was normally quiet on returning Apollo lunar missions) was also evaluated, as was the possibility of performing EVA in lunar orbit to change film and data collection cassettes.

Recovery operations for this type of mission resembled those of the C Prime and F missions (Apollo 8 and 10), so no quarantine period would have been required.

J missions (Apollo 16–20)

There were to be five J missions (J-1 through J-5) to follow on from the H series, flown by modified Apollo hardware that offered both extended duration on or around the Moon, and increased payload envelope capabilities in both to delivery to the Moon and return to Earth. Improvements in suit technology would allow far greater mobility, and the inclusion of surface transportation greatly expanded the overall capability of these missions (which became known as the 'super-science' Apollo missions). As with every manned spaceflight, mission planning allowed for contingency and alternative missions should hardware failures prevent the original mission objective, as long as there was no threat to the safety of the crew. Such alternatives for the J series were extended Earth orbital missions, or the I-type lunar orbital missions detailed above.

For these missions, hybrid translunar trajectories were necessary due to the increased gross mass of the spacecraft. It was also determined that a three-burn LOI manoeuvre might be needed to meet some landing site requirements. Lunar surface duration increased up to 54 hours, during which three periods of two-man EVAs could be performed safely. Initially, planning established that the first surface EVA would take place shortly after landing on the first day. EVA 2 would be completed on the first full day on the Moon, and EVA 3 early on ascent day. Use of the modified PLSS (6,000 BTU rating) allowed each EVA surface time to be extended and, according to preliminary planning studies, offered a maximum EVA duration (LM depressurisation to repressurisation) of 4 hours 30 minutes per man per EVA (at 1,250 BTU/h).

How far the astronauts ventured from the LM would depend on the level of activity which they were to undertake and the real-time capability of the PLSS that they were using. If, as on J-1, no Rover was carried, then this would be approximately 3 miles. On the Rover missions, this increased to about 7.5 miles.

While the two astronauts explored the surface, the CM Pilot in the orbiting CSM would monitor and carry out experiments from orbit. In addition, an interval of up to 61 hours was available between ascent stage rendezvous after lunar lift-off and TEI. Retrieval of the data cassettes would be by EVA (nominally 44 minutes) either in lunar orbit or during trans-Earth coast. The total mission duration would not exceed 16 days (384 hours).

Apollo 16 (J-1)

Originally targeted for a March or April 1971 launch, this mission to Descartes was assigned the scientific primary objectives of a selenological survey and science activity with a surface experiment deployment of ALSEP Array D from a modified LM. The lunar orbital science activity would be to fly the first SIM-bay package in the maiden flight of the modified Block II CSM. The primary technological objective was a demonstration of the capabilities of the SIM-bay and the modifications incorporated into the CSM, LM and EVA equipment. Three 4-hour 30-minute EVAs were planned (27 man-hours), with walking traverses of up to 2.5 miles from the LM during a landing duration of up to 54 hours. Operation of the SIM-bay was possible for 57–77 hours. The flight would also feature photography of future landing sites in the Davy Rille.

The priorities of this mission were to first deploy the scientific instruments (the

Passive Seismic and Heat Flow experiments), to continue the network of seismic instruments across the Moon, and to then proceed with the three geological traverses to the furrowed terra (EVA 1) and plain area (EVA 2) and the elongated cones and crater chains (EVA 3).

As this was to be the first mission of the J series, amended primary objectives for this flight included demonstrating the ability to perform an extended lunar surface mission that included up to three surface EVAs showing the capabilities of the SM SIM bay, and demonstrating the extended capabilities of the modified CSM and LM, the PLSS and SLSS, and that of the communications relay.

Apollo 17 (J-2)

This mission was originally manifested for a launch in either July or August 1971, aimed for the Marius Hills landing site. Scientific objectives were similar to those for the first J-series missions, and included photography of the Copernicus landing site area planned for Apollo 18. The three surface EVAs would use the first LRV on excursions ranging up to 6 miles maximum radius from the landing point. Duration on the surface and of the EVAs was similar to the previous J-series mission, with SIM bay experiment operation of 57–77 hours.

Mission priorities were primarily mobility-aided traverses to collect plain material and samples from volcanic land-forms. The deployment of the scientific instruments (PSE/HFE and Laser Ranging Retro Reflector) was a third priority, and could possibly have been deployed in the second or even the third EVA. With six ALSEP

The original Apollo 17 EVA traverses of the Marius Hills region.

scientific stations already deployed on the Moon, it was not as important to place this package early in the operations as it was to secure geological samples. Additional primary objectives for this mission included demonstrating the capabilities of the first Lunar Roving Vehicle.

Apollo 18 (J-3)

This mission was targeted for a landing at Copernicus in February or March 1972, in a duplicate of the previous J-mission profile. On this flight there would be no photography of other future landing sites, as the landings of the original Apollo series were being wound down.

At Copernicus, the mobility-aided geological traverse again had priority over deployment of the ALSEP package, but the mission would have continued to expand the seismic network with another PSE and the first use of a Questar telescope. The EVAs were to secure deep-seated materials from the peaks on the central formation of the crater, fill material from the crater floor, and volcanic (?) material from the domes, and to take photographs of the crater wall surrounding the landing site.

Apollo 19 (J-4)

The penultimate Apollo would have been the fourth J-series profile, assigned to July or August 1972, again without photography of future landing sites. As with

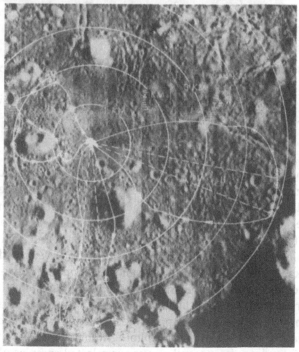

One of the original Apollo 18 primary landing sites at Copernicus, revealing the planned EVA traverses.

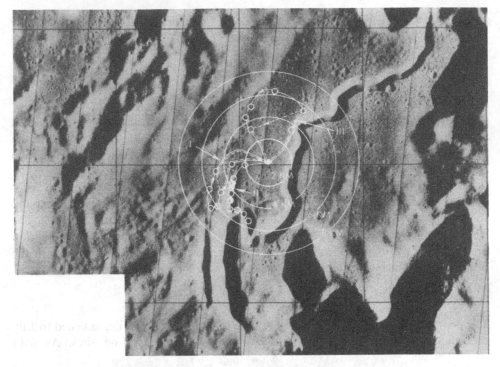

Apollo 19 EVA traverses at the Hadley Apennine site.

the previous mission, the deployment of experiments would have been of a lower priority, with a further PSE and a mass spectrometer to record any level of surface outgassing. Geological samples would have been obtained from the sinuous rille and Apennine Ridge, as well as from elongated depression materials.

Apollo 20 (J-5)

This was the shortest-lived Apollo lunar landing mission, being cancelled almost as soon as the planning document was published. Apollo 20 was originally targeted for Tycho, near the landing site of Surveyor 7. The crew would have returned part of the unmanned lander, as had the Apollo 12 crew. The mission, also not including any future landing-site photography, would have flown the fifth and final J-series profile.

Had the mission flown as planned, the initial priority was to deploy the final experiment package and complete the seismic, selenodetic network initiated by Apollo 11 and continued with the first full ALSEP on Apollo 12. In the three geological traverses, samples from highland features, lava flows and pool materials would have been retrieved. A specific primary objective for this flight, to be carried out during the third EVA, was to provide photographic records and verbal accounts of the Surveyor 7 landing site, as well as retrieve selected samples of the unmanned spacecraft for post-flight analysis.

Apollo 20 EVA traverses at Tycho.

The scope of Apollo
From this summary, it becomes clear that detailed planning for the systematic exploration of the lunar surface was under development in 1969, and was scheduled to include the flight of Apollo 20. But within weeks of the publication of the December 1969 document, the first of three Apollo missions was axed from the programme. Then, in April 1970, Apollo 13 suffered the in-flight explosion on the way to the Moon, requiring an immediate abort of the landing attempt and a dramatic and tense three-day struggle to return the astronauts safely to Earth. Five months later, two more Apollo missions were scrubbed from the launch manifest. Less than a year after the ambitious plans for ten lunar landings (classified as Phase I of the manned lunar exploration programme), and with only two landings successfully achieved, the whole programme was cut to just four more missions, stretched out to completion by December 1972 with Apollo 17.

The loss of the three Apollo missions (Apollo 15 (H-4), Apollo 19 (J-4) and Apollo 20 (J-5)) is a tale of lost opportunities and quashed dreams, and is discussed in the next chapter, in conjunction with the lost mission and lunar EVAs of Apollo 13. There were still those who hoped to follow the advanced J-series of missions with further upgraded Apollo hardware, but the events of 1970 firmly grounded those desires. These ideas – most of which never progressed further than the drawing board or scale models – would have seen the dream of Apollo journeying far beyond the Moon and deep into space.

1   *Definition of Management Responsibilities for the Apollo Lunar Exploration Program*, MSC Announcement No. 69–61, 6 May 1969 signed by Robert R. Gilruth, MSC Director.
2   *Assignment for Responsibility for Advanced Lunar Missions*. MSC Announcement No. 69–138, 30 September 1969, signed by Robert R Gilruth, Director of MSC.
3   *Future Apollo Lunar Exploration* press briefing (transcript), 9 October 1969, given by Joseph P. Loftus Jr., Manager, Program Engineering Office, MSC, Houston, Texas. NASA JSC Apollo collection.
4   *Recommend Lunar Exploration Sites (Apollo 11 through Apollo 20)*, a presentation by Farouk El-Baz/ MAS, to the Apollo Site Selection Board, 10 July 1969, NASA JSC Apollo collection.
5   *Manned Space Flight Weekly Report – July 28, 1969,* copy dated 29 July 1969, memo from G. Mueller to the Manned space Flight Management Council, NASA HQ, Washington DC. From the Apollo History Archive Collection, formerly (1988) at the NASA JSC History Office, Houston Texas; currently (2001) being relocated at the University of Houston at Clear Lake.
6   *Landing sites for Apollo 12 through Apollo 20*, NASA MSC memo, 27 August 1969, from C.H. Perrine, Assistant Chief, Systems Engineering Division (Code PD) to various MSC directorates. From the Apollo History Archive Collection, formerly (1988) at the NASA JSC History Office, Houston Texas; currently (2001) being relocated at the University of Houston at Clear Lake.
7   *Apollo Program (Phase 1 Lunar Exploration) Mission Definitions*, prepared by TRW for the NASA Apollo Program Office and the Advanced Missions Programme Office, MSC-01266, 1 December 1969. From the Jerry Carr collection (CB dated 16 December 1969).
8   *America's Next Decade in Space: A Report for the Space Task Group*, NASA, September 1969.
9   For example: *Mission Requirements, SA-506/CSM-107/ LM-5, G Type Mission, Lunar Landing* SPD9-R-038, 17 April 1969; or *Mission Requirements, SA-507/CSM-108/ LM-6, H-1 Type Mission, Lunar Landing* SPD9-R-051, 18 July 1969.

Changing the mission

This book has so far detailed the step-by-step approach of Apollo development leading to the two lunar landings in 1969 that met President Kennedy's goal. In addition, a review of what missions were to have followed these initial landings through Apollo 20 has been supplemented with details of the further application of Apollo hardware and procedures in Earth orbit, as well as around and on the Moon. In these schedules, Apollo hardware should have been flying regular missions in Earth orbit and towards the Moon at least a decade after Apollo 11. So what happened to change these plans?

From the mid-1960s there were numerous influences both inside and outside of NASA and the space programme that were to have a dramatic effect on what was hoped and planned for in manned space exploration during the 1970s and 1980s, and on what was actually budgeted for and flown. The reasons of how and why the focus of the manned programme shifted from comprehensive exploration of the Moon and Mars to more restricted missions in Earth orbit are summarised in the closing chapter, but it is worth recalling how the planning dramatically changed over a short period in the early 1970s.

WANING MOON

In the summer of 1969, Apollo was at its peak of success. The Moon had been reached, and appeared to be the main focus of space exploration for the next decade. But by November, even going to the Moon for only the second time appeared routine to the general public, and with the loss of television pictures of surface activities the media coverage waned.

In January 1969 a new President, Richard M. Nixon, had replaced Lyndon B. Johnson in the White House. Nixon had called Apollo 11 the 'greatest week since creation' as he talked to the astronauts on the surface; but he was faced with more terrestrial concerns, and the fulfilment of Kennedy's goal did not necessarily mean a blank cheque for NASA to fund its programmes. In a desire to review the future of the space programme before authorising unchecked funding for the space agency,

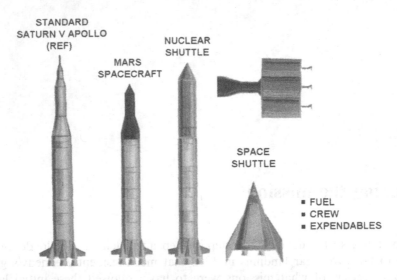

This October 1969 artwork shows various components of new hardware compared with the 'standard' Saturn V. These vehicles were based on the Apollo Saturn technology, and were proposed for manned planetary missions. A Mars spacecraft would be joined in EOR by three nuclear Shuttle stages, and the crew and supplies would be delivered by a new vehicle called the Space Shuttle.

Nixon appointed a Space Task Group that reported after a six-month study in September 1969. The STG indicated that the future of America in space would best be served by a balance of unmanned and manned activity that included paced progress to planetary spaceflight capability by the end of the century. A fast-pace programme would see a fifty-man space station, a lunar base, and Americans on Mars by 1985; a medium-pace programme suggested extensive unmanned martian exploration before committing to a manned programme, or development of a space station and Space Shuttle vehicle to create Earth orbital infrastructures and reduced costs of launching payloads into space, and then, perhaps, commit to going to Mars.

Costs ranged between $8–10 billion per year by 1980 for the fast pace programme, to $4–6 billion a year for the slower option. Nixon's response to the report was not immediate, and even the idea of sending men to Mars was not, at that time, attractive to the public, with growing inflation, social tensions, and the war in southeast Asia presenting them with more personal challenges. In 1970, President Nixon was not about to follow Kennedy's example and commit the nation 'before the decade is out', to a manned landing on Mars and return to Earth, and authorise what was an estimated price tag of $54–78 billion over the following ten years.

Funding for NASA had been a continual battle for some years, and in President Johnson's final budget submittal in early 1969 a requested NASA budget of $3.878 billion was almost 25% lower than the peak of agency funding in 1965. When the budget was resubmitted under the new administration in April, it had shrunk further to $3.833 billion, causing NASA Administrator Tom Paine to state that NASA faced 'difficult programme adjustments,' but which did not mean ending a scientifically

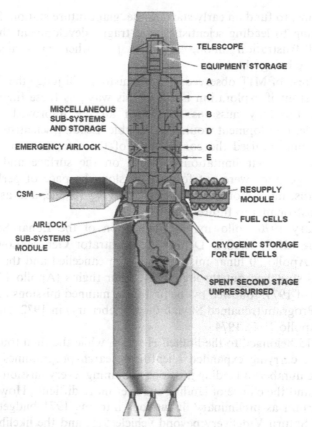

TELESCOPE

EQUIPMENT STORAGE

A

C

B

D

G

E

MISCELLANEOUS
SUB-SYSTEMS
AND STORAGE

EMERGENCY AIRLOCK

CSM

RESUPPLY
MODULE

FUEL CELLS

AIRLOCK

SUB-SYSTEMS
MODULE

CRYOGENIC STORAGE
FOR FUEL CELLS

SPENT SECOND STAGE
UNPRESSURISED

This 1967 artist's impression shows an advanced six–nine-man space station concept
based on converted Saturn V second and third stages resupplied by the Apollo CSM.
(Courtesy NASA via the British Interplanetary Society.)

active manned lunar exploration programme with Apollo, although anything
beyond that seemed uncertain. When the final budget was signed in November,
NASA would receive only $3.697 billion, making those 'difficult programme
adjustments' even more certain.

APOLLO 20: A DELETED MISSION

Rumours of a cut in the lunar flights had been circulating in scientific circles for
some months as protests were lodged against plans to curtail Apollo in favour of
freeing funds for commencing the larger space station and Shuttle programmes. The
Washington Post for 28 December 1969 reported that as many as four missions
would be scrapped, ending the programme at Apollo 16, and probably releasing $1
billion to begin what was expected to be a $50-billion development of space station
and Shuttle programmes. Another option would be to not fly the OWS and three

manned missions, to fund an early start on the 'giant future station' Such a sacrifice would, according to leading scientists, be a 'tragic' development that could cause bitterness and frustration among scientists just when vital answers in lunar exploration were being obtained.

Dr Frank Press of MIT observed: 'I think history will judge the American space program as a stunt if exploration is cut in this way', as these final four (J-series) missions were the only missions for which all the improved equipment and procedures under development would be available. Leading scientists involved in the Apollo programme praised the contribution of the astronauts of Apollo 11 and Apollo 12, despite their limitations of time on the surface and restrictions in equipment. Suggestions were put forward to slow the pace of perhaps one flight every six months, more mobility in a surface roving vehicle, and assigning 'one or two' scientist-astronauts to future flights.

On 4 January 1970, following the dedication of the Lunar Science Institute adjacent to the MSC, NASAs Deputy Administrator George Low informed the press that the Apollo 20 lunar mission had been cancelled and that the remaining seven missions stretched out through 1974. Four flights (Apollo 13–16) would fly before the end of 1971, followed by the first three manned missions under the Apollo Applications Program (renamed Skylab on 17 February) in 1972, and by Apollo 18 in 1973 and Apollo 19 in 1974.

Apollo 13–15 belonged to the limited H series, while the final four missions were of the J series, carrying expanded scientific research programmes and hardware. With a definite number of landing missions remaining, every mission became equally as important, and the choice of landing site ever more difficult. However, the future was still uncertain as preliminary discussions into the 1971 budget confirmed the suspension of Saturn V delivery beyond vehicle 515, and the likelihood that funds would not be available to resume production. America had lost its large lift capacity, but in the same meetings, funding for the next programme – the Space Shuttle – was authorised. The end of Apollo had been suspected for some time – and now it was certain.

On 7 January, disposition of Apollo 20 hardware was issued by MSC, with the decision to use the first and second stages to launch the Skylab space station, while the third stage would be converted as the back-up Orbital Workshop that might possibly be launched as a second workshop in 1976 – depending, of course, on adequate funding. Requirements for the Apollo 20 pressure suits and PLSS were deleted, and production of LM-14 was stopped pending its disposition, while CSM-115A would be retained, for the present, to be used possibly on the second OWS programme. Any experiments planned for the surface or SIM-bay would be revaluated and possibly included on the remaining flights. The remaining Apollo flights were rescheduled, and landing sites were re-evaluated, with the launch readiness dates dependent on AAP schedules.

To accommodate the AAP/Skylab series of missions (one unmanned and three manned launches), NASA was considering halting Apollo lunar operations after flying Apollo 16 in 1972, resuming the flights with Apollo 17 and Apollo 18 in 1974, and completing the approved lunar programme with Apollo 19 in 1974. The

Apollo mission planning, January 1970

Mission	Saturn vehicle	CSM serial	LM serial	ALSEP array	Mission type	Launch readiness	Revised
Apollo 13	508	109	7	B	H-2	1970 Mar 12	1970 Apr 11
Apollo 14	509	110	8	C	H-3	1970 Jul 8	1970 Sep 30
Apollo 15	510	111	9	A-2	H-4	1970 Oct 30	1971 Feb 24
Apollo 16	511	112	10	D	J-1	1971 Mar 29	Under review
Apollo 17	512	113	11	E	J-2	1971 Jul 20	Under review
Apollo 18	513	114	12	F	J-3	1972 Feb 19	Under review
Apollo 19	514	115	13	G	J-4	1972 Jul 14	Under review
Apollo 20	515	115A	14	H?	J-5	1972 Dec	Deleted

cancellation of the Apollo 20 flight also meant that the first of four LRVs could be carried on Apollo 16.

Sharpening the budget axe
On 13 January the top management at NASA gave a press conference in Washington to outline NASA's future plans.[1] Tom Paine stated: 'This week I am taking action to redirect portions of our space program to bring NASA's total operations in line with the budget which we will work with in FY 1971. Within a short time, President Nixon will make an important statement on the future of America's space program, setting forth his strong support of a vigorous and forward-looking program. We recognise that [there are] many important needs and urgent problems we face here on Earth. America's space achievements in the 1960s have rightly raised hopes that this country and all mankind can do more to overcome pressing problems of society. The space program should inspire bolder solutions and suggest new approaches. A strong space program continues as one of this nation's major national priorities. However, we recognise that under current financial restrictions, NASA must find new ways to stretch out current programs and reduce our present operational base.'

Paine explained that the goal was to move NASA forward, and in strength, but still achieving greater economy into the new decade. This, then, was the challenge that NASA management faced in 1970, and Paine expressed confidence in its ability to meet that challenge, hoping that a reduction would not mean a sacrifice in ensuring major achievements in the 1970s. Paine also stated: 'While we will be reducing our total effort, we will not dissipate the strong teams that sent men to explore the Moon and automated spacecraft to observe the planets.'

A few days afterwards, the media began reporting that NASA was to axe 50,000 jobs over the following eighteen months because of the reduced budget. By early February, President Nixon outlined his 'important statement on the future of America's space program' by slashing the FY 1971 NASA budget to $3.4 billion – 12.5% lower than for FY 1970, and the lowest for a decade. The proposals threatened the closure of two or three NASA field centres, postponed unmanned flights to the planets, threatened Skylab, and suggested even more cuts in the Apollo

This Surveyor 7 view of a region near Tycho shows the terrain which Apollo 20 would
have to have negotiated to achieve a landing near the unmanned spacecraft.

lunar programme. However, in the same statement there was an $88-million increase
for the development of the larger space station and Space Shuttle, with the first space
station placed in orbit 'by 1978'. In addition, the first manned Mars landing would
not occur before the '1980s or later'. President Nixon openly supported the
unmanned 'grand tour' missions (which became Voyager) to the outer planets,
beginning in the late 1970s, and applauded the programme's new emphasis on
unmanned planetary exploration, as well as the forward-looking development of the
Space Shuttle and the space station.

Apollo 20 crew assignment

The termination of Apollo 20 also had an effect on the Astronaut Office, with three
seats into space being lost. At the time of cancellation, Apollo 20 was still more than
two years away, and no crew had officially been assigned to the mission, although
several astronauts had been associated with such an assignment. In early 1970, teams
of astronauts were in training for Apollo 13–15, and their back-up crews were in line
to fly Apollo 16–18. Official naming of the crews for Apollo 16 through 18 was
expected after the flight of Apollo 13 in the spring, and that would have identified
back-up assignments that would reveal the assignments through Apollo 20. Based on
Slayton's planning and actual assignments, the Apollo crewing from 1970 suggested
the following Apollo 13–20 crew planning:

Apollo   Back-up crew rotation

13	Back-up crew rotate to prime crew Apollo 16
14	Back-up crew rotate to prime crew Apollo 17
15	Back-up crew rotate to prime crew Apollo 18
16	Back-up crew rotate to prime crew Apollo 19
17	Back-up crew rotate to prime crew Apollo 20

Back-up crews for Apollo 18, 19 and 20 were essentially dead-end assignments, with no rotation to later flights (Apollo 21, 22 or 23). In addition, there was a requirement for three Skylab prime crews and at least two back-up crews to be announced in 1971 to fly the missions in 1972 (see the author's *Skylab: America's Space Station*, pp. 303–305).

An Apollo 20 prime crew (original Apollo 17 back-up crew) of Conrad (CDR), Weitz (CMP) and Lousma (LMP) is based on various rumours that they were to have been assigned to the flight. The origins of such a potential crewing assignment came from the Astronaut Office and Conrad himself in the early 1980s, and later comments from Deke Slayton suggest that it would have resulted in a good crew; but there were also others in contention. Both Weitz and Lousma have stated that they have heard of such suggestions, but nothing was ever made official or mentioned in private, as when the flight was cancelled it was far too early to have identified the crew, even within the Astronaut Office. However, Conrad was known to have wanted a second trip to the Moon, and as the later landings would have been in extremely challenging areas, past experience on the Moon would certainly have been advantageous. But there were only so many seats to the Moon, and more than enough astronauts available to fill them many times over.

It appears that the 'writing was on the wall' in the Astronaut Office, and that the Apollo 20 flight was to be scrapped long before Apollo 12 flew. When Stafford informed Conrad that he would never be assigned a second landing, and that he was needed as the first Skylab crew Commander, Conrad enlightened his fellow crewmates Gordon and Bean: 'Guys, if you want to fly again, follow me to Skylab.' Bean took the advice, and became the second Skylab crew Commander. Gordon decided to wait for a landing and a command seat, and indeed was assigned to the back-up position on Apollo 15 and was in line to rotate as prime Commander of Apollo 18.

Paul Weitz had worked on the Apollo 12 support crew, but also performed significant assignments on the AAP. It is interesting to note that when Conrad moved to Skylab after his Apollo 12 flight, Weitz went with him and became his Pilot on Skylab 2, and Lousma served as Pilot to Bean on Skylab 3.

In April 2001, Jack Lousma wrote: 'Frankly, I have heard rumours about my assignment to the last mission, but I can't confirm them. Nobody in a position to act, including Deke, ever told me eyeball to eyeball or in writing, that I would fly on one of those last three missions. I can say that I was positioned to be seriously considered along with others. As a support crew-man for Apollo 9 and 10, my prime responsibility was as the prime crew's rep to work with the Cape test and check-out team to get LM-3 and LM-4 through its flight validation testing in the Manned

Space Operations Building, the Vehicle Assembly Building, and on the launch pad. By the time I had done all that, I had spent about two years and accumulated about 4,700 hours in the LM simulator just by being on hand when sim time was available. I was also directly involved in developing onboard malfunction procedures for LM-3. As support crew-man for Apollo 13, I was the test and check-out guy, this time for the CSM, but I worked with the prime crew on lunar surface exploration training for my shifts as CapCom. Further, I worked with Jerry Carr to get the LRV through development. My collateral duties were focused on lunar operations for 2–3 years.' And as Lousma aptly pointed out: 'If I was being groomed for a lunar landing, *I was ready!'*

In trying to suggest a crew for Apollo 20, it is worth recalling the rotation pattern of former CM Pilots from Apollo 7:

Apollo	CMP	Rotation
7	Eisele	Back-up CMP Apollo 10 pending resignation from NASA, never intended to rotate to Apollo 13, and replaced by Shepard back on flight status
8	Lovell	Back-up CDR Apollo 11; planned Commander Apollo 14 reassigned Commander Apollo 13
9	Scott	Back-up CDR Apollo 12; planned Commander Apollo 15
10	Young	Back-up CDR Apollo 13; planned Commander Apollo 16
11	Collins	Offered back-up Commander Apollo 14 and command of Apollo 17; declined, and replaced by Cernan (Apollo 10 LMP) as no other CMP available
12	Gordon	Back-up Apollo 15; Commander Apollo 18
13	Swigert	Replaced Mattingly as prime CMP on '13'; in this system, Swigert would therefore have been in line to receive the command of Apollo 19
14	Roosa	In reality, became back-up CMP Apollo 16 and 17; on this rotation he may have been a leading contender for the command of Apollo 20

There is also the question of the scientist-astronauts, many of whom had mixed feelings about making a lunar flight, with at least one (Henize) stating that he became an astronaut to fly a mission to an AAP space station in order to perform scientific research in his field (astronomy), and flying to the Moon was not a consideration – at least not for him. However, if a scientist-astronaut had been offered the chance to fly to the Moon, it is doubtful whether many would have declined. Don Lind firmly believes that he would have been the second scientist on the Moon, and applied for, and apparently received, helicopter training (part of the LM crew training programme). With his previous piloting background and support assignments in developing ALSEP and Apollo 11 EVA timelines, there seems little doubt that this could have been achieved, but probably not on main-line Apollo. A logical place for him would have been as back-up to geologist Jack Schmitt on

Apollo 18, and to then rotate to fly as LM Pilot on Apollo 21 – but that was under AAP, and was by no means certain even at the time of Apollo 11. Slayton reassigned Lind to AAP (Skylab) Branch Office in August 1969 when it was becoming apparent there would not be a flight assignment for him under Apollo.

The Apollo 20 crew could therefore equally have been Roosa–Weitz–Lousma. Then there was the question of experience in the last two or three difficult landings that could have led to the reassignment of former LMPs as CDRs. This is supported by the known assignment of Haise as back-up Commander Apollo 16 to rotate to Apollo 19, and suggestions that Mitchell may also have been rotated from LMP Apollo 14 to back-up Commander to Apollo 18, to return to the Moon on Apollo 20 – an assignment which Conrad had been told was not possible. This raises the question: had Apollo 13 landed, would Haise have received the Apollo 16 back-up crew assignment and flown Apollo 19? Who knows what might have been.

Flight planning, April 1970
On 6 April 1970, Ken Kleinknecht, Manager of the Skylab Program Office, presented an overview of programme scheduling at MSC in light of these changes. The overhead presentations indicated that the planning was for Apollo 13 and Apollo 14 to fly in 1970, Apollo 15 and 16 in 1971, and Apollo 17 in early 1972. Skylab 1 (unmanned OWS) would launch in the autumn of 1972, followed by the first manned flight of 28 days, then two 56-day missions into the spring and summer of 1973. There was also an indication of a fourth Skylab manned launch (or SL Rescue availability) for a 1973 launch, followed by Apollo 18 and 19 in 1974, Skylab B (second unmanned OWS) with one manned flight launched at the end of 1974, and two further manned visits to the second Skylab in the first half of 1975.

Spacecraft allocations for these missions were listed as:

Apollo	CSM	Skylab	CSM
13	109	2	116
14	110	3	117
15	111	4	118
16	112	X (Rescue?)	119
17	113	Skylab B	115A
18	114	Skylab B	Not assigned
19	115	Skylab B	Not assigned

Any decision on Apollo 18/19 would impact a 1972 Skylab launch and planning of manned flights. MSC favoured moving Apollo 18 ahead (before Skylab). At the Cape, the teams at KSC would have to maintain six-month manning intervals for each launch, and staff would eventually have to be coordinated to work both on Apollo and Skylab hardware processing.

In an effort to compensate for the loss of Apollo 20 EVAs (and rumours of further cuts), MSC evaluated three periods of surface activity on Apollo 14 (H-3) under the assumption that the third surface period could be easily dropped if it proved unworkable. Indeed, evaluation of the proposal recommended retaining the H series as two EVA missions, and only the J missions to plan for three surface periods.

APOLLO 13: THE LOST EVAS

Despite the cancellation of one lunar mission on Apollo 20, NASA was still faced with further cuts in the programme in order to fund other activities. However, the next mission, Apollo 13, was funded, and was preparing to take a third team of astronauts to the Moon in the spring of 1970. By now even the media was becoming disenchanted with the thought of yet another (but only the fifth) trip to the Moon, and coverage of the mission was not as extensive as that for earlier missions. Even the prospect of Jim Lovell making a record fourth spaceflight, and the expected landing in the highlands of the Moon instead of the relatively flat areas of the two landings, did nothing to heighten the coverage – but all that was to change after the mission was launched.

Plans for Apollo mission H-2, 1969
On 3 February 1969, a little over a year before Apollo 13 flew, NASA released a forecast for the next five Apollo missions (Apollo 9–13), scheduled to be flown over the ensuing twelve months. At that time, Apollo 13 was without an assigned prime crew or a target landing site, as it was the second back-up mission for Apollo 11 should that flight and Apollo 12 fail in their attempts at achieving the first lunar landing. Launch of '13' was manifested for 1 December 1969, by Saturn V 508, on what was expected to be the third lunar landing mission of eleven days, using CSM 109 and LM-7.

Following the success of Apollo 9, by late March it appeared that the Apollo 10 mission in May would probably clear the way for Apollo 11 to attempt the first landing in July. Apollo 12 would then make the second landing, in September, and so Apollo 13 was advanced to a launch on 10 November.

Providing Apollo 11 succeeded, the type of mission that Apollo 13 was to fly had been designated H-2 – the second in a series of four landing missions (Apollo 12–15), with the capability of pinpoint landing on the surface within a 0.6-mile target circle. During the 36-hour maximum stay time there would be two planned EVAs, both on foot and relatively close to the landing craft. The first EVA would include the deployment of the second and more complex Apollo Lunar Surface Experiment Package (ALSEP) and the collection of a few samples. The second EVA featured an expanded sample collection programme during a geological traverse away from the landing area to a prominent surface feature. The mission would also feature limited research from lunar orbit by the CM Pilot – mainly taking the opportunity to photograph potential follow-on landing sites and targets of opportunity.

But where was Apollo 13 going to land?

A place to go

During final preparations for the Apollo 11 launch from Cape Kennedy in June 1969, the flight dynamics teams were quietly confident that the mission would be accomplished. If Armstrong and Aldrin achieved the landing, a more accurate targeting would be planned as early as the second landing, on Apollo 12, or perhaps the third, on Apollo 13. Reduction of the elliptical landing footprint would result in a more precise landing in rougher terrain, allowing the astronauts to explore more geologically interesting sites than the relatively flat mare, and thereby increasing the scientific return from each flight.

As the more interesting sites were outside the equatorial zone for which the early Apollo missions were targeted, it would be necessary to relax both the propellant margins and the choice of back-up landing sites. Propellant usage in the inclined orbit required to reach these higher latitudes would be much higher than required for the equatorial zone. The option of aiming for a back-up landing site if the primary landing area could not be reached was acceptable, if restricted to similar mare terrain, but it would become more difficult to train a crew for landing at two different sites as the geology became more complex. It was decided that although back-up sites would not be assigned to the landing missions, there would be alternative sites available to allow some contingency in the event that the primary site could not be reached due to operational constraints. Although this plan still prevented very-high-latitude sites from being visited by Apollo crews, it offered a broad range of flexibility for the H-series of missions, and removed the requirement that the 'H' landing sites should be free of surface relief.

During the meeting of the Apollo Site Selection Board on 10 June 1969, the subject of a landing site for Apollo 13 was discussed. The targeting of Apollo 13 would be based on a successful landing of Apollo 11 on the relatively flat Sea of Tranquillity in July, and the demonstration of a pinpoint landing by Apollo 12 near the unmanned Surveyor 3 lander, in the Ocean of Storms. At the time this was planned for September, but on 29 July, following the success of Apollo 11, Apollo 12 was moved to November as part of a more relaxed flight schedule. Apollo 13 slipped to March 1970.

Some of the geologists wanted to send an Apollo to a large crater as soon as possible, and some consideration was given to the craters Hipparchus and Censorinus. At the meeting, however, the consensus of the Group for Lunar Exploration Planning was that the primary landing site for Apollo 13 should be the Fra Mauro formation, some 112 miles east of the intended Apollo 12 site. The whole area was generally accepted to be ejecta from the Mare Imbrium basin, and was long considered an important target for an early Apollo landing. Samples from this region would allow the dating of the geological event that created the basin. This could also be linked to similar events on Earth and other planets during what would have been a very dynamic period in the formation of our Solar System.

The LM of '13' was not to be targeted for Fra Mauro itself, but to an area just to the north. This would allow the LM crew to fly a clean line of approach from the east to a tight 0.6-mile landing circle, and also remain near a blocky rim. The first area chosen was of exactly the type of terrain rejected when sites were first selected for the first landing attempts. When detailed high-resolution photographs of the area were requested for precise mapping, it was found that the only photographs available were four single frames from the 1967 Lunar Orbiter 3 spacecraft – and these were taken more for scientific interest than for actual site planning for a subsequent landing attempt.

During 16–17 October, the Group for Lunar Exploration Planning held another meeting to refine targets for Apollo 12 through 20, based upon the experiences of Apollo 11. Fra Mauro was confirmed as the primary landing site for Apollo 13, and although back-up sites were not allocated, two alternatives were offered in case of operational constraints over launch time and lighting conditions, which might change over the months. For Apollo 13, the alternative sites offered were the crater Alphonsus and the Hyginus Rille.

On 30 October, as a result of recommendations by the Site Selection Board, the Manned Spacecraft Center recommended that Apollo 13 should be targeted for Fra Mauro, with the Hyginus Rille as alternative site. During the November flight of Apollo 12, Command Module Pilot Dick Gordon made use of the close proximity of the intended Apollo 13 site to the ground track he was flying over the Ocean of Storms. With the bonus of advantageous lighting conditions, he secured a collection of photographs of the Fra Mauro landing area. By December 1969, Fra Mauro was the official prime target for Apollo 13.

The actual landing site was narrowed down to a 1,200-foot pit, about 25 miles north of Fra Mauro. The main objective for the Apollo 13 crew was to land near a small 'drill-hole' crater, given the name Cone because of its shape. The LM landing area was moved to a smoother site to the west of Cone when the first site selected was thought to be too close to the debris field for a safe landing. It was important, however, that the LM touched down close enough to the crater to allow a geological traverse by the crew within the safe walking distance constraints.

By the time the LM approached the surface at the Cone crater site, and with a pinpoint landing having already been demonstrated by the crew of Apollo 12, the skills and training of the Commander would be such that, given the availability of extra fuel to hover, a closer landing site might be found in the landing area.

By using the hybrid trajectory on the way out to the Moon, and the Service Propulsion System on the Service Module in lunar orbit to place the LM in its descent orbit trajectory prior to undocking, there would be a greater margin of fuel on the LM. This additional fuel reserve would enable the crew to hover for a longer time than was available during Apollo 11, and would allow the Commander to select a suitable site in what was expected to be much rougher terrain.

Apollo 13 and Apollo 14 crew selection

By March 1969, America's first man in space, Al Shepard, had been restored to full flight status. After being medically grounded since 1963, Shepard wanted an early flight assignment, having missed all of Gemini and half of Apollo. Deke Slayton also wanted to assign his friend to an early mission. When Slayton told the eighteen astronauts of the prime and back-up crews for what became Apollo 7, 8 and 9 that the first landing crew would be selected from this group, he had no idea that, with

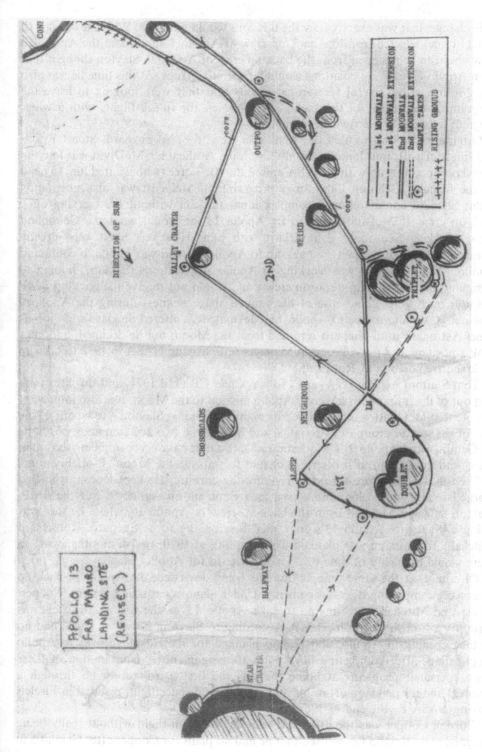

A NASA sketch of the two Apollo 13 EVA foot-traverses planned for Fra Mauro on 15–17 April 1970.

little change, that was exactly how the missions would be flown. With the Apollo 10 and 11 crews selected from the back-up crew of Apollo 7 and 8, and the Apollo 12 crew about to be assigned from the back-up crew of Apollo 9, Slayton thought that from Apollo 13, crewing positions would be open to change. At this time he was also being informed by several veteran astronauts that they were looking to leave the programme and move on to other goals. This left the future flights with a wider range of astronauts from which to choose.

Originally, Slayton wanted the Apollo 13 crew to be Al Shepard, Stuart Roosa and Jim McDivitt, but that idea did not last long. At the time, McDivitt was looking at several options both within NASA and in the Air Force (who wanted him to head up the Manned Orbiting Laboratory programme). McDivitt was also not happy about Shepard commanding a complex lunar mission without first serving on a back-up crew. If McDivitt were to fly Apollo 13, he would want the Command himself, taking Scott and Schweickart with him. However, Scott was already assigned to his own crew as the back-up to Apollo 12, and was in line to command Apollo 15. Schweickart was working in Apollo Applications (Skylab), hoping to command one of the long-duration crews, and would not receive the back-up LMP position on Apollo 12 because of his bout of space sickness during the Apollo 9 mission. As an alternative to Apollo 13, McDivitt was offered Shepard's old job as Chief Astronaut until Shepard returned from the Moon; but he declined, and later took a position as Apollo Program Manager until leaving NASA in 1972 to take up an executive position at Rockwell.

Tom Stafford assumed Shepard's role as Chief CB until 1971, and this also took him out of the running for a second Apollo mission to the Moon. It is also important to note that McDivitt was considered by many – including Slayton – to be one of the best of the second group of astronauts, and he received two key command positions on Gemini 4 and Apollo 9. The question was: could he have accepted being second in command to Shepard, if it offered a chance to walk on the Moon? Probably not.

Slayton was therefore forced to re-think the crewing. He kept Roosa, who had worked solidly since 1966, and served as a CapCom on Apollo 9. For the LMP, Slayton assigned Mitchell from the back-up crew of Apollo 10, where he too, was doing a good job. Apollo 13's crew was therefore set to be Shepard, Roosa and Mitchell. The mission was planned for the spring of 1970, twelve months away, so there would be plenty of time to prepare Shepard for Apollo.

Of course, at the same time as this was being discussed, the Apollo 10 back-up crew were completing their assignments. Under Slayton's rotation system, Cooper, Eisele and Mitchell were in line to take Apollo 13 to the Moon. Mitchell was certainly a firm favourite for LMP. According to Slayton, however, Eisele had no further opening in Apollo, and he had planned for him to move over to Apollo Applications after completing his Apollo 10 assignment. A combination of these plans, personal problems at home (he was the first astronaut to go through a divorce) and no real support in his making a second spaceflight, resulted in Eisele's leaving NASA by the end of 1969.

Gordon Cooper was the astronaut who lost a Moon-flight without really being assigned to it. Slayton has written that he had no plans for Cooper after Gemini 5 in

1965. He helped fill out dead-end assignments on the back-up crews for Gemini 12 and Apollo 10 before Slayton began reassessing crewing from Apollo 13 onwards. To secure a seat as Commander on Apollo 13, Cooper would have had to do a good job in the back-up role on Apollo 10, indicating a dedication to training, which according to Slayton he never did. When it was rumoured that Shepard was to be offered Apollo 13 over Cooper, without first serving as back-up, Cooper was not very happy, especially since he had filled two back-up roles as well as flying two 'duration' missions. Shepard had had 15 minutes of flight experience in 1961, and was back-up to Cooper on his Mercury flight in 1963, but he had had no other crew assignments. With no suggestion that he would be offered the command of Apollo 14 or 15, Cooper also left NASA, and retired from the USAF within twelve months.

Slayton then considered the flight crew just off Apollo 10 to take the role of the back-up crew. Stafford took Shepard's old job, Young was promoted from CMP to Commander, and Cernan would remain as LMP. To fill Young's place as CMP, Slayton brought in Jack Swigert, who had done a good job on the support crew of Apollo 7, and was doing an equally good job on the support crew of Apollo 11. Thus, the original back-up crew for Apollo 13 (Young–Swigert–Cernan) would expect to rotate to the prime crew for Apollo 16. Yet again, this was not to be. Cernan turned down the role as back-up LMP on '13', telling Slayton that he wanted to wait for a command of his own – which was a gamble, but that one he was prepared to take. Slayton accepted this, and assigned Charlie Duke as LMP on Young's crew, giving Cernan his own crew as back-up on Apollo 14.

Slayton therefore forwarded the crewing selection for Apollo 13 and 14 to Washington as:

Mission	Type	Position	Prime	Back-up
Apollo 13	H-2	Commander	Shepard	Young
		CM Pilot	Roosa	Swigert
		LM Pilot	Mitchell	Duke
Apollo 14	H-3	Commander	Lovell	Cernan
		CM Pilot	Mattingly	Evans
		LM Pilot	Haise	Engle

This was promptly turned down, as it was thought that Shepard's crew did not have enough experience and that there was insufficient time to train them before Apollo 13 flew. To solve this, Slayton simply exchanged the prime crews, asking if Lovell thought his crew would be ready in time to take Apollo 13, instead of Apollo 14, to the Moon. Lovell's answer was 'yes'. It would be a push, but the experience was there in Lovell's crew to make it happen; after all, the target was still the Moon, and they would be going three months early. Slayton therefore resubmitted the plan to Washington, and this time it was accepted. Lovell obtained an earlier flight to the Moon, and Shepard was also going – if a little later. Young was in line to command Apollo 16, and Cernan's gamble paid off as he was in line for Apollo 17 (which turned out to be the last of the series).

Finally, on 14 August 1969, the crews for Apollo 13 and Apollo 14 were named:

Mission	Type	Position	Prime	Back-up	Support
Apollo 13	H-2	Commander	Lovell	Young	Brand (CM specialist)
		CM Pilot	Mattingly	Swigert	Lousma (LM specialist/CM duties)
		LM Pilot	Haise	Duke	Kerwin (scientist-astronaut)
					England (Mission Scientist)
Apollo 14	H-3	Commander	Shepard	Cernan	Chapman (Mission Scientist)
		CM Pilot	Roosa	Evans	Pogue (CM specialist)
		LM Pilot	Mitchell	Engle	McCandless (LM specialist)

The Mission Scientist position was a new role in which a scientist-astronaut on the support crew served as an interface between the flight crew and the scientific community. This continued with all remaining Apollo missions and into the Skylab programme.

Preparing for flight
From August 1969 until launch in the spring of 1970, Lovell's crew trained for a mission to Fra Mauro, and on 8 January 1970 they gained a little extra preparation time when NASA announced that the launch had been slipped one month to 11 April, to provide a more detailed analysis of the flight plan and intended landing site.

All progressed well until just five days before launch, when Mattingly was found to have been exposed to rubella by coming into contact with one of Charlie Duke's children, whose friend had the measles. Lovell, Haise and the back-up crew were all immune, as they had all had the illness in their childhood. Mattingly, however, had not. Rubella has an incubation period of ten days, and if Mattingly were to develop the illness when alone in lunar orbit, and with his colleagues on the surface, the mission would be placed in great jeopardy. It was therefore decided to remove Mattingly from the flight.

Other options open to flight planners included delaying the flight for a month or two. Once Mattingly recovered – if, indeed, he developed the measles – they could fly the intact crew. This, however, also meant a significant change in lighting conditions at the landing site and recovery sites, and the keeping of the crew in peak condition for several weeks, which would be very difficult.

Alternatively, the back-up crew could fly the mission, and Lovell's crew would rotate to Apollo 16. With that option it has to be remembered that Young's crew had not received priority in the training programme for Apollo 13. Also, Lovell and Haise had trained extensively as a team for their specific landing area, and had

Apollo 13 astronauts Jim Lovell (left) and Fred Haise simulate the lunar traverse at Kilauea, Hawaii, in December 1969. Both astronauts carry cameras and communication equipment to talk to astronauts who were to have served as CapComs during EVA. The crew also trained with the lunar tools which they planned to use at Fra Mauro.

worked with the flight hardware longer than had Young and Duke, who were already beginning to focus their training on assignment on Apollo 16.

It was therefore decided that the best and easiest option would be to exchange Swigert for Mattingly and meet the 11 April launch date. For the next few days, Swigert was put though his paces by simulation teams, trainers and controllers, and by Lovell and Haise, before he was allowed to fly the mission. Mattingly would join Young and Duke in supporting the Apollo 13 mission on the ground, and then be assigned to fly Apollo 16.

Back to the Moon
Apollo 13 left Earth on 11 April 1970, carrying Lovell, Swigert and Haise. It was a record fourth trip into space for Lovell, and at the same time he became the first man to make a second trip to the Moon. It was the first flight for Swigert and

Haise, but it would be a new adventure for all three, in a way that none of them suspected.

Events leading up to the evening television broadcast of 13 April had progressed so smoothly that if the rest of the flight went as well, there was a chance of the flight controllers and support teams becoming bored. It is now etched in the pages of history what happened after the television show had finished, and boredom certainly played no part in the new mission of Apollo 13.

But what if Apollo 13 had not suffered its in-flight explosion, and had gone on to achieve its goal of landing at Fra Mauro and returning safely to Earth? What was the lost mission of Apollo 13, and what had Lovell and his crew trained to accomplish?

In February 1971, some of what was planned for Apollo 13 was demonstrated by the crew of Apollo 14, who were targeted to the Fra Mauro formation after the return of Apollo 13 – such was the importance of the landing site. (Some of the images used here come from Apollo 14, and could easily have been taken by Lovell and Haise had they landed as planned.

In one respect, Shepard actually managed to fly '13', as he wanted to do back in 1969. On Earth, Lovell watched Shepard carry out some of the tasks he had trained to accomplish at Fra Mauro. It was ironic that by accepting a quicker route to the Moon on Apollo 13, Lovell lost his Moon-walk, and by being delayed to Apollo 14 for more training, Shepard *kept* his Moon-walk.

Outward: 21.00, 13 April–15 April 1970

During the planned flight out to the Moon,[2] Lovell and his crew were to conduct further inspections of the LM to prepare it for landing at Fra Mauro, and continue with regular television broadcasts from space. The outward flight to the Moon was planned to last 73–75 hours, with Apollo 13 arriving at the Moon around 18.00 Houston Time, on Tuesday, 14 April. Just prior to midnight Houston Time, the Service Propulsion System of the Service Module was to perform a 5-minute 57-second burn, to place the docked spacecraft in a lunar orbit of 170 x 60 miles inclined at 5.3° at about 77 hours 25 minutes.

On previous Apollo lunar missions, the spacecraft had a near-circular orbit at 70 miles altitude, and then the LM had undocked and used its descent engine to lower its orbit to 10 miles before committing itself for landing. On Apollo 13, because of the terrain in the landing area, the need for as much fuel as possible to hover meant that the docked spacecraft would perform the descent orbit insertion on the second revolution. The SPS engine would be used for 23 seconds at GET 81 hours 45 minutes to lower the orbit to 60 x 8 miles above the surface. This manoeuvre would give Lovell an extra 14 seconds of hovering time to look for a safe landing site.

Two revolutions later, during Rev 4, came the first opportunities to photograph the three candidate landing sites. Swigert was to take photographs of Censorinus from this low orbit, and then later a second set from a higher orbit. At GET 86 hours 10 minutes, a crew rest period was to have begun, to last for approximately 8½ hours.

By Rev 10, at GET 96 hours 30 minutes, Lovell and Haise would have begun a check-out of the LM *Aquarius* prior to undocking on Rev 12 at GET 99 hours 16

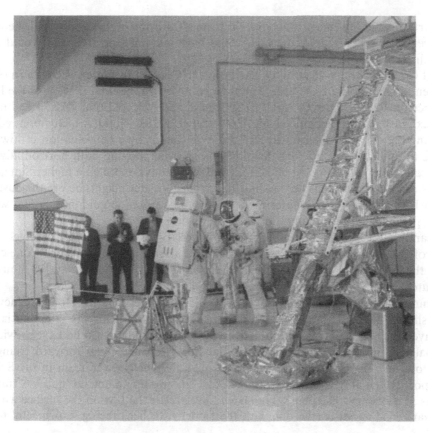

Jim Lovell (*left*) and Fred Haise participate in a 1-g suited EVA simulation at KSC during February 1970. Here they are practising the deployment of equipment near the LM forward strut and MESA facility on the descent stage.

minutes. When the two spacecraft undocked, Swigert would have visually inspected *Aquarius* and Lovell and Haise would have looked over *Odyssey* prior to the separation manoeuvre that began the LM descent to the surface.

Lovell and Haise would have begun the descent to the Moon on Wednesday, 15 April, taking two revolutions and four hours. While the two astronauts in the lander set up for landing, Swigert would complete landmark tracking from the vantage point of just eight miles above the surface. This was the first time that this would be attempted on an Apollo mission, and Swigert's task was to determine whether it was possible. At such a low orbit, the velocity was dramatically increased, and the objective was to determine whether such a low pass was feasible to track landmarks on or around the planned landing site.

After completing this task and ensuring that the LM was ready for descent, Swigert would have fired the SPS engine for 4 seconds at GET 100 hours 35 minutes to circularise the orbit at 59 x 62 miles for the duration of the landing phase. As the CSM came around to fly over the landing area again, on Rev 13, there came a second opportunity for landmark tracking – this time from the higher orbit, and therefore 'slower'. During this time, the LM would have remained passive, and would have visually tracked the CSM.

At GET 103 hours 21 minutes, the Power to Descent Initiation (PDI) burn of the descent engine would have been commanded, to begin the final descent to the lunar surface on Rev 14. During Apollo 12, Conrad had reported that dust kicked up by the exhaust of the descent engine had created a dust cloud, which obscured the landing area. It was determined that this was caused by a combination of low Sun angle at landing, and the fact that LM *Intrepid* was moving both horizontally and vertically at landing. For Apollo 13, Lovell would have expected a higher Sun angle, and would have selected ther landing area from a greater height, before the dust was disturbed. Bringing *Aquarius* straight down, the automatic guidance system would have prevented any horizontal drift of the vehicle.

Landing was targeted between a pair of overlapping craters called Doublet and a row of three overlapping craters called Triplet, about 2,000 feet west of Cone crater (see the EVA map, p. 279). The landing was planned to occur at 103 hours 42 minutes GET (20.55 Houston Time) on Flight Day 5, 15 April 1970.

Immediately upon landing, the crew would have prepared for an emergency lift-off, should the occasion arise. After verification that the LM was safe, and having received the 'Go' for a stay on the surface, the two astronauts would have provided a verbal description of the view of the landing site out of the forward triangular windows of *Aquarius*. This description would have helped the team in the Science Support Room at MCC-H identify the exact landing site to amend the EVA traverse plans as necessary. The plan was to have Lovell settle the LM on the surface with the forward landing strut (the one with the ladder attached) facing 'down Sun' to the west.

The first EVA was planned for about 4 hours after landing, giving the crew a short rest and meal break prior to exiting the LM. It was planned that Apollo 13 would have the two astronauts spend 8 hours each (16 man-hours) divided into two 4-hour periods, outside *Aquarius*. These would be separated by a 14-hour rest period of housekeeping, meal times, and sleep.

EVA-1: the first steps

Any stay on the surface by Apollo crews was open-ended, and depended upon real-time events and planning, up to the planned maximum stay for the mission. For Apollo 13 this would have been 33.5 hours.

In order of priority, the EVAs were planned to accomplish:

- Photography through the LM cabin windows.
- Contingency sample collection.
- EVA mobility evaluation.
- LM external inspection.
- Deployment of surface experiments.
- Selected sample collection.
- Lunar field geology.

Following suit donning and cabin depressurisation, Lovell was to have exited first in the normal way, facing away from the hatch, kneeling down, and backing out of the hatch onto the porch in the prone position, guided by Haise. After checking that he was clear of the hatch, Lovell would have dropped the Lunar Equipment Conveyor (LEC) to the surface. This would be used to lift items for return back into the cabin at the end of each EVA.

He would have then dropped a Jettison Bag, which contained the Lunar Sequence Camera storage bag, and four arm-rests, used by the Pilots during the descent, and no longer needed. This would be the first of a series of dumps of unwanted equipment during the stay on the surface, as the astronauts made room and saved weight by replacing the unwanted equipment and hardware with the Moon-rocks which they were to bring home.

Descending the ladder, Lovell would have pulled a D-ring lanyard to deploy the Modular Equipment Stowage Assembly (MESA) located on the descent stage quadrant under Haise's window. He would then have descended the nine ladder rungs towards the forward footpad. While Lovell descended to the surface, inside the LM, Haise would have photographed his Commander's actions with the 70-mm electric data camera that Lovell was planning to use on the surface later in the EVA (the March 1970 Apollo 13 Lunar Surface Procedures document does not state that a television camera was fixed to the MESA to record Lovell's first steps on the Moon), and the 16-mm Surface Sequence Camera.

Once Lovell felt comfortable on the surface, Haise would have passed down (via the LEC) the 70-mm electric data camera to allow Lovell to photodocument the area around the landing site in a panoramic sweep. This would have been followed by the retrieval of the contingency soil sampler out of his EVA suit to collect 1–2 lbs of material. This was to cover the possibility of an early termination of the EVA prior to any further collection of samples. Lovell would then have temporarily stowed the bagged sample on one of the forward secondary landing struts. Meanwhile, Haise, leaving the forward hatch slightly ajar (during the Apollo 11 EVA, Armstrong told Aldrin that this was a particularly good thought), would have exited the LM, descended the ladder (guided and photographed by his Commander), and joined Lovell on the surface.

As Lovell unpacked and deployed the S-band umbrella-like antenna on the surface, to improve communications with Earth, Haise would have photographed the operation using Lovell's 70-mm data camera. (Apparently, as with Apollo 11 and 12, only one of the crew (Lovell) was assigned to have the chest-mounted camera, but both astronauts would have taken turns in using it.) Haise would then have set up the television camera, which up to this time had been stowed. During camera operations on Apollo 12, the camera was inadvertently pointed at the Sun during its move from the LM to the tripod on the surface. The resulting damage prevented transmissions from the Moon. On Apollo 13, there was a colour television camera in both *Odyssey* and *Aquarius*, and an extra black-and-white camera carried down to the surface in the LM – just in case.

Haise would have moved the television camera (with a 100-foot cable link) back to about 50 feet in front of the LM to the right side, pointing down Sun to cover

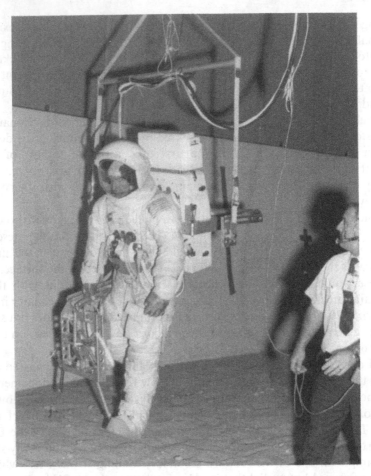

During EVA-2, Lovell and Haise were to perform a geological traverse to Cone crater. They would not have had the Modular Equipment Transporter (MET) planned for Apollo 14, nor the LRV to assist in moving their equipment, and so everything had to be carried by hand or on the suits. Here, Lovell uses a ⅙ gravity simulator and carries the lunar tool kit during EVA training at MSC, Houston, in January 1970.

activities around the lander, and then towards the proposed experiment deployment area. One of the new features on Apollo 13 was the addition of identification markings on one of the EVA suits. On the first landings, identification between the astronauts on movie and television film was almost impossible; sometimes even the astronauts themselves could not remember who was doing what as they helped each other in their tasks. Therefore, on Lovell's suit, red stripes were added to his EMU elbows and knees. This was expected to aid in identifying him in the photographs (a trend that continues into the current Shuttle EVA operations). Interestingly, during the EVAs on later missions, the dust problem made the suits grubby and tended to mask the stripes.

After setting up the television camera, Haise was to move back to the MESA and prepare for the transfer of equipment back into the LM cabin using an Equipment Transfer Bag (ETB). From the MESA, Haise would have loaded two Portable Life Support Systems (PLSS) lithium hydroxide (LiOH) canisters and two spare PLSS batteries into the bag. Retrieving the contingency sampler from the secondary strut where Lovell had earlier placed it, he would have removed the sample bag from the handle, and also put that in the ETB.

If at any time Lovell required help in the deployment of the S-band antenna, Haise was on hand to either help connect the 30-foot cable or to steady the structure. (Experience on Apollo 12 showed that set-up could be a two-man job.) When completed, Haise would have returned up the ladder to the LM cabin and flipped the LM antenna switch to the S-band erectable antenna position, checking with Mission Control on the quality of reception.

Lovell would then have passed up the ETB to Haise in the LM cabin using the LEC. When Haise had the bag in the LM cabin he would have emptied its contents and temporally stowed them in the cabin. He would then have transferred the 16-mm camera, two extra film magazines, the black-and-white television camera, and the EVA-1 traverse map, down to Lovell waiting at the bottom of the ladder, before descending the ladder once more to the surface.

The next task planned for both astronauts was the deployment of the US flag, after which each man would have moved to different sides of the lander to begin combined verbal and photographic documentation on the external condition of *Aquarius*.

They would then have gone to the Scientific Equipment Bay (SEQ) located at Quad II in the rear of the LM descent stage, to remove the Apollo Lunar Surface Experiment Package No. 3 assigned to their mission.

ALSEP-3 deployment

Whoever reached the SEQ bay first would have opened the compartment by manipulating lanyards (the door opening similar to an overhead garage door), revealing the two ALSEP packages inside. To protect the astronauts from the Radioisotope Thermoelectric Generator (RTC) power pack on the left of the open bay, a small door would have swung horizontally out as they opened the bay door. The two packages could have been removed either manually (like taking suitcases off a shelf) or by using an extending boom and a ratchet device to lower each one to the surface by cables (the preferred option).

It was planned that Haise would have lowered Package 2 first, and then taken it to the vicinity of the RTG fuel cask. From Package 2 he would then have unpacked and set up the Hand Tool Carrier (HTC), which would also have been used to carry the ALSEP tools. These also had to be transferred from a stowed position on Package 2. Package 2 also contained the Apollo Lunar Surface Drill (ALSD), which would have been moved to a temporary position on a convenient foot-pad.

Lovell, meanwhile, would be off-loading Package 1, which contained all the surface experiments which they were to deploy: the Passive Seismic Experiment (PSE); the Cold Cathode Gauge Experiment (CCGE); the Charged Particle Lunar

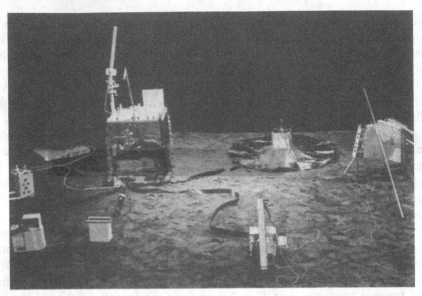

The Apollo 13 ALSEP laid out during a 1-g simulation on Earth, similar to the proposed configuration for deployment on the lunar surface.

Experiment (CPLE); the Heat Flow Experiment (HFEE); and the Central Station. He would also have been available to help Haise stow the required tools on the carrier frame.

Haise was assigned the task of removing the RTG fuel capsule, by lowering the cask horizontally, removing the dome using special tools, and transferring the fuel capsule to the RTG on Package 2, where the generator would have started heating up. Following this, Haise would have transferred the HTC and the ALSD to the MESA on the other side of the lander.

Lovell would have closed the SEQ bay door again to protect the thermal environment inside the LM descent stage, and would then have joined two mast sections out of the tool stowage on Package 2 to form a carrying handle for the two ALSEP packages. This would have then enabled them to transfer the ALSEP barbell fashion to the deployment site. The carrying handle would also have formed the antenna mast for the central station. Lovell would have left the packages on the surface to complete a television panoramic sweep and three still photographic panoramas.

As Lovell completed his tasks at the back of the LM, Haise, now at the MESA would have unloaded the Lunar Sample Return Container (SRC) No. 1 and opened it to remove the six core sample drill stems, sealing caps and supplies for the 'selected' sample which they were to collect after deployment of ALSEP. The LMP would have placed all of these on the HTC, and would follow this with the deployment of the Solar Wind Composition Experiment. He would then have placed the Close-Up Stereo Camera (to be used on EVA-2) in the Sun to conserve its battery charge.

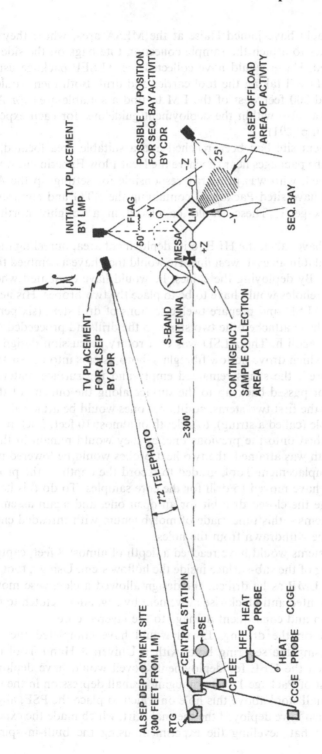

A sketch of the proposed layout of equipment near to the landed Apollo 13 LM at Fra Mauro, including the television camera locations, the deployment site plan for the ALSEP experiments and location of the flag, S-band antenna, and preferred contingency sample collection area.

By now Lovell would have joined Haise at the MESA area, where they would have helped each other to attach the sample collection tote bags on the side of the PLSS. This completed, Haise would have collected the ALSEP package using the barbell handle, with Lovell taking the tool carrier and drill. Both men would have traversed the required 300 feet west of the LM to find a suitable area for ALSEP deployment, which was also within the deployment guidelines for each experiment (see the illustration on p. 291).

Once the deployment site had been reached and a suitable area located, Haise would have lowered the packages near to where the Heat Flow Experiment was to be set up. Assisting Lovell, who was primarily responsible for setting up the ALSEP package, they would have sited Package 2 containing the RTG and connected the power cable to Package 1. These would be placed in a roughly north–south alignment.

Haise would then have taken the HFE to its deployment area, unreeling the cable as he went, and placed it in an east–west line. He would then have assembled the heat flow probe packages. By deploying the cables, he would have evaluated where the drilling of the two boreholes would have to be to place the two probes. His next task was to assemble the ALSD and prepare twelve sections of drill stem (six per hole).

Haise would then have attached the two stems to the drill and proceeded to bore down into the lunar regolith. The ALSD was of a rotary percussion design with a quick release chuck, which drove hollow fibreglass–boron stems into the surface. As the bit head was closed, the stems remained empty and the surface material was either pushed aside or passed back up to the surface along the outside of the drill thread. After sinking the first two stems, additional ones would be attached one at a time, up to six per hole (called a string), to a depth of almost 10 feet. Each new stem would simply be pushed onto the previous one, as they would remain in the hole. Once the correct depth was attained, the two heat probes would be lowered into the holes by using an Emplacement Tool, graded to record the depth of the probe.

Haise would then have moved to drill for the core samples. To do this he would have had to exchange the closed drill bit for an open one, and again assemble the initial core sample stems – this time made of molybdenum with threaded ends – as they would have to be withdrawn from the holes.

This string of six stems would have reached a depth of almost 8 feet, capturing a sample of the layering of the sub-surface inside the hollow stem. Using a foot treadle held in place by the LMP as he drilled, the design allowed a clockwise movement downwards, but prevented anti-clockwise movement by engaging a clutch to prevent inadvertent extraction and consequent damage to the layered core.

As Haise worked on the drilling, Lovell would have completed the ALSEP deployment by releasing the securing bolts with a Universal Hand Tool (UHT). After Haise had taken the HFE for deployment, Lovell would have deployed the PSE some 10 feet east of Package 1. After gouging a small depression in the surface, he was to set up a small stool above this hole on which to place the PSE, aligned to the east. He would then have deployed the thermal skirt, which made the experiment resemble a sombrero hat, levelling the experiment using the built-in spirit level indicator.

At the Central Station, Lovell would then have removed the CCGE, taking it 60 feet south-west, as it contained a delicate atmosphere sensor and required deployment as far away from contamination from the other experiments as possible. He would then have removed the CPLEE from the Central Station, and placed the solar particle sensor ten feet due south. The dust detector experiment would have remained on the Central Station structure.

Next came the deployment of the Central Station itself by releasing hold-down bolts and then extending the structure to reveal its Sun-shield. This would have been followed by mounting the antenna on its stem (the former ALSEP carrying handles) and aiming it at the Earth by using the inbuilt aiming device. By pushing a button, a dummy load would be released across the RTG leads for the initial power-up of the Central Station. It would have taken several minutes to power up the station to transceiver operation levels. Finally, at the request of Mission Control, Lovell would have used the UHT to initiate a switch to begin experiment operations. In the event of a failure of this system there were back-up auxiliary switches for the transmitter, and each experiment had the facility to be cycled on individually.

With the ALSEP operations completed, Lovell would have performed panoramic photodocumentation of each experiment and the general layout area. If Haise was having difficulty with the HFEW he could have assisted at this point, or filmed him operating the drill and core sample equipment.

EVA-1 geological sample collection
With ALSEP in operation, Lovell would then have joined Haise to remove the core sample from the borehole. This would be done by first tugging upward on the drill. If it became stuck, they would have used the power of the drilling action to break soil cohesion, while at the same time pulling. When the two stems were clear of the surface, the drill could be disengaged and discarded at the site, as it was no longer needed. A cap would then have been placed on the top of the open stem to preserve the contents.

Lovell would then have used a wrench to loosen, but not separate, the two sections. Then, using the still attached treadle and ratchet, he would have raised the stem to below the next joint, where it could also be loosened but not separated. This would be repeated until all six stems were raised and the five joints loosened.

The astronauts would then have used sealing caps to separate each stem carefully, marking each one as they did so, and verbally recording the sequence to Mission Control. This would allow the six sections to be reassembled in sequence at the Lunar Receiving Laboratory, Houston, at the end of the mission, so that the scientists could split the core tubes revealing an intact, layered core inside.

As the crew obtained a final visual and photographic record of the work site, the EVA plans would be revised according to the timeline. The two astronauts were then to return to the LM. An EVA extension was possible to send the two astronauts towards Star crater if they could reach it in the time allowed.

The return to the LM would have been via Doublet crater to collect further samples, documented prior to and after selection and deposited in sample bags (or, if too large, in the tote bags on their PLSS). Lovell would have filled Haise's bag, and *vice versa*, as each man would have been unable to reach his own tote bag.

Back at the LM, the television would have been repositioned to show LM ingress and EVA-2 egress. Some of the samples were to be taken into the LM cabin to allow the astronauts to perform preliminary advanced geological analysis. The information from this would have allowed the Science Support Team to revise the planning for EVA-2 while the men rested.

The samples from the tote bags would have been placed in two Weigh Bags (from SRC-1) and placed on the scales, which were part of the MESA. If short of the maximum, the crew would have filled them with random rocks and fine soil from around the LM. They would then be sealed and placed with the six core tubes, and the SRC sealed. Haise would then have removed the second sample return container and placed it in the Sun, covering it with the thermal shroud salvaged from the S-band antenna to maintain proper thermal balance and the integrity of the indium seal on the box.

Using a special brush, the two astronauts would then have cleaned the dust from each other's suit before Haise climbed back into the LM. Lovell would then have sent up the sealed rock box, one of the tote bags containing the samples to be examined, exposed film cassettes, and both cameras, using the ETB. This completed, Lovell would have ascended the ladder to re-enter the cabin and, in closing the hatch to repressurise the vehicle, end the first EVA of about 4 hours duration.

EVA-1 contingency plans

EVA1 one man In the event of only one EMU being available to support an EVA (due to system failure on the second unit, forcing the second astronaut to remain on the LM ECS umbilical), there were plans to allow one astronaut to perform an almost normal four-hour or more solo EVA. Should any LM sub-system become degraded enough to require constant monitoring or manual intervention to maintain system integrity, again a one-man EVA was planned.

For Apollo 13, EVA training was structured so that Lovell was the ALSEP specialist and Haise was the ALSD specialist. As the ALSEP deployment was more important than the drilling operations, the CDR was the prime candidate for the EVA-1 one-man contingency EVA. The HFE would not be attempted in this scenario. If anything was wrong with Lovell's EMU, he could have used parts of Haise's unit to accomplish the one-man EVA task.

This abbreviated EVA would also omit core-sample collection and deployment the S-band antenna, and would have reduced television coverage. The SRC No. 1 would also not have been used, since it mainly carried the six core sample tubes. Haise would have performed verbal and visual observations from the LM cabin, and photographed (still/movie/television) Lovell's activities on the surface. Lovell would have tried to remain in view of the LMP in front of the LM during this type of EVA.

EVA-1 one man – minimum time If, for any reason, only a limited but safe surface-stay was available, there could have been time to accomplish a short one-man EVA. The objectives of this EVA would be the deployment of the MESA, the use of the LEC, an evaluation of surface mobility, and the collection of the contingency sample. In addition, a visual/verbal and, where possible, photographic documentation of the landing site from the surface out to the horizon would be included. Where possible, selected documentary photographs would be taken for analysis during post-flight crew debriefing. The selected crew-member – again, probably Lovell – would remain in visual sight of the other astronaut in the LM (around the forward strut area), so that all activity could be recorded on film and television. This short EVA would probably have lasted no longer than an hour.

EVA-2: a walk on the wild side

Following the rest period and preparations for their second EVA, Lovell would have exited the LM first and descended to the surface. Haise would then have used the LEC and ETB to send down the cameras, extra magazines, the tote bag taken inside at the end of EVA-1, and the EVA-2 traverse map. He would then have exited the LM to join his Commander on the surface.

The first task would have been to prepare the equipment needed for the geological traverse to Cone crater. As with Apollo 12, this would have been accomplished on foot (and unlike the later Apollo 14 trip to Fra Mauro, without the Modular Equipment Transporter (MET)). Lovell would have moved the second sample box from its shaded position on one of the foot-pads onto the MESA table, to be opened. From this, he would have removed a new sample-bag dispenser, three core tubes and three special sample containers, a new pair of Weigh Bags, and the Solar Wind

This Apollo 14 scene, with Al Shepard, should have appeared in the Apollo 13 photographs, when Jim Lovell would have been pictured examining the lunar terrain.

Apollo 11 astronaut Buzz Aldrin described the lunar panorama as 'magnificent desolation'. Here Apollo 14's LM *Antares* sits at the Fra Mauro landing site. This scene could instead have shown Apollo 13's LM *Aquarius* sitting in the sunlight as Lovell and Haise explored the surface.

Experiment Bag. The dispenser and core tubes would have been snapped onto the HTC, and the rest of the equipment would be placed on the MESA until the end of the EVA. The astronauts would have loaded the geological sampling tools onto the HTC (including the vertical photographic reference device known as the gnomon, and the two-piece hoe/shovel to be used for digging trenches).

Each astronaut would then have stowed equipment in his partner's tote bag. Stored in a pouch on Haise's bag would be the 100-foot safety tether, camera filters, and special sample containers. A similar pouch on Lovell's bag would have contained the extra film canisters. Haise would then have checked the Lunar Surface Close-Up Camera (CSC) by taking some trial photographs near the LM, before passing it to Lovell, who would have carried it during the EVA. Haise would then have placed the 16-mm camera in a special pouch in the HTC, which he was to carry.

The next task would have been to attach a large scoop to the extension handle and move to a representative sampling area near the LM. Using a 2-inch sieve attachment, one of the astronauts would have spent five minutes gathering samples and depositing the collected rocks and chips in a sample bag, which would then be stowed in the ETB. After calibrating their cameras and taking a series of photographs of the special contrast chart on the HTC, they would have collected a further sample before moving away from the LM area.

The so-called 'contaminated sample' from under the LM would have been a fine scoop of material collected by Lovell, photodocumented and again bagged and stowed in the ETB. This would have revealed the effect of the LM exhaust and outgassing contamination of a vehicle while on the surface, providing information that might be of use in designing future landing and surface vehicles.

The next three hours of the EVA would have been dependent upon the review of where the LM had landed and any amendments to the timeline that this required, or on timeline constraints on the surface. Coordination between the Science Support Room at Houston and Mission Control would have been combined with input from the crew and handled through mission scientist and Chief EVA CapCom astronaut Tony England.

Because the crew would have been out of the field of view of the television camera for most of the traverse, air-to-ground commentary would have been essential in relaying information to the scientists on the ground about what the astronauts were actually seeing and doing. The crew would have had to report all movements between sampling stations, noting the direction in which they were heading and the distance and direction from the LM. Occasionally, the astronauts would have read the film counters to the CapCom to allow the ground team to keep track of the photographic frames used and the remaining frames available, and to update their records. Every time the crew changed direction or advanced on a new leg of the traverse, they had to record 12–14-frame photodocumentation of the surrounding panorama, including feature location and LM direction.

To prevent Sun glare, each EVA helmet had a central pull-down sun visor, for the first time. Lovell's suit also held the main telecommunications relay back to the LM, so he was required to remain in direct line-of-sight link to the LM communications system during the traverse. If a large object (such as a rock or boulder or fairly deep crater depression) was encountered, it was planned that Haise would move behind or into the obstruction to check communications with Mission Control via Lovell's EMU Extra Vehicular Communication System (EVCS).

At selected sites on the EVA, the two astronauts would have taken core samples by using a hammer to drive a hollow tube into the soil. Using the attached extension handle, the tube would then have been pulled out, capped and reattached to the HTC. Throughout the EVA, small trenches would have been dug and photodocumented, to evaluate surface structure and soil mechanics across the terrain traversed.

When a candidate sample site was identified by either of the astronauts, a well-practised procedure would have been followed, photographing and recording the samples to be taken. Lovell would have placed the gnomon in close proximity to the sample, and would then have taken stereoscopic pictures cross-Sun at a distance of five feet. Haise would have approached the sample from 15 feet, either before or after taking the sample (or both), to obtain pictures of the background with the camera focused at 74 feet, while trying to be within 45° of cross-Sun orientation. He would then have taken a second set of photographs down Sun, with a focus of only 5 feet.

Either astronaut would then have retrieved the sample by scoop, tongs or gloved hands. Small samples or fine soil material would have been bagged, and their number recorded. Large unbagged samples would have been placed in one of the tote bags. Lovell would then have completed the sampling sequence by taking a cross-Sun photograph of the site from 5 feet. If the situation had justified it, a CSC would have been used, with the frame number and orientation of the camera reported to CapCom.

The crew would then have retrieved their equipment and moved on to the next site. If the sampling area was larger than the 5 feet that could be covered in one frame of film, a stereo pair would have been taken at 15 feet to record the general area. One extra down-Sun frame would also have been exposed, as well as individual documentation of each sample, to provide *in situ* comparison of the selection area and the relationship to the other nearby samples.

Some of the special equipment to be used by the astronauts included a special polarised camera filter attached to Lovell's camera. Using this, he would have performed a photodocumentation programme of different kinds of rock in different locations, using a variety of Sun angles and filter settings. Some of these photographs would have been of rocks to be gathered, and others would be the general nature of an area not sampled.

At the most distant location from the LM (which would have been at the rim of Cone crater), the crew would have rested for a short time before completing their tasks at the site and starting back to the LM. Cone crater was the primary target for the second EVA. It was thought they would have ascended a gentle 5–6-degree slope from the base of the mound (where some of the unwanted equipment would have been discarded before the climb was begun). At the crater rim, Lovell and Haise would have had a spectacular view of the undulating lunar surface, both down into the crater itself and back across the landing area towards *Aquarius* on the valley floor below them. They were expected to be 8,700 feet from the LM at this point. (It is interesting to recall the difficulties, due to the featureless terrain, encountered by Apollo 14 astronauts Shepard and Mitchell in trying to reach the same Cone crater. The MET helped support equipment, but in the end the astronauts had to carry the MET with them, as it dragged in the soil.)

During the traverse, the crew would have collected a variety of samples for the Gas Analysis Sample, and two or three micro breccias and crystalline rocks would have been selected for the Magnetic Sample. Both of these sample containers were small, hand-held, can-like mini-SRC storage tubes, with their own special sealing features.

The astronauts would also have used a hoe/shovel (developed from a USMC implement) to dig a 2-foot deep trench. They would have carefully photodocumented the progress of the digging for the soil mechanics experiment. Samples would have been taken from the top, bottom and sides, and where dissimilar layers met. When completed, Haise would have made a boot-print in the top of the outfill, and would have recorded comments. (There was some discussion about having Haise perform a 'run' at the mound to jump on the top, to evaluate how the soil broke down in reduced gravity.)

There were also plans to retrieve a small sample of finely graded material from the base of the trench in a third Special Environment Sample (SES) container. Lovell was tasked to dig the trench, while Haise would have recorded the action with the sequence camera. This 'movie' camera had three magazines available, totalling about 23 minutes of film at 12 fps. In addition to the trench-digging experiment, it would have been used to record rock-rolling experiments down craters, locomotion across the surface, sampling techniques, and anything else the astronauts decided to capture on film. At the end of the third magazine, the camera would have been discarded to save carrying it further, with the exposed film magazines stored in the pockets of the tote bags. The CSC would have been used as required, and when its capacity of 100 pairs of frames was completed, its magazine would also have been stored in a tote bag pocket and the camera discarded.

During the trip back to the LM, further geological sites of interest would have been sampled and, subject to available time, an EVA-2 extension was planned to Triplet crater. During Apollo 12 surface traverses, Conrad had complained that while the work was not fatiguing, it was hot and drying. For Apollo 13, an 8-ounce water-bag for drinking during EVA was provided in the suit. It was accessed through a tube on which the astronaut sucked to quench his thirst. According to Haise, it was just enough to 'wet your whistle.'

Back at the LM, the HTC would have been deposited at the MESA, and each astronaut would have helped his partner remove the filled tote bags. The first Weigh Bag would have been filled with documented and bagged samples, topped up with unbagged samples if there was any room remaining. The second Weigh Bag would have been filled and placed with the first in the SRC (along with the SES containers), and then sealed.

Any remaining samples would have been left in a tote bag and transferred into the cabin, where the selected samples examined between the EVAs were also to be stored. While Lovell sealed the rock box, Haise would have retrieved the solar wind experiment, rolled it, bagged it, and stored it in the ETB, along with all exposed film magazines, the cameras, sieved contaminated samples, and any other items of equipment that were to make the trip home.

After again dusting off each other with the brush as much as possible, Haise would have returned to the LM cabin while Lovell ferried up the rock box, ETB and other items to him using the LEC. Once this was done, Lovell would have left the lunar surface, after four hours, for the last time – a job well done.

Inside the LM, the crew would have completed several hours of discarding unwanted equipment, including the removed PLSS, lunar overshoes, and filled urine bags, as well as repressurising the cabin, cleaning and stowing the gear, and preparing for lift-off.

EVA-2 contingency plans

As with the first EVA, there was a plan to perform a second one-man EVA (if it was safe enough to proceed) in the event of restrictions on a full two-man traverse.

This time, either Lovell or Haise could be outside for at least four hours, and possibly more. In the event of an EMU sub-system not working properly, or the LM requiring continuous monitoring, contingency would prevent a two-person excursion. In the latter case, relating to LM systems fluctuations, it could have meant that a second fully operating EMU was available but the crew were unable to leave the LM for long. This would have allowed some sort of limited operational, contingency or emergency support exit by the second astronaut, staying in the immediate vicinity of the LM forward strut.

The timeline presented for a solo EVA-2 revealed a fairly extensive geological traverse, without going too far from the LM and staying in line of sight of the crew compartment windows. Again, still and motion picture documentation by the second crew-member would be attempted. Prior to exit, a revised plan would be discussed and explained to the crew via the CapCom. This time no tote bag would be carried, all samples being carried on the HTC, as this would occupy one hand and the stereo camera the other. The astronaut would also have had to handle all photodocumentation as well as sample collection, but SRC 2 would have been used as it would on a normal EVA-2 operation. Hand-driven core samples may have been attempted if time allowed. The tubes would also be stored in SRC 2. SRC 1 would have been used for overfill and the quick samples collected during EVA-1. Trench digging and panoramic views would also feature in this type of EVA.

Lunar orbit operations, 15–17 April 1970

While his two colleagues were exploring the lunar surface, CMP Jack Swigert would have been kept busy in *Odyssey*. Continuing the work of past CM pilots, he would have had the task of carrying out visual observations and photography of the lunar surface. As this was an H mission, there was no SIM-bay carried on the Service Module. For communications with Earth he would have had his own CapCom, independent of Tony England, who would have worked with the surface crew.

After tracking the descent of the LM and trying to locate it on the surface after landing, Swigert had three targets for photography. These were candidate landing sites for later missions (Apollo 18 and 19 were still on the manifest when Apollo 13 flew). The second opportunity to photograph Censorinus would have occurred during Rev 27, with a Sun angle of 73° compared to that of 50° during the earlier photo-opportunity.

Both Descartes and the Davy crater chain (not the rille, as some documents listed) would also have been target sites later in the mission (Orbit 42). Swigert was also to take photographs of the impact crater produced by their S-IVB third stage, which would have been deliberately crashed on the surface prior to their arrival in lunar orbit, its impact recorded by the Apollo 12 instruments. (In fact, the crash actually occurred – the only successful element of science from the Apollo 13 mission as flown, apart from some of the best photographs of the lunar far side ever taken).

Homeward bound, 17–21 April 1970

Following the end of a nominal EVA-2, Lovell and Haise would have eaten and rested before preparing for a planned lunar lift-off at GET 137 hours 09 minutes (06.22 Houston Time), on Friday, 17 April, after 34 hours on the surface.

Ascent would have been during Rev 31, rendezvous during the next orbit, and docking finally accomplished on Rev 33, the same as during Apollo 12. Following the transfer of the astronauts, the planned 95 lbs of samples, exposed film, cameras and other equipment, the unmanned LM would have been prepared for undocking, which was to be achieved on the far side of the Moon during Rev 35 (GET 144 hours 30 minutes). Following a 75-second impact burn, the LM would have swung around to the front of the Moon and descended to impact about 30 minutes later, some 30–40 miles from the ALSEP set up at Fra Mauro. The Apollo 12 ALSEP would also have recorded the impact.

Five minutes after the planned impact of the LM on the Moon, the crew would have begun their eight-hour rest period. Following this, there would have been a plane change at GET 154 hours 13 minutes on Rev 40 to set up for the photodocumentation for Descartes (62°–72° Sun angle) and the Davy crater chain (47°–52° Sun angle) during Revs 41 and 45. Other photography and television coverage from lunar orbit would have been carried out as events dictated.

After 90 hours in lunar orbit, the Service Module would have performed the Trans-Earth Injection (TEI) burn during Rev 46 (18 April), to begin a three-day (71–73-hour) flight home. Splash-down was planned for the Pacific Ocean, some 200 miles from Christmas Island, at 12.17 pm Houston Time, on 21 April 1970, for a mission planned to last between 238 and 241 hours.

The recovery would have been similar to that of Apollo 12, with the crew wearing biological face-masks and commencing a period of quarantine before being allowed to fully celebrate their triumphant return as the third lunar landing crew. Jim Lovell and Fred Haise would have had to adjust to life after spaceflight as the fifth and sixth men to walk on the Moon.

MISSION CANCELLED

Apollo 13 did not, of course, achieve the landing at Fra Mauro, and the events of the mission from 13 April until recovery on the 17 April have been well documented. The intended landing site for Apollo 13 was ranked high on programme objectives, and with the loss of the samples and scientific information such a landing would have provided the scientists, the Site Selection Board recommended changing the landing site of Apollo 14 (Littrow) to that of Fra Mauro. On 7 May 1970 the suggestion was accepted, and Apollo 14 would head towards the landing site planned for Apollo 13 – but after the results of the investigation by the Apollo 13 Review Board, which recommended a number of changes to the spacecraft to ensure that such an accident would not occur again. By the time Apollo 14 was launched in January 1971, two more lunar missions had been axed.

By the end of April, and with the Apollo 13 astronauts safely back on Earth, the media soon began to pick up on the news that NASA was thinking of cancelling two more lunar missions – not solely as a result of the Apollo 13 accident, but in the desire to use the hardware for other objectives. Neither the President nor the Administrator at NASA wanted to end the programme with an accident like that during Apollo 13, and they supported a continuing effort for at least four more landings through Apollo 17 in 1972. The hardware was available, astronauts trained to fly the missions, and NASA desired to place several more packages of instruments on the Moon to establish a network that would continue to return scientific data over the next five to ten years.

Apollo 20 was terminated for Skylab, and the idea of using Apollo 19 for Skylab B and Apollo 18 as an initial launch for a huge 100-crew 10-year space base would be a wise application of Apollo hardware faced with reduced budgets. It would also show that by ending the programme with Apollo 17, NASA did not think that Apollo 11 and 12 were gambles that paid off in light of Apollo 13. The majority of scientific objectives set by Apollo would have been met, and using former Apollo hardware would generate a major start for new programmes.

Mission planning, summer 1970

During the summer of 1970 there continued a re-evaluation of intended landing sites for the remaining Apollo missions. Since November, studies of the modes of mobility at the Marius Hills, Copernicus peaks and Hadley Apennine had been conducted by BellCom under NASA contract (NASW-417), using the J series mission classification and hardware (A7L suit/Uprated PLSS/LRV/extended LM). From the studies, the contractor concluded that, apart from ALSEP deployment, the use of the LRVs doubled the scientific activity on each flight over a walking mode of exploration. Driving at an average of 6 mph a typical rover traverse could cover 6–10 miles and visit from seven to ten sampling stations, each about 0.75 miles apart, depending on scientific objectives and where the LM landed close to the science targeting landing area.

By 10 June 1970 a meeting of the Site Selection Sub Group reviewed potential landing sites for Apollo 15 through 17.[3] The loss of Davy crater area landing site photography from Apollo 13 to designate it as a primary target for Apollo 15, and the lack of time to take photographs on Apollo 14 and develop them in time for Apollo 15, eliminated that site from the mission.

Censorinus was still considered to be the favourite site for sampling highland material, but it was about to be rejected due to lack of detailed photography of the landing area to add to the approach terrain photographs taken during Apollo 10. Littrow was a strong contender for good science both on the surface, from ALSEP, and on lunar orbit of Apollo 15, but the selection group felt the selection of this site could result in losing the Marius Hills site on a later mission. The group also reviewed a further dozen sites, some of them being rejected due to the similarity of mare landings on Apollo 11 and 12 (or duplicating more of the interesting site finally selected).

The meeting recommended that the Marius Hills be the candidate for Apollo 15 (H-4), with Littrow considered as an alternative site, and that the candidate site for Apollo 16 (J-1) be Descartes. The group also noted that *if* the H-4 mission was cancelled, then the Marius Hills would become the primary site for Apollo 16, and Descartes for Apollo 17. Options under further evaluation for Apollo 17–19 included the Davy crater region, Copernicus peaks and Hadley Apennine, depending on further photography of these sites from Apollo 15 onwards before definitive landing site planning could be detailed.

Studies continued into the Marius Hills site comparing a walking mission over an LRV mission. In summary, the results indicated that a reduction of station stops on each EVA would be the result of the distances between the scientific targets of interest. These were originally selected without strict operational constraints, when more detailed studies and simulations revealed that the site required accurate landing, and if a 6-mph LRV traverse were to be achieved, then the confidence to meet all objectives was high, although less than that would reduce the chances of achieving all objectives from the stretched out sampling sites. In addition, it was important that the ALSEP be set up early in the exploration to supplement the recovered samples. The time taken to deploy ALSEP reduced the time for traverse on the first EVA, and 'subtracts for the confidence in achieving the Marius Hills Mission objectives'. In August 1970, a CB memo from astronaut Tony England stated that instead of a proposed seventeen sampling stations on EVA 2, and thirteen planned for EVA 3, only ten per EVA would be possible due to the ten-minute contingency time required at each major stop, and four minutes at each minor stop. One of the conclusions therefore indicated that at this site the LRV would not increase science time, but would actually decrease it by about an hour.[4]

Meanwhile, work continued on planning the J series of missions from Apollo 16 through 19. Two 80-lb sub-satellites were planned for Apollo 16, and eight to be ejected into lunar orbit to study solar wind particles and magnetic phenomena from 60 nautical miles altitude. From Apollo 17 there were plans for an Extended Life ALSEP (ELLSEP) costing $29.1 million for the three packages. The decision to negotiate a contract (not yet fully approved) with Bendix was dated 18 August 1970, and included the design, development, test, delivery and supply of an ELLSEP for Apollo 17, with an option to purchase two more units for Apollo 18 and 19.

The proposed experiments – which were to operate for at least two years, doubling the earlier design life of ALSEP – included a *Passive Seismic Experiment (PSE)* to measure natural seismology and lunar interior properties; *Heat Flow Experiment (HFE)* measuring net outward heat flux from the lunar interior; *Lunar Seismic Profile (LSP)* recording reflective energy from man-inducted shock waves; *Lunar Ejecta and Meteorites (LEAM)* measuring particle velocity, mass and angle of impact and the ejecta particles; *Lunar Surface Gravimeter (LSG)* detecting and measuring free oscillation and variations in tidal gravity at low frequencies; *Mass Spectrometer (MS)* obtaining soil composition measurements on the surface; and *Lunar Surface Magnetometer (LSM)* measuring the magnetic field and temperature variations on the surface.

These experiments were provisionally assigned to Apollo 17 as Array E, consisting of LEAM, LSP, MS, PSE, HFE, LSM (one-year design) and ASE.

Those for Apollo 18 (Array F) and 19 (Array G) consisted of the LEAM, LSP, MS, PSE and HFE, plus the LSG and a two-year LSM. Final definition of the Apollo 18 and 19 packages was based on development of the Apollo 17 array.

Apollo J-mission crewing, summer 1970
While politicians, administrators and scientists debated the pros and cons of flying to the Moon or creating a space station, the astronaut group assigned to Apollo

continued their training, but kept one eye on the 'policy makers and mission breakers', aware that their flights were by no means certain.

Although no crews had been officially assigned to Apollo 20, there were astronauts in training as back-up crews who were expected to rotate to Apollo 15 through 19.

The Apollo 12 back-up crew (Scott–Worden–Irwin) had been officially named to Apollo 15 on 26 March 1970, and their back-up crew (Gordon–Schmitt–Brand) was planned to rotate to Apollo 18. Jack Schmitt was the first scientist-astronaut assigned to a flight crew, and probably remained the only one assigned to Apollo. All the other active scientist-astronauts were by then in Apollo support roles or were eligible for assignment to AAP/Skylab.

Following the Apollo 13 mission, the back-up crew of Young and Duke were joined by Mattingly to form the crew of Apollo 16. Slayton assigned Fred Haise on a second chance at landing with Bill Pogue and Jerry Carr, the Apollo 16 back-up crew, who would, under normal circumstance, expect to rotate to Apollo 19 as prime crew. No official announcement concerning Apollo 16 was expected until after Apollo 14 flew later in 1970.

Mission	Type	Position	Prime	Back-up	Support
Apollo 15	H-4	Commander	Scott	Gordon	Joe Allen (mission scientist)
		CM Pilot	Worden	Brand	Karl Henize (CM duties/scientist)
		LM Pilot	Irwin	Schmitt	Bob Parker (scientist)
Apollo 16	J-1	Commander	Young	Haise	Tony England (mission scientist)
		CM Pilot	Roosa	Pogue	Hank Hartsfield (Ex-MOL)
		LM Pilot	Mitchell	Carr	Don Peterson (Ex-MOL)

During the summer of 1970, the crews intended for Apollo 16 were just commencing their training programme. According to comments from Haise, Carr and Pogue, their training as back-ups to Apollo 16 lasted for only about six months, and focused on the development of the mission and training plan for that mission, with very little discussion concerning Apollo 19. Preliminary training in mission simulators depended on availability after the Apollo 14 and 15 crew's priority, and in Haise's checking out in the Lunar Landing Training Vehicle. Despite the designation, an LM Pilot did not fly the vehicle, and only the Commander of the prime or back-up crews participated in LLTV training. Therefore, neither Cernan, Gordon nor Haise had flown this vehicle before they were assigned as back-up Commanders on Apollo 14–16. Other training involved CSM and LM upgrade presentations during classroom lectures, and briefings on ALSEP and SIM-bay experiments.

Two astronauts participate in evaluation tests of the Apollo LRV concept in 1970. Astronauts Jack Lousma (probable LMP Apollo 20) drives the early one-man systems evaluation version of the LRV. At the rear is Apollo 19 LMP Jerry Carr. With the cancellation of their missions, both men moved to Skylab and flew to that space station in 1973.

In correspondence, Haise revealed that he and Carr (with Pogue) had participated in a field geology exercise using the LRV with the Apollo 16 prime crew in the summer of 1970, and remembered seeing a photograph, at some point, of the intended 'Apollo 19' trio during this training. Three decades later the exact dates were unclear, but further research revealed that Young and Duke completed geological field trips during July (San Juan Mountains, New Mexico and Medicine Hat, Alberta, Canada) and September (Colorado Plateau). (A search for photographs from these field trips was still being conducted when this book went to press, and although no stills were found of these specific activates, the July dates would presumably be the strongest candidates for Haise's recollection.)

However, after less than six months' preliminary training as a crew of Haise, Pogue and Carr and their intended Apollo 16/Apollo 19 mission, training dramatically changed once more. During September 1970, what scientists feared, and what astronauts expected, happened. NASA – faced with even tighter budget restrictions – was forced to cancel two more Apollo flights to the Moon, although

Astronaut Jim Irwin uses a scoop to make a trench in the lunar soil during one of the Apollo 15 EVAs in 1971. In the background, eight miles away, is the 15,000-foot high Mount Hadley. Apollo 19 should have landed in this region, and if it had done so, a similar scene would have featured Jerry Carr instead of Jim Irwin.

Apollo 15 Commander Dave Scott on the slope of Hadley Delta. If Apollo 19 had landed here as planned, it could have been Fred Haise that climbed the lower slopes of this mountainous region.

A photograph of Hadley Rille taken by Jim Irwin, showing Dave Scott near the LRV – a scene that might have been recorded by Fred Haise or Jerry Carr during the Apollo 19 mission.

exactly which ones was subject to more heated debate between scientists and programme managers.

The axe falls

On 15 June, AWST reported that NASA was undergoing an internal argument between continuing the Apollo programme and using the hardware for more than one Skylab space station.[5] The argument centred on cancelling either Apollo 19 or 15, with the clearest indication to cut Apollo 15 – the last of the H missions. This would allow flying all four of the remaining J series with expanded capability of surface and orbital science, while the relocation of Apollo 15s hardware would allow hardware for a second space station in the mind-1970s. Then, in July the *Washington Post* reported not one mission but perhaps four Apollo flights might be cut from the programme.[6]

Three flights – Apollo 15, 18 and 19 – were definitely considered, and would release sufficient hardware to launch an intermediate space station in 1976, partly as a celebration of the United States' 200th birthday, the paper reported – although NASA dismissed that notion completely, with the media misinterpreting a general comment that if NASA could launch a space station during 1976 it would be coincidental with the bicentennial celebration and not part of it. The plan was to launch Skylab A in 1972 (AS-515), Skylab B in 1975 (AS-510?), and a larger six-man station (Skylab C?) in 1976 (AS-513/514), using the eight Saturn 1Bs in storage for carrying astronauts to and from the stations before the Shuttle was available. The paper quoted a 'space source' as saying: 'A large Skylab makes sense, because then we have an instant use for the Space Shuttle in 1978. The Space Shuttle first mission could be a trip to the six-man Skylab.' The *Washington Post* also pointed out the two hurdles on this approach: the scientific community involved with Apollo, and the fifty or so astronauts who reportedly viewed going to the Moon more rewarding than long flights in Earth orbit.

Further cuts seriously threatened the loyalty of many senior employees. The desire of MSC was to fly all Apollos to the end of 1972 and to then fly Skylab, and not split the landing programme, with the space station programme avoiding the loss of momentum and performance levels to prepare and fly the lunar missions. A delay would also help Skylab hardware preparations and crew training.

By September 1970, again, what scientists feared and astronauts expected happened. On 2 September 1970, the outgoing NASA Administrator Tom Paine issued a statement.[7] After reviewing the considerations and recommendations across several areas, three options were considered:

- Fly Apollo 14–17 on six-month centres, complete Skylab A, and then fly Apollo 18 and 19
- Fly Apollo 14, 16–18 on six-month centres, terminate Apollo, and then fly Skylab A.
- Fly Apollo 14, 16–19 on five-month centres, terminate Apollo, and then fly Skylab A.

The recommendations from the National Academy of Sciences Space Science Board and the Lunar and Planetary Mission Board was that all the Apollos should be retained; but NASA faced hard decisions, and with fiscal and programme pressures, the second option seemed to be the only option. Cutting Apollo 15 and 19 effectively cancelled out the Apollo Budget deficit of $42.1 million, and would be assisted by a rapid phase-down in manpower levels at all Apollo facilities, with the remaining missions renumbered Apollo 14 through 17. The major scientific effects were the chance of six sites, instead of eight six ALSEPs instead of eight (Apollo 17 retaining a revised ELLSEP), three flown SIM-bay missions instead of four, and operation of three rovers instead of four. With the hardware, all work would stop on CSM 111 and LM 13, and the crews for the missions would once again undergo changes. In September 1970, mission planning was as follows:

Mission	*Old schedule*	*New schedule*
Apollo 14 (H-3)	Fourth Quarter 1970	Launch early 1971
Apollo 15 (H-4)	First Quarter 1971	Cancelled Saturn V used to launch new Apollo 15
Apollo 16 (J-1)	Third Quarter 1971	Renumbered Apollo 15 launch, third quarter 1971
Apollo 17 (J-2)	Second Quarter 1972	Renumbered Apollo 16 launch, second quarter 1972

Apollo 18 (J-3)	Second Quarter 1974	Renumbered Apollo 17 launch, third quarter 1972
Apollo 19 (J-4)	Third Quarter 1974	Cancelled
Skylab A	End 1972	Launch end 1972

Disposing of the hardware
The Saturn Vs for the last three missions were reassigned to other roles:

AS-513 (Apollo 18) The first and second stages and instrument unit were used to launch Skylab OWS in May 1973, instead of the Apollo 20 hardware originally planned. The third stage was converted for static display at JSC, Houston, Texas. The OWS (S-IVB 212) was already under fabrication for the flight hardware.

AS-514 (Apollo 19) First stage on display at JSC; second and third stages on display at Apollo–Saturn V Center, KSC, Florida. The first two stages, together with S-IVB 515, could have become Skylab B.

AS-515 (Apollo 20) First stage on display at Michoud Assembly Facility; second stage on display at JSC. The S-IVB stage was converted as the back-up OWS Skylab B. When that programme was terminated it was cut up and displayed in the National Air and Space Museum as a representation of Skylab A.

Arguing the case
On 15 September, Tom Paine resigned as NASA Administrator, stating that the decision was personal, and not based on recent budget cuts. Paine thought that the time was right to leave the agency now that post-Apollo programming had been completed.

The next consideration was where to send the remaining missions now that the site opportunities had diminished to just three. Scientists argued that defining the major questions on the origins and make-up of the Moon was planned over at least ten landings; but now, with the cancellation of three missions and the loss of the Apollo 13 landing, and most of what was planned to follow Apollo to the Moon, these would be at the most only six sites. The losses were not exactly catastrophic, although to some scientists they appeared so. Then, in December 1970 there were fears that Apollo 16 and 17 might be scrapped to saved more money, and that samples from some of the more scientifically interesting regions might be gathered in minute samples by Soviet unmanned Luna sample return missions, as demonstrated by Luna 16 in September 1970, or via extended journey by unmanned rover, as currently being exhibited by the Soviet Lunokhod 1. Discussion as to whether Russia was ever in a 'Moon race' were fed by comments from the Soviet Union that their programme direction was the creation of permanent manned space stations and unmanned scientific exploration of the Moon, and that they never intended sending men to the Moon. With hindsight and new information, we now know that that was not exactly true, and that there was an active Soviet manned lunar programme in the late 1960s and early 1970s, although it did not help the case for sustaining Apollo.

Studies of potential landing sites during the next site selection sub-group meeting in August and September changed the landing sites available for Apollo 15, narrowed from Littrow, Descartes, Hadley Apennine and the Marius Hills. The consensus for Apollo 15 (which was always almost impossible to obtain from every participant) at the site meeting was to aim for a landing that would provide a major advance in lunar studies whether flown as an H or a J mission, offering the capability of significant walking or LRV transverses. From this, the argument for either Hadley or Marius continued until Dave Scott said that he favoured Hadley, although he could land at either. On 24 September, the new J-series Apollo 15 was targeted for Hadley Apennine, the former landing site of the recently cancelled Apollo 19.

In the same meeting, Apollo 16 was assigned to the Descartes site, but the decision concerning Apollo 17 would be reviewed following the Apollo 14 and Apollo 15 missions. The Marius Hills and Copernicus remained favourites, but in February 1972 it was finally decided that Apollo 17 would head for the Taurus Littrow valley. It would be the last Apollo mission... or would it?

On 12 August 1971, Dave Scott, recently returned from the Moon on Apollo 15, made an unexpected suggestion to reinstate Apollo 18 and 19, which prompted the media to suggest that NASA might request funds to re-manifest the two missions, although in reality this was highly improbable. At the Apollo 15 post-flight press conference, Scott was quoted as saying: 'I think we ought to reinstate them right now. I think we should explore the Moon to a far greater extent than we're planning now... because I can guarantee you that people will never get tired of finding new things up there.'

Then, in September 1971, as the Soviets flew Luna 18 to the Moon, more rumours in the press indicated that consideration was being given to possibly flying Apollo 18 and Apollo 19 on revised lunar missions – perhaps in view of recent discussions with the Soviets on space cooperation – in an international lunar mission with the Soviets. It was equally acknowledged, however, that these plans were nothing but 'a gleam in the eye' at NASA, although serious consideration was also was being given to fly a fourth crew to Skylab on a mission of about 21 days, depending on consumables and the state of the station when the third crew came home after 56 days. But there was one more Apollo mission that was more than a 'gleam in the eye'.

LUNAR POLAR MAPPING MISSION

In September 1971, an in-house study at NASA evaluated sending Apollo 18 on a trip to map the Moon from polar orbit.[8] The study proposed several possible launch dates for Apollo 18 to evaluate scheduling with Apollo 17 and Skylab, and these depended on hardware availability, processing flows, test chamber use, and contractor support. (Grumman's LM team was being scaled down to a minimum level from January 1972.) Costs varied, and also depended on there being no other flights between this and the beginning of Shuttle operations, although the memo indicated that if there were 'additional missions' authorised in the 1974–1975 time-frame using Saturn 1B CSM-only-type missions, the overall costs would be reduced in processing the hardware. These studies briefly revitalised the calls for an Apollo CSM/LM Lab in polar orbit, which was discussed during the 1960s under Lunar AAP programming, and which would have seen a 28-day mission allowing favourable lighting conditions over the whole lunar globe.

The decision to proceed would have to have been made by 1 December 1971 to achieve the launch dates in 1973, although launch of the lunar polar orbiter could be slipped to April 1974. Unfortunately, the option of sending Apollo 18 to polar orbit was not pursued.

So who would have crewed such a mission? That depended on whether the Apollo–Soyuz Test Project would have been authorised, funded and flown. If not, then perhaps Stafford might have opted for the mission, together with Brand (or Swigert?) and Slayton. Then again, perhaps there would have been a case for flying a scientist-astronaut who was not assigned to Skylab (Joe Allen?). Once again, we will never know.

THE LAST MEN OF APOLLO

The cancellation of Apollo 15 and Apollo 19 resulted in changes to the crewing assignments. Although they would not be assigned to Apollo 18, Gordon, Brand and Schmitt continued as the back-up for the new J-class Apollo 15 mission. The crew for Apollo 17 still had to be announced, and though Cernan, Evans and Engle were the normal rotation choice, the Gordon–Brand–Schmitt team was not exactly out of the running. Indeed, Gordon has said that when Cernan crashed a helicopter in January 1971, during Apollo 14 training, he thought he might after all be reassigned to replace Cernan on the last Apollo, and was a little disappointed when this was not the case. The assignment of the back-up crew for Apollo 16 certainly was a dead-end assignment, but Haise was prepared to remain as back-up Commander. Slayton reassigned Carr and Pogue to Skylab, and they were replaced by Roosa and Mitchell. The announcement was made official on 3 March 1971.

The reason for selecting Schmitt over Engle on Apollo 17 has since been discussed many times: how the argument for an experienced geologist over a test pilot was required on at least one Apollo landing – and especially the last one.[9] Early in August, Engle took compassionate leave due to a family bereavement, and after returning to Houston on 10 August he telephoned the CB office to ask whether there were any messages for him. There was just one – from Al Shepard, Chief of the CB. When he contacted Shepard, Engle was told that he was to be replaced on Apollo 17 by Schmitt. Although bitterly disappointed, he received the news positively and determined to use his two-year training experience to help to make '17' the best mission ever.

Engle has commented that the most difficult job was telling his young children that he would not be going to the Moon. The official announcement of crewing for Apollo 17 – Cernan–Evans–Schmitt, with Scott–Worden–Irwin as back-ups – came on 13 August 1971. Engle, it was revealed, would move with Gordon over to Shuttle development, while Brand would move over to Skylab. By 28 August, Engle was on

The original Apollo 17 lunar landing crew undergoes geological training in the Pinacate volcanic area of north-western Sonora, Mexico. Here they are training as the Apollo 14 back-up crew in February 1970. Gene Cernan (Commander) is followed by Joe Engle (LMP) pulling the MET. In August 1971, Apollo 15 back-up LMP Jack Schmitt replaced Engle on the last Apollo to the Moon.

Apollo 15 back-up crew Dick Gordon (Commander, *left*) and Jack Schmitt (LMP) during EVA traverse training at Taos, New Mexico, during March 1971. This should have been the landing crew for Apollo 18 before the budget cuts. Schmitt reached the Moon on Apollo 17, but Gordon was reassigned to early Shuttle development after completing his Apollo 15 back-up role.

a prearranged hunting trip with several astronauts, including Jack Schmitt. Apparently, during the preparations for the hunt, Schmitt refused to hold sighting targets for Engle in case the former Apollo 17 LMP suddenly had the idea of trying to regain his lost Moon seat!

In May 1972, Scott's team was removed from the back-up position on '17' in the aftermath of unauthorised postal covers flown on Apollo 15, and was replaced by Young, Roosa and Duke in a dead-end assignment to close out Apollo.

Apollo crew assignments, 1971–1972

Mission	Type	Position	Prime	Back-up	Support
Apollo 15	J-1	Commander	Scott	Gordon	Allen (Mission Scientist)
Landing, Aug 1971		CM Pilot	Worden	Brand	Henize (CM scientist-astronaut)
Hadley Apennine		LM Pilot	Irwin	Schmitt	Parker (scientist)
Apollo 16	J-2	Commander	Young	Haise	England (Mission Scientist)
Landing, Apr 1972		CM Pilot	Roosa	Roosa	Hartsfield (ex-MOL)
Descartes		LM Pilot	Mitchell	Mitchell	Peterson (ex-MOL)
Apollo 17	J-3	Commander	Cernan	Young	Fullerton (ex-MOL)
Landing, Dec 1972		CM Pilot	Evans	Roosa	Parker (Mission Scientist)
Taurus Littrow		LM Pilot	Schmitt	Duke	Overmyer (ex-MOL)

The flight operations of Apollo 14–17, Skylab and the ASTP have been well documented and are not expanded upon here. The crewing assignments for what evolved from AAP (Skylab) were announced on 19 January 1972, and discussion of that evolution from 1970 is included in the author's *Skylab: America's Space Station*. The crew for the very last Apollo that would dock with a Soviet Soyuz in 1975 was named on 30 January 1973. (See the Table on p. 314.)

The ASTP crew-members were the last Americans in space until the first flight of the Shuttle in April 1981. Apollo '18' splashed down on 24 July 1975, less than seven years after Apollo 7 began the series. In just 81 months, the manned space-flight programme of Apollo entered the history books. It was all over.

Skylab/ASTP* (AAP) crew assignments, 1973–1975

Mission	Position	Prime	Back-up	Support
Skylab 2	Commander	Conrad	Schweickart	Crippen (ex-MOL)
Planned 28 days	Pilot	Weitz	McCandless	Thornton (scientist-astronaut)
Flew 28 days	Science Pilot	Kerwin	Musgrave	Henize (scientist-astronaut)
				Hartsfield (ex-MOL)
Skylab 3	Commander	Bean	Brand	*ditto*
Planned 56 days	Pilot	Lousma	Lind	
Flew 59 days	Science Pilot	Garriott	Lenoir	
Skylab 4	Commander	Carr	Brand	*ditto*
Planned 56 days	Pilot	Pogue	Lind	
Flew 84 days	Science Pilot	Gibson	Lenoir	

(The Brand–Lind–Lenoir back-up crew was for a short while considered for Skylab 5 – the 3–4 week close-out mission – until Skylab 4 was extended from 59 to 84 days)

Apollo '18' ASTP*	Commander	Stafford	Bean	Crippen (all ex-MOL)
docking mission,	Pilot	Brand	Evans	Truly
July 1975	Docking	Slayton	Lousma	Overmyer
	Module Pilot			Bobko

* Apollo–Soyuz Test Project
† Swigert would have originally received this assignment, but he was involved in the Apollo 15 postal cover incident

1 *NASA Future Plans*, NASA News (press conference transcript) 13 January 1970. Participants included Dr Thomas Paine, NASA Administrator, George Low, NASA Deputy Administrator, Dr John Naugle, NASA Associate Administrator, Office of Space Science and Applications, Milton Klein, Manager of the AEC-NASA Space Nuclear Propulsion Office, and Julian Scheer, NASA Assistant Administrator for Public Affairs.
2 Documents used in the preparation of the EVA planning of Apollo 13 were: Apollo 13 Press Kit (70–50K) 26 March 1970, released 2 April 1970; Apollo 13 Briefing, by the Mission Planning and Analysis Division, MSC, 13 March 1970 (MPAD 70-300 S) (transcript and overheads); Fred Haise Press Conference, 25 February 1970, KSC (Apollo 13 PC-1) (transcript); Apollo 13 and 14 Mission Review, 24 February 1970, NASA MSC (overheads); Apollo 13 Flight Plan Final, 16 March 1970; Apollo 13 Lunar Surface Procedures, Lunar Surface Operations Office, MOB, FCSD, NASA MSC, 16 March 1970; and *Newsweek*, 13 April 1970 (*Aquarius* and *Odyssey*) pp. 62–63; all from AIS archives; and Apollo 13, The NASA mission Reports, ed. R. Godwin, Apogee Books, 2000.
3 GLEP Site Selection Subgroup Minutes, 5 June 1970, signed by Noel Hinners, BellCom Inc. Washington. From the Apollo History Archives, originals researched at NASA JSC History Office (1988–1994).
4 *Evaluation of Mobility Modes on Lunar Exploration Traverse*. BellCom Inc (TR-69–340–4) dated 14 November 1969; Lunar Surface Traverses, letter from TRW Apollo support Group to NASA ASPO, 24 April 1970; Memo from A. W. England, Mission Scientist, CB, re. comparison of science time available on LTRV *versus* walking mission, all filed in the NASA Apollo Collection.
5 NASA urged to cut missions to the Moon, *Aviation Week & Space Technology*, 15 June 1970 p. 15.
6 3 *Apollo Cancellation Weighed for Big Skylab, Washington Post* 9 July 1970 p. A12.

7   Statement by Dr Thomas O. Paine, Administrator of NASA, 2 September 1970. (Copy marked from Apollo 19 file.) Apollo Collection, JSC History Office.

8   Lunar Polar Mapping Mission (Apollo 18), data fax transmission from Robert Hock, Manager Apollo Skylab Programs, KSC, to 'Apollo Spacecraft Manager', MSC, 23 September 1971. From the Apollo History Archive, JSC.

9   See previous references: *Deke, American Manned Spaceflight*; *Where No Man Has Gone Before* and *The Last Man on the Moon*, Eugene Cernan and Don Davis, St Martin's Press, 1999.

Beyond the Moon

The majority of this volume has concentrated on reviewing what NASA and its contractors were proposing for the utilisation of Apollo hardware beyond the first manned landing on the Moon. Many of these plans were no more than paper studies and proposals and were never intended to progress to flight hardware, but offer a guideline to securing the funding to develop the hardware (Saturn 1B, Saturn V, CSM and LM) to attempt advanced missions in Earth orbit and on or around the Moon well into the 1970s.

There are clear examples of these plans evolving into flight operations. Apollo X became AES, and then AAP, and evolved into the OWS project (and ASTP); the studies of adapting the Service Module to carry instruments evolved into the ATM, to be launched under the Apollo J-mission SIM-bay programme; the application of the LM for other missions beyond mainstream Apollo including, for a while, the ATM, then the J series of lunar missions; the development of surface transport, begun as MOLAB, evolved into LSSM, and flew as the LRV used on Apollo 15–17; and the principle that adopting the S-IVB stage of a Saturn into an early space station was a logical step after Apollo – and this finally flew as Skylab. In many respects these ideas – grouped under the general heading 'Apollo Applications Program' – were advantageous, broad and far-reaching. But even these were, perhaps, not so far-reaching as sending Apollo–derived hardware to Mars.

APOLLO IN THE MANNED PLANETARY PROGRAMME

The idea of using Apollo hardware to support manned flights to Mars had evolved from studies in the early 1960s which, in turn, built upon years of study into sending astronauts to the Red Planet. Author David S.F. Portree has conducted an extensive survey of Mars mission planning, and his work is highly recommended for further study in this field.[1] In this brief summary, this primary source is supported by the three Congressional Staff Studies by the NASA Oversight Committee in 1966 and 1967, and by the Presidential Report of 1967 (previously detailed in the references).

One of the early leading personalities who supported manned flights to Mars was

Wernher von Braun, who had published popular books on the subject in the 1950s. In the early 1960s, von Braun was at the Marshall Space Flight Center, and was involved in the development of the Saturn family of launch vehicles as part of the newly created Apollo programme. Apollo was originally just one part of a ten-year long-range NASA plan that included the creation of a space station and a circumlunar flight before the end of the decade, which, in a sequence of progressive steps, might eventually lead to manned exploration of Mars.

After Kennedy committed America to the Moon in 1961, the Saturn vehicles were studied to support the EOR mode and to provide experience in orbital rendezvous, while the much larger Nova was evaluated for the Direct Ascent mode and to provide the means for launching the large amounts of payload into orbit using fewer launch vehicles, both of which would be useful in the subsequent manned Mars programme. The 1962 decision to use LOR for Apollo using just one Saturn per mission removed the immediate requirement for multiple production of Saturn rockets or development of the large Nova. Although it represented a cheaper and quicker way of reaching the Moon, LOR would also restrict any long-term plans after Apollo.

In addition, LOR was also seen as developing the experience and techniques of reaching Mars in orbital rendezvous out of Earth orbit, surface explorations, orbital reconnaissance and survey, and the provision of the necessary hardware to support a manned planetary programme based on Apollo and Saturn hardware. It became apparent that the Apollo–Saturn system could support the development of a post-Apollo programme that had a manned flight to Mars as its objective. The successful completion of this goal in the near term hinged on the availability of hardware, and notably the Saturn launch vehicle. It took some time to physically build the Saturns, and its early qualification to carry astronauts was critical for reaching the Moon by 1970. Once the Saturn was man-rated, then the number required operationally could be defined, and if no clear commitment to a programme beyond Apollo was forthcoming then the production flow of heavy lift launcher would slow perhaps beyond a point where it could be recovered, and the development of anything on the scale of the Nova would perhaps never leave the drawing board.

To assist in planning the manned lunar programme, a fleet of unmanned robotic explorers was dispatched to survey the Moon before any astronaut ventured there. The same approach would be required for the Mars effort, so that when the Apollo programme ended some time in the early 1970s there was already a sustained programme of unmanned planetary probes exploring the Red Planet to provide the information necessary to plan the first manned flights.

Windows of launch opportunity to the Moon occur each month. This was useful for Apollo, but with Mars everything depends on the alignment of the two planets in their respective orbits to define the best opportunity to dispatch spacecraft. This occurs approximately every two years, and does not necessarily result in the shortest flight path. One of the earliest and most favourable opportunities to reach Mars occurred in 1971, but if Apollo landed in 1970 there would be no possibility of mounting a manned Mars expedition that soon. It seemed reasonable that the 1980s would represent the first opportunity to send Americans to Mars. To fill the void, there was a requirement for an intermediate programme, between Apollo lunar exploration in the early 1970s and Mars surface exploration in the early 1980s, that retained the momentum of manned spaceflight.

What was needed was a programme that could use Apollo–Saturn technology developed for the Moon as a stepping stone for the later planetary landing programme, allowing time to develop the launch vehicle and propulsion technology required to mount manned missions to Mars. In Apollo lunar mission planning there were studies of testing Apollo in ever higher Earth orbits to eventually circumnavigate the Moon and return to Earth without entering orbit, which would be achieved on later missions. This same principle was investigated during 1963–1964 under the Early Manned Planetary–Interplanetary Roundtrip Expedition (EMPIRE).

EMPIRE

The EMPIRE study[2] consisted of three aerospace contractors (Aeronutronic, Lockheed and General Dynamics) reviewing fly-by missions (or orbital mission profiles) that used a 1970 launch as a baseline, planning the 'missions' that also involved not only Mars but also Venus. These studies also resulted in establishing a requirement to proceed with the Nova launch vehicle, expand the nuclear rocket programme, and provide continuation of Apollo hardware beyond its use in the lunar programme.

Two main flight trajectories were studied. The first, termed 'hot', took the spacecraft inside the orbit of Earth, closer to the Sun and possibly flying by Venus (depending on the launch window) and flying by Mars, returning to Earth after an 18-month mission. The second profile, termed 'cold', required the spacecraft to leave Earth and journey out towards Mars, which it flew by three months later. It would then continue on past the planet out to the asteroid belt, where it would reach its furthest point in its trajectory and head back towards Earth, arriving 22 months after departure.

While Nova and new spacecraft designs were studies, Lockheed's study featured the use of a Saturn C-5 and Apollo CSM, and replacement of the LM by a folded lightweight fly-by vehicle. A nuclear upper stage would place the vehicle on course for Mars, whereupon the CSM would separate and allow the fly-by element to 'unfold' two long booms from a central hub, docking the CSM at one boom to counterweight the habitation module on the other boom. In the centre was the weightless hub that contained the necessary course correction, solar power subsystems, automated probes for deployment at Mars, and a radiation shelter for the crew. After returning to Earth, the crew would have used the CSM for performing Earth recovery, in much the same way as at the end of the lunar mission.

The Lockheed study also used an Apollo-type CM as the Earth recovery vehicle – a design that would appear in several Mars manned mission studies of the period, as would the use of Saturn and Nova launch vehicles, essentially attempting to commit for long-term production of those vehicles beyond that of the lunar programme.

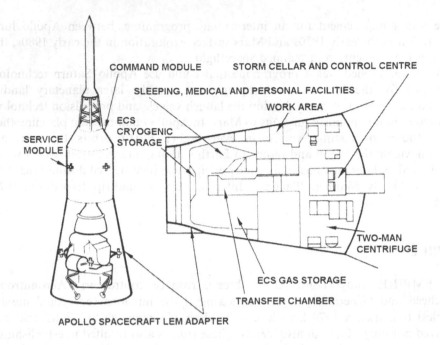

COMMAND MODULE   STORM CELLAR AND CONTROL CENTRE

SLEEPING, MEDICAL AND PERSONAL FACILITIES

WORK AREA

SERVICE
MODULE

ECS
CRYOGENIC
STORAGE

TWO-MAN
CENTRIFUGE

ECS GAS STORAGE

TRANSFER CHAMBER

APOLLO SPACECRAFT LEM ADAPTER

Standard advanced laboratory for Earth orbital, lunar and planetary missions.

An interplanetary Apollo CSM?

By 1964, the idea of mounting a Mars fly-by prior to landing was gaining support in NASA and in industry. During a June 1964 Mars symposium at MSFC, Max Faget, one of the designers of the Mercury spacecraft, observed that a manned Mars fly-by would require a commitment to develop the technology to support a crew on interplanetary flight, and if there was no commitment, to 'we will not get [to Mars] in my lifetime'.

Von Braun, too, was a supporter of Mars fly-bys to provide the information required to mount a manned landing, and in order to support this he promoted the use of the Saturn V as the logistics vehicle to orbit the required hardware. A post-Saturn rocket (Nova) seemed unlikely in light of the NASA post-1964 budget that featured reduced post-Saturn funding. In NASA's projections, post-Apollo operations would be focused on Earth orbital and expanded lunar exploration based on Apollo hardware, and von Braun stated that based on the reliability record of the Saturn in Apollo, this could lead to the establishment of a production-line-type Saturn V delivery which would increase the availability of the launch for a broader programme that hopefully would include an authorised manned Mars programme.

Lunar Exploration Systems for Apollo (LESA)

The desire to create an Apollo logistics system that encompassed the Earth–Moon system was, in 1964, a logical progression after the initial landings. In conjunction

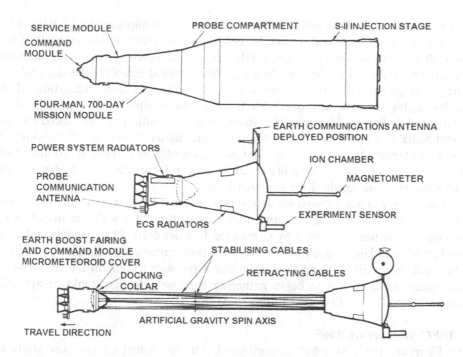

Mars fly-by spacecraft based on Apollo Saturn hardware. (*Top*) Earth orbit configuration, (*centre*) non-spinning configuration (docked), (*bottom*) spinning configuration.

with the planned AAP space station programme such a system offered a return on the huge financial and manpower investment of placing an American on the Moon by 1969, and would, it was forecast, lead to the creation of a permanent American presence on the Moon by the mid-1970s. The creation of a lunar base would also provide sustained technological and operational experience that could be used in conjunction with the space station programme in developing the techniques of long-duration spaceflight and exploration which were required in the planning the first manned flight to Mars.

Throughout the 1960s, dozens of designs and concepts for a permanent space and lunar base were put forward. In part this was to try to secure a post-Apollo programme (AAP) while the funds were still available and hardware was being produced. One example of this planning was the 1964 LESA concept study by Boeing, planned as a mid-1970s post-Apollo preliminary lunar outpost.[3] This 4–5-year programme suggested the initial use of modified Apollo spacecraft, to be gradually replaced by a new vehicle, with all initial hardware launched by the Saturn V. The operations would begin with the surface delivery of a 25,000-lb Shelter by means of a Lunar Logistics Vehicle.

The first phase continued with a modified LM delivering three astronauts near the shelter for a 28-day stay. The proposed design had a 2,800-cubic-foot volume in two chambers. The outer toroidal chamber provided storage, airlocks and connections to later shelters as the base expanded, and storage for an unpressurised LRV for local excursions, while the inner volume housed the living quarters. The transfer of lunar soil to the top of the structure would also be completed to provide additional radiation protection inside. The first expedition would establish the site and begin the local geological surveys, astronomical observations and evaluation of new technologies such as improved EVA suits and equipment for prolonged and reapeated surface exploration. This included the evaluation of 'Advanced Development Suits' that featured a construction of 'hard' materials instead of the softer suits used in Mercury through Apollo. The hard suits offered a constant volume during body movement making mobility much easier, and operated at a higher pressure, reducing the time required for pre-breathing.

The second phase allowed six astronauts to remain for six months by expanding the base with a second shelter and additional rovers, and with the installation of nuclear power units delivered by unmanned logistics craft. Phase 3 delivered a third shelter with a planned duration in excess of twelve months, while Phase 4 delivered the final shelter allowing up to twenty-four astronauts to remain on the surface for two years or more and to begin gathering baseline data that could be applied to preparations for the first expeditions to Mars.

MSFC Mars report, 1965

In February 1965, Marshall Spaceflight Center published an in-house study that featured the use of adapted Apollo hardware for a Mars programme.[4] The report established that the technology to support manned Mars missions using Saturn and Apollo hardware would be available by the late 1970s. It included the use of RL-10 engines for rendezvous and docking manoeuvres, the LM descent engine for course corrections, and of a modified CSM.

The flight profile featured six Saturn Vs and a single Saturn 1B. The first Saturn V would launch the unmanned Mars fly-by spacecraft, followed by the four Saturn Vs delivering LO tankers. The final Saturn V would deliver the Earth-departure stage based on the second stage of the Saturn, designated S-IIB, with a full tank of LH but an empty LO tank. The crew would launch on the Saturn 1B and use their Apollo CSM to rendezvous, dock and transfer to the fly-by spacecraft. The CM would be protected in a pressurised hanger that also provided a shirt-sleeve environment for the crew. After docking with the Earth-departure stage, the tankers would dock in turn, and transfer all four loads of LO. Then the Earth-departure stage would perform the Trans-Mars Injection (TMI) burn and then detach from the departing spacecraft.

A Mars fly-by armada of automated probes, each carrying 1,000 lbs of scientific equipment, would be released as the crew observed the planet. As the Earth approached at the end of the 660–690-day mission, the crew would transfer to the CM, and separate from the main fly-by vehicle by using the CSM in an Apollo lunar-type re-entry trajectory.

The MSFC report justified both the science and the engineering to mount such an expedition, and suggested a political justification as well, underlining that the next step from lunar landing 'in this decade' needed a major new undertaking to ensure

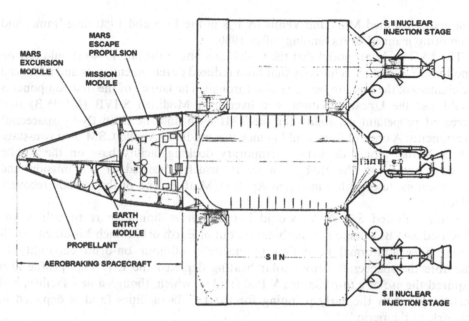

MARS ESCAPE PROPULSION

MARS EXCURSION MODULE

MISSION MODULE

S II NUCLEAR INJECTION STAGE

EARTH ENTRY MODULE

PROPELLANT

AEROBRAKING SPACECRAFT

S II N

S II NUCLEAR INJECTION STAGE

Manned Mars landing spacecraft with nuclear injection stages.

that a planetary landing in the mid-1980s would prevent a public decline in supporting a manned space programme and keep the US at the forefront of that exploration.

Planetary Joint Action Group (JAG)
In November 1963, Vice President Lyndon B. Johnson took up office after the assassination of President Kennedy. As a key figure in the decision to go to the Moon, Johnson continued to support the Apollo programme, but was unsure of where the space programme should progress after achieving the first landing. He therefore asked NASA to provide a list of suitable future goals. In reply, NASA emphasised the use of Apollo hardware in Earth orbit to create an interim space station and expanded lunar operations. Prominent scientists rejected this idea, stating that planetary exploration should be the next goal; but with increasing hostilities in south-east Asia, the federal administration was cautious of committing huge amounts of funds to NASA when it would clearly be required elsewhere. The resulting FY 1966 appropriation was the first of many reductions in the annual NASA budget that affected Apollo, the AAP, and any new-start programmes.

Meanwhile, the Mars studies continued, and in spring 1966 a second major manned planetary study was initiated that focused on piloted planetary missions using modified Apollo technology and advanced concepts such a nuclear engines.[5] The programme envisaged for Mars exploration included the AAP (1968–1973) remaining aloft for increasingly long periods to simulate the duration of manned flights to Mars; using Mariner and Voyager (1969–1973) unmanned probes as robotic scouting missions similar to the lunar robotic programme for Apollo;

conducting manned Mars and Venus fly-bys in the 1975 and 1980 time-frame; and attempting manned Mars landings after 1980.

The JAG study proposed that the fly-by missions in the late 1970s should rely on Apollo hardware and technology that both reduced development costs and advanced the chances of the mission being carried through. The launch of the first components would use the Uprated Saturn V delivering a Modified S-IVB (MS-IVB) that increased propellant capability for the TMI burn and the main fly-by spacecraft components. A crew of four would launch in a modified Apollo CSM on a two-stage Saturn V that would dock to a temporary docking facility base on the Apollo docking system on the fly-by spacecraft, which consisted of the mission and habitation modules with a modified Apollo CM at the end of the mission recovery vehicle.

Three Uprated Saturn Vs would launch twelve hours apart to deliver the Modified S-IVBs to take the combination out of Earth orbit. Each Modified S-IVB would have additional insulation to ensure a 60-hour on-orbit capability to assemble the spacecraft before solar heating depleted the LH. This profile also required the use of a third Saturn V Pad (Pad C) which, though a new facility, had been included in the early planning for the LC 39 facilities (and is depicted in artwork of the period).

Using the CSM, a series of rendezvous and docking manoeuvres would bring the three Modified S-IVB stages to the main structure and then redock to transfer inside the habitation modules. The CSM and temporary docking facility would be ejected, as it was no longer required for the mission, and one by one the Modified S-IVB stages would ignite, burn and separate to propel the vehicle out of Earth orbit and towards Mars, on a journey that would take about 130 days. The programme envisaged a range of unmanned planetary probes, including an ambitious automated sample return spacecraft that would gather a small amount of martian soil to be

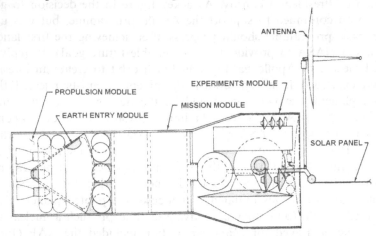

This 1966 Planetary Joint Action Group study used existing and near-term technology for its piloted Mars fly-by spacecraft design. The Earth Entry Module (*left*) was based on the Apollo Command Module.

returned to the fly-by spacecraft. The study of the samples would be conducted during the 537 days that it would take to return to Earth via the asteroid belt, during which there would also be an opportunity to study the far side of the Sun. The crew would use the modified CM – the Earth Entry Module (EEM) for end-of-mission recovery, and land using retro-rockets to cushion the landing, with the now unmanned fly-by spacecraft heading into solar orbit after, a mission of 667 days (the study used a projected time-scale of 3 October 1975 to 18 July 1977). This Apollo-based Mars fly-by was proposed as a new start for the NASA FY 1969 budget.

When the report was published in March 1967, the Vietnam War was costing thousands of American servicemen's lives, social unrest was becoming headline news in mainland USA, and there were spiralling social welfare problems, all of which placed pressure on the federal budget. NASA had also just lost three astronauts in the Apollo pad fire, and the word from Capitol Hill was that to help alleviate the financial strain, it could no longer depend on unchallenged funding from Congress.

Undeterred, and apparently not realising that 'no new starts' also applied to their study, JAG amended their plans – this time using only two Modified S-IVBs – deleted the requirement to build a third launch pad, and proposed a delay in the programme to 1975–1982; but also requested a nine-month engineering study authorised in the FY 1969 budget. The word from congress on 'no new starts' meant just that, and even the usually strong NASA supporter Congressmen Joseph Karth (Democrat, Minnesota) told NASA: 'Bluntly, a manned mission to Mars or Venus by 1975 or 1977 is now and always has been out of the question, and anyone who persists in this kind of misallocation of resources is going to be stopped.'[6] Even President Johnson was restrained in his support of NASA's aims: 'Some hard choices must be made between the necessary and the desirable. We dare not eliminate the necessary. Our task is to pare the desirable.'

Scheduling the Saturns
The launch of the first Saturn V in November 1967 was a boost for NASA, Apollo and the Mars planners, as it represented the key to all that America hoped to achieve in space over the next decade. The first fifteen launchers were on order, but from the AAP and Mars planning, far more Saturn rockets would be required to sustain even a modest programme through the 1970s, although there were beginning to emerge plans that pointed to a new smaller but reusable launch system called the Space Shuttle, which would supplement the heavy lift launch capability of the Saturn V.

Studies of improving Saturn technology by attaching four solid propellant strap-ons on the now 470 foot tall Uprated Saturn V indicated a payload delivery capability of 274 tons into a 262-mile orbit, and the addition of nuclear engines as upper stages instead of the cryogenic engines used for Apollo. During the Congressional studies by the Subcommittee on NASA Oversight (*Study of Further National Space Objectives*, and *Apollo Program Pace and Progress in 1965–1967*) as well as the President's Science Advisory Committee (1967) report (cited in the references and the Bibliography), a study had been made on the number of launch vehicles required to support the projected AAP/planetary programme. The general consensus indicated that at least two Saturn V/Apollo (CSM/LM) systems would be

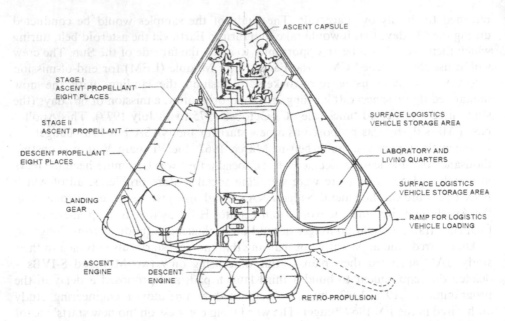

North American Rockwell's 1968 Mars lander, based on the Apollo Command Module design, for atmospheric entry at Mars.

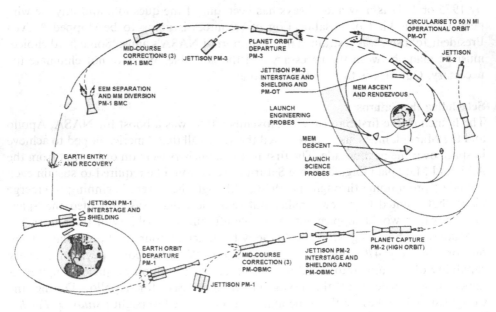

In January 1968, Boeing prepared this complex Mars expedition plan incorporating nuclear rockets and an opposition-class trajectory. (Courtesy Boeing Aircraft Corporation.)

required for the continued lunar exploration period immediately following Apollo's initial two or three landings (through 1975). The White House Joint Space Panel 1967 report also suggested (p. 36) that a third Saturn V/Apollo complete system should be provided annually as a back-up, and 'at least for a few years a fourth should be available for other probable missions, manned or unmanned.'

If this statement was applied to the projected Apollo programme, the first batch of fifteen ordered Saturn Vs would remain allocated to the main Apollo lunar programme to achieve initial landings through 1970. If not required, they could be utilised to support the post-Apollo programme as required. In addition, a suggested production rate of three or four Saturn Vs a year for the period 1971–1975 would provide for an additional fifteen to twenty Saturn V launch systems and a production line philosophy that would see a significant reduction in manpower and overheads. This would also allow a sustained flow of hardware beyond that time-scale, providing the hardware desired by von Braun and hoped for by NASA, contractors, lunar and planetary mission planners, and scientists.

The production flow of the Saturn 1B after the delivery of the initial order was, however, not so clearly defined, with the launch costs being double that of the Titan III M. Studies suggested further analyses of utilising the Titan II M carrying a four- or six-man CSM, as an alternative ferry system, to a Saturn 1B launcher, and that work on developing the larger Apollo CSMs should begin in the near future.

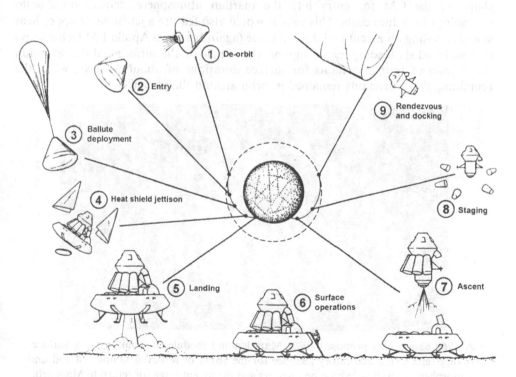

North American Rockwell's plan for landing on Mars and returning to Mars orbit.

As this report was published, the consequences of the federal restrictions in the NASA's overall budget prevented any increase in Saturn V production programme beyond the fifteen already ordered. Therefore, even though plans continued over the next three years on expanding the programme beyond Apollo on the Moon and out to Mars, in reality the funds to mount the programme were unauthorised, the hardware remained unordered, and the Congressional and public desire to support them was diminishing.

Boeing Mars study report, 1968

In January 1968, Boeing published its final report of a fourteen-month study that represented one of the most detailed planning concepts for a manned Mars mission.[7] The design featured a 582-foot long vehicle comprised of a 474-foot five-stage propulsion module and an 108-foot piloted spacecraft that itself had a mass of 140.5 tons. Each of the 33-foot diameter and 158-foot length propulsion modules housed 192.5 tons of LH and a 195,000-lb NERVA solid core nuclear thermal rocket with an engine bell in excess of 13 feet in diameter and measuring 40 feet in length. A six-person crew would ride into orbit in Apollo CSMs launched on Saturn 1Bs, with the Mars spacecraft placed in Earth orbit in sections by six Uprated Saturn Vs.

In addition, a 30-foot diameter Manned Excursion Module (MEM), designed by North American Rockwell (formerly North American Aviation), resembled the shape of the CM for entry into the martian atmosphere, relying on Apollo technology to reduce costs. This vehicle would also feature a jettisonable upper heat shield revealing an ascent and descent stage (again based on Apollo LM technology) that included six landing legs to support the vehicle on the surface and the provision for a team of four astronauts for surface durations of about 30 days, while the remaining two astronauts remained in orbit around the planet.

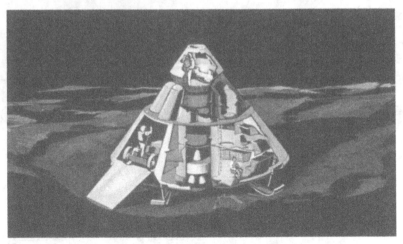

A 1969 design of a proposed 1980s Mars landing module. It would carry a surface-roving vehicle and laboratory facilities for the study of retrieved samples of soil and atmosphere, as well as habitation quarters and the ascent stage for return to Mars orbit after 30 days on the planet.

The MEM would require a 1971 new-start to the programme to support a 1982 manned Mars landing capability. However, the Boeing report indicated that the programme was to begin in 1976 with Earth orbital tests using a two-stage Saturn-MEM and Apollo CSM to test orbital rendezvous and docking, followed by Earth atmospheric entry and land landing from 1978. The first expedition could then be mounted in 1985–1986.

The cost of developing the MEM was estimated at $4.1 billion, and the whole programme at $29 billion, and generated a number of media reports when announced. But the news was dominated by other affairs. The Vietnam War was escalating, social unrest was on the increase, on 4 April 1968 civil rights leader Martin Luther King was assassinated, and then on 6 June Senator Robert Kennedy – the Democrat Party front-runner for the Presidential election – was also shot and killed.

By 1969, NASA was still forecasting extended lunar operation, Earth orbital space stations and manned Mars missions based on Apollo technology, but the new President, Richard M. Nixon, requested an evaluation of all options for the future of the space programme after Apollo, resulting in the Space Task Group report. Here the emphasis was placed on a five-year programme of integrating Earth orbital operations with lunar expeditions involving AAP space stations, extended Apollo lunar missions, and, from 1975, the development of a reusable Space Shuttle carrying standardised space station payload modules. These generic modular spacecraft would also be used as orbiter ferries, lunar logistic vehicles, and space tugs. Connected to this projected long-range planning was a plan drawn up by von Braun that featured upgraded Saturn Vs and Space Shuttles as the launching capability for manned Mars expeditions. The overall planning revealed that by 1980 up to thirty astronauts could be in space at any one time, manning space stations or lunar bases, flying missions to support those programmes, or crewing the first expeditions to Mars. Von Braun's Mars expedition was aimed for a November 1981 Earth departure date, with a landing on the Red Planet in August 1982. After a stay of one or two months, the return would be via a Venus fly-by in February 1983 and return to Earth in August 1983.

Because of inevitable funding restrictions, an alternative approach was also proposed by Robert Seamans, who suggested that NASA should continue with the step-by-step approach of AAP and tests of the Space Transportation System, establishing the feasibility and need for manned flights to Mars. In this way, the Space Shuttle and space station, rather than Mars, constituted the major focus for 1970s post-Apollo planning. The resulting budget hearing predicted further reductions in NASA funding, by cancelling one of the Moon flights and terminating the already suspended Saturn V production line.

In his State of the Union Address of 22 January 1970, President Nixon did not direct the nation towards new goals in space such as a manned Mars landing within a decade, but rather in creating an environmental Earth observation programme to repair the damage to our own planet and to explore new worlds with robotic spacecraft. By the end of 1970 the future of manned lunar exploration was limited to only another four landings over the ensuing two years, and sending humans to Mars would be at 'some undefined point in the future'.

THE SECOND MAJOR CHALLENGE OF SPACE

In summarising the lost dreams and potential of Apollo one reference seems to be the synthesis of the mixed feelings and lack of a clear direction of the American programme at the height of the Apollo programme. In 1969, shortly after Apollo 11 achieved the national lunar goal, a collection of specialists in the field of physical and social science interpreted their own different views on the impact of not only Apollo's achievement but on the general view of the direction of the space programme at that time and in the immediate future.[8] Three decades after publication, it still provides interesting reading, in light of more recent developments with the International Space Station.

Professor Sir Bernard Lovell, at Jodrell Bank Radio Observatory, argued (in 1969) that the annual cost of Apollo was only 15% that of the Vietnam War budget, and that the science budget was only a fraction of the gross national product, and yet represented 'tremendous accomplishments... for the future of man'. He also indicated support for continued lunar and planetary exploration, but predicted that even the success of Apollo would not be enough to reverse the decreasing NASA budget.

Freeman Dyson, of the Institute for Advanced Strategy at Princeton University, considered that the programme consisted of two unequal parts: the scientific programme using unmanned vehicles and only absorbing about 10% of the budget, and 'the unscientific program including manned flights [that took] nine-tenths of the money.' He remarked that the general view among the public was that spaceflight was always going to be so expensive that it would never be available to large masses of people, and that the development of a reusable launch vehicle was the key to greater access to space. 'The policy has been to do things fast rather than cheaper,' and a slow-down in the programme would, he predicted, generate more work in long-range programmes to create reusable systems, that would allow an increase in space traffic 'within a few decades after it is achieved'.

Sidney Hyman, of the Adlai Stevenson Institute of International Affairs, recalled that the political–military–economic reason of Apollo's commitment to the Moon in 1961 was in response to the apparent Soviet supremacy in science and technology. Furthermore, if America could place a man on the Moon in the chosen period, it would be a trademark of American skill and technical power. By 1969 the political and social climate had changed, and there was a 'chorus of sobering voices [saying] on all sides that the Moon-landing... was the supreme emblem of America's wrong priorities' in spending billions of dollars to put a man on the Moon while social deprivations haunted the Americans cities.

Charles S. Sheldon II, Chief of the Science Policy Research Division at the Library of Congress, Mose Harvey, Director of Advanced International Studies, University of Miami, and Philip M. Smith, Programme Director, Antarctic Program Office, National Science Foundation, presented views of the impact of the Soviet programme on the direction of the American space effort, and in looking to the future how cooperation in space, in light of a warming of relations during the Cold War, was a key to future success on Earth and in space, based to some degree on experiences in the Antarctic programme.

William Leavitt, Senior Editor of the *Air Force Space Digest*, recalled the cancellation of the USAF Manned Orbiting Laboratory programme as a result of duplication with the NASA Skylab programme, and as a direct financial implication of the growing military effort in Vietnam and recent advances in automated reconnaissance techniques. Citing the MOL as an example of the developing argument between manned and unmanned space exploration, only months before Apollo was severely cut back, Leavitt observed that the manned programme must be justified in its own right and not advertised to the public as an immediate and direct application. There was, Leavitt observed, rising public and political pressure for 'more practical pay-off from space and less emphasis on beating the Russians.' Leavitt's comments clearly implied that Apollo, the AAP and Mars would not deliver and return personal benefits to the homes of Americans fast enough. 'While a great many Americans might stay up far into the might to monitor on television such an historic event as man's first walk on the Moon, the enthusiasm needed to put up the monies to pay for a continuous undertaking of this kind comes somewhat harder in a period plagued by inflation, war abroad and frightening ferment at home.'

As an enthusiastic teenager who watched Apollo 11 on television, I learned of the worldwide millions who followed the flight and applauded the achievement, but I was unaware that the apparent wave of support for the programme would not continue for very long. Information supplied by correspondence from NASA Chief Historian Roger Launius in February 2000 revealed the results of a polling analysis of public opinion of the programme. Apparently, throughout the 1960s only 35–45% of Americans polled favoured Apollo, and the peak of 50% in favour lasted only about three weeks during Apollo 11 in the summer of 1969. That figure rapidly fell way after the landing had been achieved, and took with it any sustained or improved political and public support for further landings. Roger Launius commented in 2000: 'What is really interesting about this is not that the last three landings were cancelled, but that there were any missions after Apollo 11 at all. A whole lot of people were saying 'been there, done that, next problem'.'[9]

In the 1969 collection of observations into the status of the space programme, the next three authors – Harold Urey, Professor of Chemistry at the University of California at San Diego, Irving Michelson, Professor of Physics at the Illinois Institute of Technology, and John O'Keefe, Assistant Chief, Laboratory of Theoretical Studies, NASA Goddard Spaceflight Center, Maryland – surveyed the future of lunar exploration and research after Apollo, while Thornton Page (scientific consultant, NASA MSC) reviewed the advances in automated programmes. At this time, the Apollo 8 images of Jim Lovell's 'Grand oasis in the vastness of space' was spurring an emerging environmental movement to protect the Earth's fragile environment, and not, as NASA had hoped, increasing the desire to further space exploration. Even within NASA the addition of the new science of Earth resources was gaining support, to be included in Skylab as a major objective with the astronomical and biomedical research programmes.

Franklin A. Long, VP Research and Advanced Studies, Cornell University, Wernher von Braun, Director of NASA MSFC, Huntsville, Alabama, (who still enthusiastically supported the advanced used of the Saturn system), Ernst

Stuhlinger, Associate Director for Science at Marshall, and Sidney Sternberg, Division VP and General Manager, RCA Electromagnetic and Aviation Systems, examined the technological impact of Apollo. An interesting observation on the impact of the aerospace industry on the space programme was commented on by Franklin Long.

Long observed that reported annual figures of sales in the aerospace industry had risen from $20 billion in 1960 to $27 billion in 1968, while sales to NASA represented $4.5 billion, and those to the military, $1.2 billion. Thus the total space manufacturing effort represented only 25% effort in the total aerospace industry and only 15% to NASA. There was obvious expansion in the civilian air industry, and no sign of a slow-down of military expenditure, except in space, but no forecast of a significant increase in the civilian space programme, with an expected decline rather than growth. Long also estimated that the workforces recruited for special skills to support the space programme would in part be transferable to other aerospace programmes, but 'production cutbacks will... cause turmoil and difficulty for individual companies and employees.' The tantalising prospect for NASA and the aerospace industry – the manned exploration of Mars – would require extensive participation by the entire aerospace industry, but unfortunately an 'early decision to go ahead does not appear bright. The current national mood is to focus on domestic programmes... greatly increased capability for exploration of the planets by unmanned space problems will... be developed.' As a result, unless the cost of manned space efforts could be lowered in a major way, 'the national interest in another large and expensive manned space effort may remain low.'

Ernst Stuhlinger also commented that Apollo had been a tremendous training ground for young engineers and scientists, who could move to less adventurous programmes which were equally demanding of their skills. The space programme that culminated in Apollo was the cutting edge of 1970s technology, but was abandoned at its peak.

A new challenge?

In 1973, George Mueller was interviewed[10] on the decisions to curtail the Moon programme in 1970, and stated: 'People in Houston were concerned about the safety of flying out to the Moon, and Bob Gilruth in particular wanted to limit the number of flights... but I think the real forcing function was the scientific community's desire to stretch the time intervals between flights, and you found you were going to start flying equipment that was old, and the cost of maintaining that equipment over long periods of time was quite clearly going to be high, so we tried to develop a program that would provide a continuing manned space programme and bring on new equipment as early as possible.'

Mueller continued to say that the plan was to combine the desires of the scientists and to ensure that there were no significant gaps in the manned programme. The programme would end at Apollo 17 in 1972, but he added that with hindsight he thought this was a mistake, and that greater frequency would have used up all the hardware already available, as the actual cost was, in comparison, very little once the hardware had had been developed. Storing it over time took the money that could have instead supported flying the hardware. However, Mueller also realised that reaching the Moon quickly by flying as few missions as possible would keep the general enthusiasm high, both inside and outside NASA, 'and reduce the risk to absolute minimum... the more times you fly out there the more probability there is you won't come back.'

Accurately forecasting the fate of Apollo, Wernher von Braun was quoted in the *Los Angeles Times* (19 June 1966): 'The first successful lunar landing – a visit of thirty-six hours or less – will just scratch the surface. Its greatest achievement will be the demonstration of the ability to travel a quarter of a million miles from the Earth, land on another heavenly body, and return safely to Earth. Other journeys must follow. We must use the Saturn rockets, the Apollo spacecraft and the launch facilities built up in Project Apollo over and over again to gain the fullest return in our investment. To make one [landing] on the Moon and go there no more will be as senseless as building a locomotive and a transcontinental railroad, and then make one trip from New York to Los Angeles. I firmly believe our traffic to the Moon will increase when our railroad in space has been completed.' Von Braun also quoted President Johnson: 'We expect to explore the Moon – not just visit it or photograph it. We plan to chart the planets as well'; and he then added his own commitment for 'investigating the planets, especially Mars, [which] is the second major challenge of space. We have accepted that challenge too.'

Apollo had the capacity to create von Braun's 'railroad in space', but was only able to surpass his 'one landing' by a fraction of what was planned and hoped for.

In 2002 we are still awaiting for that accepted challenge to be fulfilled. During the 1960s and early 1970s there were dozens of plans and proposals for post-Apollo lunar bases and space stations, as well as a growing interest in exploring Mars. These ideas have never been abandoned by NASA, and are often replaced with new forecasts; but these still remain as studies, with no defined timetable to return to the Moon or depart for Mars.

Don Lind's comments in the Foreword of this book, in relation to the exploration of the Pacific basin, present an interesting observation. Indeed, the same could be said of the exploration of the American West during the 1800s. However, as Ernst Stuhlinger also observed in 1969: 'In times of affluence, and of mental saturation by movies and TV, our young are badly in need of a stimulant that makes them aware of the high challenge of science... [which is] so highly specialised that they do not appear too exciting to youngsters unless there is a visible and easily undeniable goal to which these sciences open up the gates.' Space exploration offered that goal in 1969, as it continues to do today – but is there still a desire to take up that challenge?

When I was a teenager, the Apollo missions enthralled me, and left me wanting for more, and I was amazed and disappointed to read, in 1970, that Apollo would end just two years later, after it seemed to have achieved the most challenging hurdle of all – reaching the Moon. At the time I felt let down that so much was promised but never materialised although I never gave up on the programme. Years later, further understanding of how and why those plans did not evolve produced a bigger picture of the complexities of the space programme, and eventually led to my setting

Mission achieved. Apollo 14 Commander Al Shepard – the first American in space, and the fifth man on the Moon – holds with pride the third US flag to be deployed on the lunar surface. This scene was repeated only another three times before the Apollo programme was terminated in 1972. Thirty years later we still await the ceremony of unfurling the seventh flag on the Moon.

down in print what I had hoped for three decades earlier. I still missed the excitement of watching men walk on the Moon, but at least I understood why no other Apollo astronauts journeyed to the Moon. However, I did not expect that thirty years later we would not yet have returned there.

This book was planned as a view through that 'gate' of what Apollo hardware could have and might have achieved in the 1970s and 1980s (and further discussion via the Astro Info Service website is welcomed and encouraged). In 2002 we do not yet have the grand dreams and plans of the post-Apollo era for beyond the ISS, which could excite the imagination enough to take us back to the Moon and out to Mars. This, therefore, is a small supplementary record of missions that are neither lost nor forgotten for those who remember Apollo. It is also for those who are too young to remember the magic of Apollo and dreams of a permanent manned lunar base, but who yet dream of one day watching the first humans explore Mars. For all the advancements in science, engineering, technology, spin-offs, environmental studies and international cooperation, the exploration of space must also be promoted as a great adventure full of plans, studies and dreams, in order to provide future generations with inspiration to push the space frontier towards that next, long-awaited step onto the sands of Mars.

1   *Humans To Mars: Fifty Years of Mission Planning 1950–2000*, David S.F. Portree, NASA Monographs in Aerospace history 21, NASA SP-200104521. This is an excellent overview of American plans and concepts designed to send humans to Mars over a period of five decades.

2   *EMPIRE: A study of Early Manned Interplanetary Expeditions*, Aeronautical Division, Ford Motor Company, NASA CR-51709, 21 December 1962. Detailed in Portree, *Humans to Mars*, NASA SP-2001-4521

3   *Initial Concept of Lunar Exploration Systems for Apollo*, NASA Contractor Report 39, Boeing Company Aerospace Division, 1964. Detailed in the *Romance to Reality* web site.

4   *Manned Planetary Reconnaissance Mission Study. Venus/Mars Flyby*, NASA TM X-53205, 1965. Detailed in Portree, *Humans to Mars*.

5   *Planetary Exploration Utilizing a Manned Flight System*, Planetary JAG, NASA Washington DC, 1966. Detailed in Portree, *Humans to Mars*.

6   Priority Shifts Block Space Plans, William Normyle, AWST, 11 September 1967.

7   *Integrated Manned Interplanetary Spacecraft Concept Definition*, Volume 1, Summary, Boeing Aerospace Group, Seattle, Washington, NASA CR-66558, January 1968. Described in Portree, *Humans to Mars*.

8   *Men In Space: The Impact on Science, Technology and International Cooperation*, ed. E. Rabinowitch and R. Leis, Basic Books, 1969

9   Correspondence from Roger Launius, via Praxis, 21 February 2000.

10 *'Before This Decade Is Out': Personal Reflections of the Apollo Program*, Chapter 5, pp.104–105, ed. Glen Swanson, NASA SP-4223, 1999.

Bibliography

In addition to the references cited at the end of each chapter, the following publications were used for reference, and supplement those in the author's *Skylab: America's Space Station* and *Gemini: Steps to the Moon*, both published by Springer–Praxis (2001).

1962–75 *Aviation Week and Space Technology.*

1963–79 *Spaceflight*, British Interplanetary Society, London.

1963–75 *Aeronautics and Astronautics*, NASA SP-4000 annual reference series.

1966 *Future National Space Objective.* Staff Study for the Subcommittee on NASA Oversight of the Committee on Science and Astronautics, US House of Representatives 89th Congress, 2nd Session Serial 0, US Government Printing Office. (From the library and archives of the BIS, London).

1967 *Apollo Program Pace and Progress.* Staff Study for the Subcommittee on NASA Oversight of the Committee on Science and Astronautics, US House of Representatives 90th Congress, 1st Session Serial F, US Government Printing Office. (From the library and archives of the BIS, London).
 The Space Program in the Post Apollo Period. A Report for the President's Science Advisory Committee, Prepared by the Joint Space Panels, The White House, February 1967. (From the library and archives of the BIS, London).

1969–78 *The Apollo Spacecraft: a Chronology*, Ertel, Morse, Bays, Brooks, Newkirk, NASA SP-4009, Volume 1 (1969), Volume 11 (1973), Volume 111 (1976), Volume IV (1978).

1971 *Pioneering in Outer Space*, Bondi, Hage, James, Mueller, Oliphant, Heinemann Educational Books, London.

1974 *Carrying the Fire*, Michael Collins, Farrar Straus Giroux, New York.

1977 *Skylab: a Chronology*, Roland Newkirk, Ivan Ertel with Courtney Brooks, NASA SP-4001.
 The All American Boys, Walter Cunningham, Macmillan.

1978 *Moonport: a History of Apollo Launch Facilities and Operations*, Charles D. Benson, William B. Faherty, NASA SP-4204.

The Partnership: a History of the Apollo–Soyuz Test Project, Edward Ezell and Linda Newman Ezell, NASA SP-4209.

1979 *Chariots for Apollo: a History of Manned Lunar Spacecraft*, Courtney Brooks, James Grimwood and Lloyd Swenson, NASA SP-4205.

1981 *Stages to Saturn: a Technology of Apollo–Saturn*, Roger E. Bilstein, NASA SP-4206.

1983 *Living and Working in Space: a History of Skylab*, David Compton and Charles Benson, NASA SP-4208.

1988 *NASA Historical Data Book*, Volume II, Programs and Projects 1958–1968; Volume III, Programs and Projects 1969–1978, Linda Newman Ezell, NASA SP-4012.

Countdown, Frank Borman with Robert Sterling, Silver Arrow Books.

1989 *Where No Man Has Gone Before: a History of Apollo Lunar Exploration Missions*, William D. Compton, NASA SP-4214.

1993 *To a Rocky Moon: a Geologist's History of Lunar Exploration*, Don E. Wilhelms, University of Arizona Press.

1994 *Deke! The US Manned Space Program from Mercury to the Shuttle*, Donald K. Slayton with Michael Cassutt, Forge Books.

US Space Gear. Outfitting the Astronaut, Lillian D. Kozloski, Smithsonian.

1995 *Exploring the Unknown*, Volume 1, Organizing for Exploration, ed. John Logsdon, NASA SP-4407.

1999 *Exploring the Moon: The Apollo Expeditions*, David M. Harland, Springer–Praxis.

2000 *Apollo by the Numbers: a Statistical Reference*, Richard W. Orloff, NASA SP-2000-4029.

2001 *Humans to Mars: Fifty Years of Mission Planning, 1950–2000*, David S.F. Portree, Monographs in Aerospace History No. 21, NASA SP-2001-452 1.

Web sites
David Portree *(Romance to Reality)*: www.dsfportree/explore.htm
Mark Wade: www.astronautix.com

Index